AF779051

Polymer Materials in Environmental Chemistry

Polymer Materials in Environmental Chemistry

Editor

Vinod V.T. Padil

MDPI • Basel • Beijing • Wuhan • Barcelona • Belgrade • Manchester • Tokyo • Cluj • Tianjin

Editor
Vinod V.T. Padil
Technical University of Liberec
(TUL)
Czech Republic

Editorial Office
MDPI
St. Alban-Anlage 66
4052 Basel, Switzerland

This is a reprint of articles from the Special Issue published online in the open access journal *Polymers* (ISSN 2073-4360) (available at: https://www.mdpi.com/journal/polymers/special_issues/poly_mater_environ_chem).

For citation purposes, cite each article independently as indicated on the article page online and as indicated below:

LastName, A.A.; LastName, B.B.; LastName, C.C. Article Title. *Journal Name* **Year**, *Volume Number*, Page Range.

ISBN 978-3-0365-3471-8 (Hbk)
ISBN 978-3-0365-3472-5 (PDF)

© 2022 by the authors. Articles in this book are Open Access and distributed under the Creative Commons Attribution (CC BY) license, which allows users to download, copy and build upon published articles, as long as the author and publisher are properly credited, which ensures maximum dissemination and a wider impact of our publications.
The book as a whole is distributed by MDPI under the terms and conditions of the Creative Commons license CC BY-NC-ND.

Contents

About the Editor

Vinod V.T. Padil is an Assistant Professor and Senior Researcher at the Institute for Nanomaterials, Advanced Technologies, and Innovation, Technical University of Liberec (TUL), Czech Republic. He received his Ph.D. in Biotechnology from Acharya Nagarjuna University, India. He also worked as a scientific officer at the Board of Radiation and Isotope Technology (BRIT), Department of Atomic Energy (DAE), India, as well as a postdoctoral fellow at the National University of Singapore. He visited research fellows at Wrexham Glyndwr University, UK, and Macromolecular Chemistry II, Bayreuth University, Germany. He has authored/co-authored more than 85 research articles, 10 book chapters, and many research proceedings in international conferences (with a H-Index of 26). His research interests include green synthesis of nanoparticles, environmental bioremediation using nanostructures, 2D/3D polymer nanocomposites, electrospinning, and bioplastic fibers/films for food packaging application.

polymers

MDPI

Review

Microscopic Techniques for the Analysis of Micro and Nanostructures of Biopolymers and Their Derivatives

Abhilash Venkateshaiah [1], Vinod V.T. Padil [1,*], Malladi Nagalakshmaiah [2,*], Stanisław Waclawek [1], Miroslav Černík [1,*] and Rajender S. Varma [3,*]

1 Department of Nanomaterials in Natural Sciences, Institute for Nanomaterials, Advanced Technology and Innovation, Technical University of Liberec, 461 17 Liberec, Czech Republic; abhilash.venkateshaiah@tul.cz (A.V.); stanislaw.waclawek@tul.cz (S.W.)

2 IMT Lille Douai, Department of Polymers and Composites Technology and Mechanical Engineering (TPCIM), 941 rue Charles Bourseul, CS10838, F-59508 Douai, France

3 Regional Centre of Advanced Technologies and Materials, Department of Physical Chemistry, Faculty of Science, Palacký University in Olomouc, Šlechtitelů 27, 783 71 Olomouc, Czech Republic

* Correspondence: vinod.padil@tul.cz (V.V.T.P.); nagalakshmaiah.malladi@imt-lille-douai.fr (M.N.); miroslav.cernik@tul.cz (M.Č.); Varma.Rajender@epa.gov (R.S.V.)

Received: 26 January 2020; Accepted: 24 February 2020; Published: 27 February 2020

Abstract: Natural biopolymers, a class of materials extracted from renewable sources, is garnering interest due to growing concerns over environmental safety; biopolymers have the advantage of biocompatibility and biodegradability, an imperative requirement. The synthesis of nanoparticles and nanofibers from biopolymers provides a green platform relative to the conventional methods that use hazardous chemicals. However, it is challenging to characterize these nanoparticles and fibers due to the variation in size, shape, and morphology. In order to evaluate these properties, microscopic techniques such as optical microscopy, atomic force microscopy (AFM), and transmission electron microscopy (TEM) are essential. With the advent of new biopolymer systems, it is necessary to obtain insights into the fundamental structures of these systems to determine their structural, physical, and morphological properties, which play a vital role in defining their performance and applications. Microscopic techniques perform a decisive role in revealing intricate details, which assists in the appraisal of microstructure, surface morphology, chemical composition, and interfacial properties. This review highlights the significance of various microscopic techniques incorporating the literature details that help characterize biopolymers and their derivatives.

Keywords: biopolymers; microstructures; nanostructures; surface morphology; filler dispersion; chemical composition; optical microscopy; scanning electron microscopy; transmission electron microscopy; atomic force microscopy

1. Introduction

The ever-rising concern about the environmental impact of synthetic polymers has stimulated a great deal of research interest in the polymers of biological origins. The combination of sustainability and biodegradability of biopolymers is the reason they are gaining precedence over the fossil fuel-derived polymers [1]. Historically, biopolymers had commanded a lot of importance in various applications such as food, clothing, ropes, and furniture [2]. However, since the discovery of synthetic polymers, these biopolymers were replaced. With the increasing public awareness of global warming and the environmental toll as well as the rapidly depleting fuel sources, these biomaterials are emerging as a potential replacement for conventional polymers; during the last two decades, this interest has been the driving force in the development of sustainable biopolymers from renewable sources [3–5].

Biopolymers encompass a wide range of materials derived from biological sources like plants, animals, trees, microorganisms, and also materials synthesized from other sources including sugars, proteins, vegetable oils, fats, resins, and exudate [6]. The common distinguishable property with all these biomaterials is that, given a definite time, they break down to form simple molecules such as carbon dioxide and water under the enzymatic action of microorganisms. In comparison with conventional polymers with a simpler structure, nature-derived biopolymers have a wide variety of structural complexities, often reliant on many factors including the source, their species, the age of the plant, and the method of extraction.

Biopolymers, owing to their versatile nature have tremendous potential to replace conventional polymers in a wide range of applications including packaging, textiles, cosmetics, food technology, drug delivery, and structural materials [6]. However, the commercialization of biopolymers is hampered by shortcomings of economic and engineering aspects. Competition with their cheaper synthetic counterparts, lower mechanical strength, and high hydrophilicity impede their industrial use. Several techniques have been developed to overcome these drawbacks including chemical modification [7,8], blending with other biodegradable polymers [9,10], and by using additive fillers [11,12] and plasticizers [13]. Recently, biopolymer research crossed paths with nanotechnology and the resulting materials have been making tremendous progress. The advent of nanotechnology in the biopolymer avenue has opened many potential doors. In combination with the nanomaterials, biopolymers display superior properties when compared with unblended ones. The intersection of the fields of biopolymers and nanotechnology has been in more than one form; biopolymers blending with nanomaterials offered nanocomposites [14,15], besides being made into nanoparticles [16,17] and nanofibers [18–20]. Additionally, biopolymers have also been used as reducing and stabilizing agents in the generation of nanoparticles [21–25]. The fusion of these two fields has paved the path for a whole class of materials, which not only have the potential to replace conventional fossil fuel-based polymers but also are environmentally benign.

Contemporary research ensued regarding the emergence of various classes of materials; hence, proper characterization is necessary to gain insights into their complex structure and morphology to determine their ideal applications. Microscopy techniques represent a class of multifunctional techniques often used for the in-depth analysis and understanding of materials over a large-scale magnification range. Microscopy is the observation of materials and their properties in the range of millimeters to nanometers, which are not visible to the naked eye. Observation of the materials at such high magnifications can provide physical, chemical, and structural information, which are associated with the performance of the material. Microscopy techniques can provide the necessary information by acquiring an image of the material and the morphology of the sample is analyzed with the help of this image at micro or nanoscale. Different microscopic techniques have been developed over the years ranging from optical microscopes to electron microscopes and scanning probe microscopes for the analysis of material morphology on varying length scales.

Microscopy techniques have been a prominent part of the material research over decades and have aided in numerous discoveries. The unprecedented importance of their contributions in the field of material research can be recognized by the fact that most research articles provide a direct or indirect reference to their use. With the increased focus on biopolymers in the recent past, the use of microscopic techniques for their analysis has been an invaluable asset in aiding researchers' innate need to understand these natural wonder materials. The aim of this article is to accentuate various microscopic techniques often used to investigate the bio-polymeric systems and their influence on current research. This article provides a compilation of the assortment of microscopic techniques used in the field of biopolymers along with an introduction to their mechanism of operation. Furthermore, the article emphasizes the significance of microscopic techniques in the elucidation of various micro and nanostructures of biopolymers. Additionally, several leading articles have been cited in the present review, which will aid the researchers to comprehend the recent development of various microscopic techniques deployed.

2. Biopolymers

Polymers generated from renewable resources or polymers that degrade into carbon dioxide and water under certain conditions in a limited time are termed biopolymers and are classified into distinct types as shown in Figure 1 [26–28]. Since different criteria are followed in the literature for their classification, herein, we have briefly discussed assorted biopolymers, their classification, and applications.

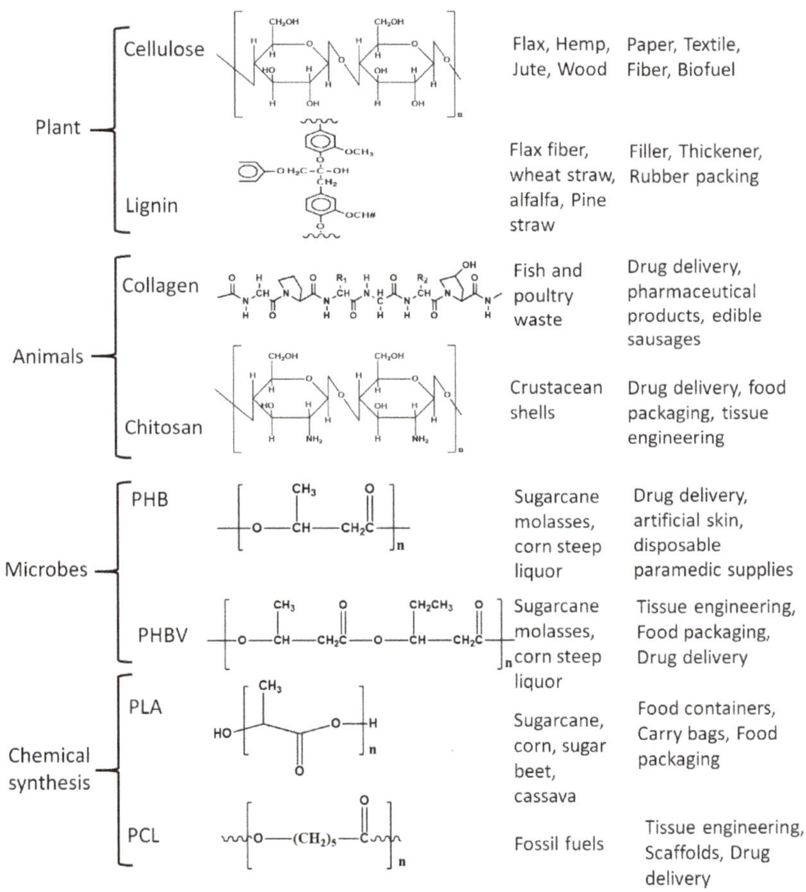

Figure 1. Common biopolymers with their structure, source, and applications.

Biopolymers are mainly divided into two types as natural or synthetic/manmade polymers; further, they are divided into subsections depending on their manufacturing process and inherent functional groups. Natural biopolymers are further sub-classified into two types based on the sources, as plant-derived or animal-derived biopolymers; plant-based biopolymers are those which are obtained from plants, trees, or biomass and are called agro-polymers, such as cellulose, starch, hemicellulose, and lignin, among others [27,28]. Cellulose is the most abundantly available polymer in today's world with around 1.5×10^{12} tons of annual production and is considered an inexhaustible source of biopolymer [29], often extracted from various agricultural and other natural sources [30,31]. Cellulose contains long chains of unbranched β (1→4) linked D- glucopyranosyl units and the chain length of β (1→4) glucan depends on factors including the species of the plant, growth environment, and maturity. Cellulose has found major applications in paper, textiles, and fiber industries, and even as a biofuel source. Lignin is the second most naturally available biopolymer after cellulose with a highly branched structure bearing various oxygenated functional groups like hydroxyl, carboxyl, carbonyl, and methoxyl groups [32,33]. Lignin comprises 15%–40% of dry matter in woody plants and consists of three phenyl propane units also called monolignols namely, coniferyl-, sinapyl-, and *p*-coumaryl alcohol [34]. Lignin has been used as a rubber intensifier, filler in composite materials, in rubber packing, thickener for paints and coatings [34]. Starch is another abundant polysaccharide extractable from agricultural raw materials, main crops being potatoes, corn, and rice, where starch is produced as granules and stored in seeds and swollen stems. The composition of starch varies from plant to plant but mostly consists of amylose and amylopectin. Amylose is a linear molecule bearing (1→4)-linked α-D-glucopyranosyl units with a very few branched links, while amylopectin has a highly branched structure with (1→4)-linkages of α-D- glucopyranosyl residues and (1→6) linkages at the branch points [35]. Applications of starch include disposable food service wares, food packaging, carrying bags, and loose fill products [36].

Biopolymers derived from animals significantly differ from those derived from plants and include proteins (e. g. collagen) and polysaccharides (e. g. chitosan). Collagen is an abundant protein constituent present in the connective tissues of vertebrate and invertebrate animals with various types identified to date, but Type I collagen is the most common and explored. It consists of three polypeptide subunits, each composed of amino acids containing glycine, proline, hydroxyproline, and lysine [37] and is used in pharmaceutical products, drug delivery applications, and edible casings [38]. Chitosan is a random copolymer derived from chitin, a polysaccharide found in the crustacean shells [39] which upon alkaline deacetylation yields chitosan and consists of β-1,4 glycosidic linked D-glucosamine and *N*-acetyl-D-glucosamine units [40]. Chitosan, because of its non-toxic nature, has been used as a food additive, in drug delivery, and owing to its film-forming properties has been deployed in tissue engineering and food packaging [41–43]. Additional applications for chitosan are in the beverage industry, as a support for enzyme immobilization, and also as a reinforcing filler [44,45].

Man-made or synthetic biopolymers are derived with human intervention and can be further divided into the biopolymers generated from microorganisms and chemically synthesized polymers, e.g., polyhydroxyalkanoates (PHA), are produced from microorganisms at certain biotechnological (pH and temperature) conditions and specific nutrients [46,47]. Renewable PHA is synthesized by bacteria from such renewable resources as carbon source, and owing to their plastic-like properties combined with biodegradability is viewed as a potential replacement for polyethylene and propylene, with applications in food packaging [48,49]. However, this potential is being encumbered by its poor mechanical properties and brittle nature [48,49]. PHB is the most popular member of the PHA family acquired via fermentation of sugars by the bacteria *Alcaligenes eutrophus*. PHB is non-toxic and degrades in vivo into d3-hydroxybutyric acid, which is commonly found in human blood. Not surprisingly, PHB has been used in heart valves, controlled drug release, artificial skin, and disposable paramedical supplies [50,51]. PHB has inherent poor mechanical properties and efforts are being made to improve those to compete with the conventional polymers [52,53]. Blending is one such endeavor, wherein, PHB family blends are found to be compatible and have enhanced co-crystallization as exemplified by poly (3-hydroxybutyrate-co-3-hydroxyvalerate), (PHBV) which is a blend of PHB and hydroxyvaleric acid. This copolymer has less crystallinity, improved flexibility and processability, and is commercially available under the name Biopol.

Chemically synthesized biopolymers include polymers made from monomers (e.g., polylactic acid), from biobased materials, and biopolymers obtained via petro-based monomers (e.g., poly-ε-Caprolactone). Polylactic acid (PLA) is a biodegradable aliphatic thermoplastic polyester prepared from starch as a major source [54,55]. The lactic acid monomer is extracted by enzymatic hydrolysis of starch from corn, tapioca, and sugarcane and is polymerized by polycondensation reaction to get low molecular weight PLA and ring-opening polymerization of lactides to obtain high molecular weight PLA [56]. PLA has zero to low toxicity, is biodegradable, and has mechanical properties comparable to those of conventional commercial polymers [10,57,58]. Among the petro-derived biopolymers, poly-ε-caprolactone (PCL) is the most common and popular biodegradable polyester and is obtained by catalytic ring-opening polymerization of ε-caprolactone [59,60]. Owing to its biodegradability combined with nontoxicity and biocompatibility, it is used in the preparation of scaffolds for tissue engineering and in controlled drug delivery [61–63].

3. Microscopic Techniques

Currently, a plethora of characterization techniques are available to analyze and characterize materials, and they assist in recognizing the end-use applications of the materials by providing a thorough knowledge of the structure and property relationships. Among these, microscopy techniques command a unique position in analyzing various features such as morphology, chemical composition and structure, topology, interfacial properties, molecular, microstructure, and micromechanical properties.

Several microscopy techniques used for the characterization and analysis of diverse biopolymers and their derivatives are discussed in this article with a brief introduction to the mechanism of the microscopic techniques followed by their usage in research (Figure 2 and Table 1).

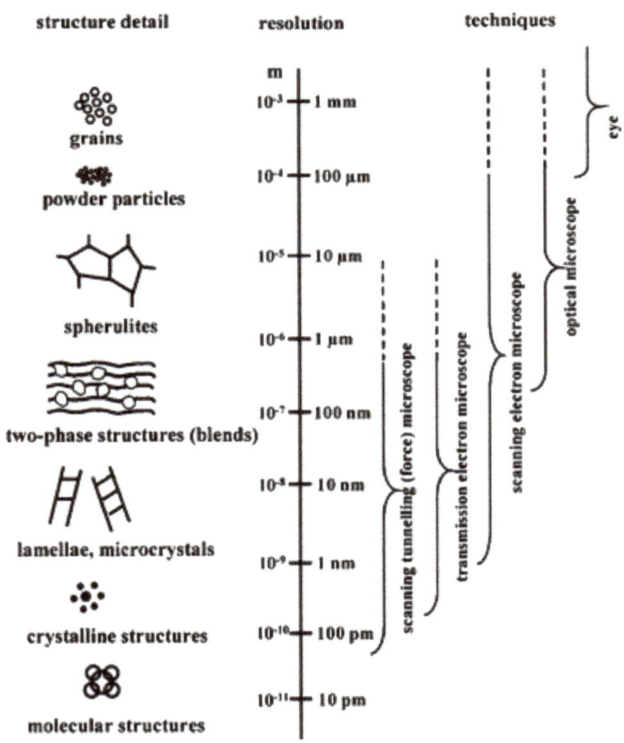

Figure 2. Assorted polymeric structures and the attainable resolutions with various microscopic techniques [64]. (Reprinted with permission from Michler et al., 2008.)

Table 1. Summary of various microscopic techniques, their applications and relevant biopolymers illustrated in the current review.

Technique	Application	Biopolymer	References
Optical Microscopy	Fiber diameter	Poly(ε-caprolactone)/chitosan blend	[65]
	Size and ahape	Starch granules	[66,67]
	Filler dispersion	starch/ Gum Arabic/nanocellulose	[68]
Scanning electron microscopy (SEM)	Particle size	Chitosan	[69]
	Particle shape	Starch granules	[66]
	Fiber diameter and surface modification	Gum Arabic, Gum Karaya, Kondagogu gum	[19]
	Crystal alignment	Cellulose nanocrystals	[70]
	Failure behavior	Gelatin/maltodextrin	[71]
SEM + energy-dispersive X-ray spectroscopy	Elemental composition	Cellulose	[72,73]
Transmission electron microscopy (TEM)	Particle dispersion	Cellulose nanofiber	[74]
	Particle Size	Kondagogu gum biopolymer assisted Pt nanoparticles	[24]
	Core shell structure	Chitosan/PEO	[75]
TEM + selected area electron diffraction	Crystallographic analysis	Biopolymer assisted nanoparticles	[21,24]
Atomic force microscopy	Molecular structure and conformation	Xanthan gum	[76–78]
	Nanomaterial topography	Nanocellulose	[79,80]
	Particle size and shape	Nanocellulose	[81]
Chemical force microscopy	Chemical interactions	Chitosan	[82]
Magnetic force microscopy	Magnetic properties	Chitosan based magnetic nanohydrogels	[83]
Scanning tunneling microscopy	Molecular structure Particle Size Surface modification	Bacterial polysaccharides Cellulose Cellulose	[84,85] [86] [87]

3.1. Optical Microscopy

Optical microscopy is probably the simplest and oldest among the microscopy techniques [88]. It is a two-dimensional imaging technique and since its time of inception, there have been incessant research upgrades to suit modern needs [89–92]. A simple optical microscope comprises of two converging lenses, an objective, and an eyepiece and utilizes the optical theory of lenses to operate, wherein light emerging from the sample will be collected by the objective and directed towards the eyepiece [93]. Usually, the sample is illuminated by two methods—episcopic (reflected) or diascopic (transmitted). Generally, the reflected techniques provide the essential information, and transmitted light optical microscopy is opted to gain insights into the microstructures, while in some cases etchants, stains, or dyes may be required for entrenched analysis of morphology [94–96]. Light detecting devices, namely charge-coupled camera, photodiodes, photomultiplier tubes, and other optical sensors are widely employed to collect the image of the sample. Modern optical microscopic imaging systems are equipped with electron scanning systems like galvano-mirrors, acousto-optic deflectors, or fast confocal illumination systems. It is the simplest of all the other microscopic techniques, as the samples can be directly viewed at a magnification up to 1500× with a theoretical resolution of 200–300 nm in lateral resolution, and 500–700 nm in axial resolution [97].

Optical microscopy has been a part of biopolymer research for a long time to analyze samples, due to its simplicity and little-to-no sample preparation. Optical microscopy assists in observing different features such as size, shape, uniformity, void content, failure analysis, and quality control [98–100]. An optical microscope can discern the filler dispersion in composites on a larger scale, thus providing a wider perspective of filler distribution globally in the sample than other techniques. It has been successfully used in determining the uniformity of poly(ε-caprolactone)/chitosan blend fibers to be used in tissue engineering [65], wherein the dry fiber diameter as a function of total polymer concentration was analyzed and the average fiber diameter was determined by utilizing the microscopy images with a computer image analyzer [65]. Similarly, optical microscopy has been used to analyze the microstructure and morphology of starch granules assisting in determining the performance of the composites as a function of the size and shape of the granules [66]. Results acquired via an optical microscope with 50× magnification revealed that the granules had a variety of morphology from oval, spherical, polygon, to irregular shapes depending on their botanical origin [66]. Govindaraju et al. made similar observations, where they observed varied morphology of starch granules obtained from different sources (Figure 3); besides evaluating the granule sizes and observing starch degradation after hydrolysis [67], the granules were found to be polyhedral in the case of rice starch, while corn starch showed spherical and polyhedral structures.

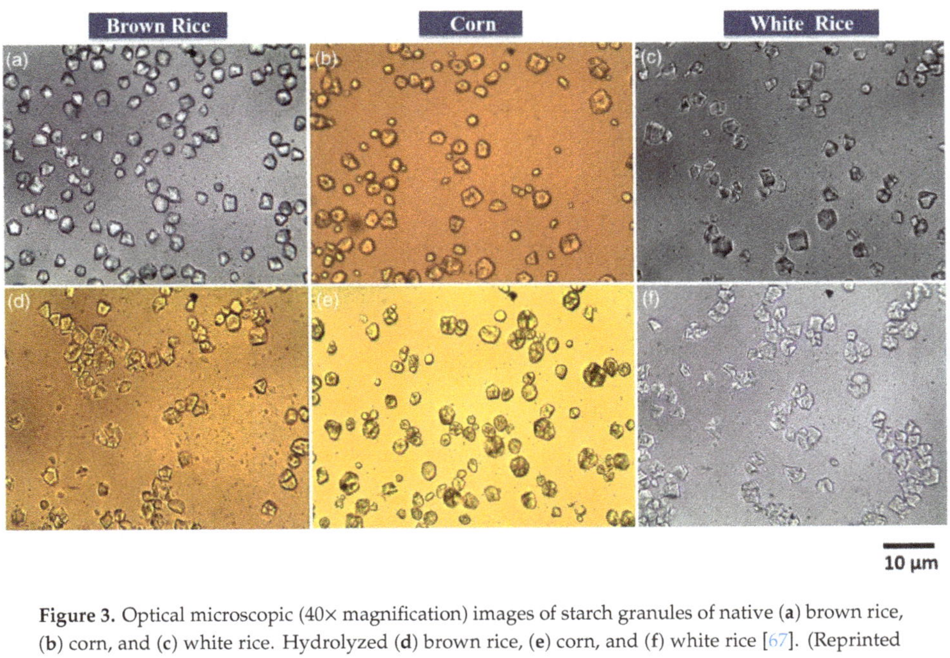

Figure 3. Optical microscopic (40× magnification) images of starch granules of native (**a**) brown rice, (**b**) corn, and (**c**) white rice. Hydrolyzed (**d**) brown rice, (**e**) corn, and (**f**) white rice [67]. (Reprinted with permission from Govindaraju et al., 2020.)

The optical microscopy technique has also been able to discern the phase separation between different components and the crystallization behavior of different biopolymers [101,102]. Herein, microscopes equipped with a temperature controller have been used to study the effect of temperature and blend ratio on the phase structure of the ensuing biopolymer blends. Micrographs of the blend systems were obtained in situ to study the crystallization behavior of the blends [101,102]. Optical microscopy is not only limited to solid samples, and has been used to analyze emulsions of biopolymers to determine the size shape and uniformity of droplet size [103,104]; images are obtained by placing a drop of emulsion on the microscopic slide and covering it with a coverslip before observing under the microscope.

When the dimensions of the samples fall into the nanometer scale like in the case of cellulose nanocrystals and cellulose nanofibers, the use of optical microscopy might seem impractical. However, as mentioned earlier there have been massive developments in the field and many variations of optical microscopy are available nowadays; fluorescence microscopy technique has broadened its horizons by extending the diffraction-limited resolution to nano dimensions and achieving super-resolution. The resolution is extended to smaller values by employing far-field imaging methods like confocal, multiphoton, 4Pi microscopy, and structured illumination and spatially patterned excitation to achieve super-resolution in florescence spectroscopy [105–107]. To use fluorescence spectroscopy on the samples, they must contain chromophores and there have been different approaches to label either the matrix or the filler with chromophores. An example of such a study is where cellulose nanocrystals of varying charge contents were fluorescently labeled with 5-(4,6-dichlorotriazinyl) aminofluorescein (DTAF) and analyzed [108]; labeled nanocrystals were fluorescently active and were comparable with the unlabeled counterparts in surface chemistry and behavior. The labeled nanocrystals were used as optical markers to determine the dispersion quality of cellulose nanocrystal loaded polyvinyl alcohol composites [108].

Laser scanning confocal microscopy (LCSM) is another variant that presents certain advantages over the conventional wide-field optical microscopy, as it has the ability to minimize or eliminate the background noise from the focal plane and is capable of taking a series of optical sections in case of thick specimens. It uses spatial filtering techniques to remove any out-of-focus light from specimens with a thickness exceeding the immediate focus plane. This technique can provide better images than conventional fluorescence microscopy from the samples prepared for the same, but LCSM cannot provide nanoscale resolution [109,110]. LCSM, in combination with Förster resonance energy transfer (FRET), has been successfully used in the study of DTAF labeled cellulose nanofibrils (CNF) incorporated into coumarin 30 (C30) labeled polyethylene (PE) matrix [111]. FRET can be defined as a phenomenon wherein an energy transfer occurs between a donor chromophore and an acceptor chromophore when certain conditions are met. FRET enables certain nano features, which are not in the resolution limit of the optical microscopy, and the FRET/LCSM combination can provide information at nanoscale while scanning the entire sample at macroscale. This technique has been used to extract information about the interface of the CNF and the PE matrix.

From Figure 4, one can see images with the fluorescence of C30 [(A), (D)] and CNF [(B), (E)] obtained by donor and acceptor filters while the energy transfer efficiency maps can be seen in Figure 4 [(C), (F)], which were calculated by applying algorithms to the confocal images. The inset images evidently show that FRET does not occur when there is no distribution of CNF in the PE matrix, which can be explained by the inability of the C30 to penetrate the CNF agglomerates. All this information suggests that information in nanoscale related to the interface of the filler and matrix can be obtained by LCSM/FRET.

From all the above-mentioned examples and studies, it is evident that even though the optical microscopy technique is one of the oldest techniques it can still provide a significant amount of information; variants of optical microscopy are competent with the newer techniques and will prove to be useful for a very long time.

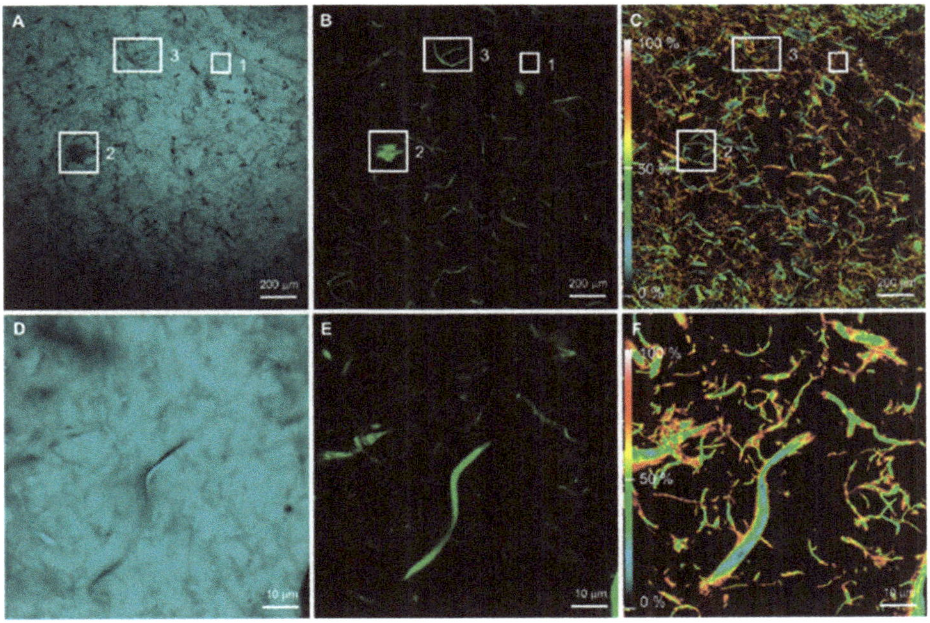

Figure 4. (**A**) Coumarin 30 (C30) fluorescence (donor filter set) at 5× magnification; (**B**) cellulose nanofibril (CNF) fluorescence (acceptor filter set) at 5× magnificationl (**C**) energy transfer efficiency map at 5× magnification; (**D**) C30 fluorescence at 100× magnification; (**E**) CNF fluorescence at 100× magnification; (**F**) energy-transfer-efficiency map at 100× magnification [111]. (Reprinted with permission from Zammarano et al., 2011.)

3.2. Scanning Electron Microscopy

Scanning electron microscopy (SEM) is one of the most versatile, distinguished, and popular techniques in research as well as industrial sectors. It is a class of electron microscopy which uses a high-energy electron beam to scan the samples and provide a high magnification and resolution image. The electron beam from the electron gun interacts with the electrons on the sample and produces certain signals about the surface topography. The SEM images are obtained by analyzing the signals from the secondary and backscattered electrons, which contain information regarding the sample. This technique provides abundant information about the samples being analyzed, including but not limited to surface morphology, crystallinity, elemental composition [112]. In theory, when the electron wavelength and their energy are taken into account, resolutions smaller than the radius of an atom could be obtained by this technique. However, limitations arising due to the lenses and the sample preparations restricts its working resolution to the order of 1–2 nm [112]. The SEM analysis can be carried out under high vacuum or low vacuum and even under wet conditions [113]. Samples for the analysis of SEM are typically frozen under liquid nitrogen and coated to avoid charging and metal shadowing or common deployment of negative staining for contrast enhancement [114].

SEM has been widely used for decades in the study and analysis of biopolymer systems to gain information pertaining to the structure, morphology, size, shape, surface modifications, wear and tear, etc. SEM studies have played pivotal roles in the analysis of biopolymers [115–117], biopolymer nanoparticles [118,119], biopolymer assisted nanoparticles [21–23], biopolymer nanocomposites [120–124], and biopolymer fibers [125–128]. SEM images can be helpful in observing and analyzing different particle structures varying from fibers to microparticles and nanocrystals [129,130]; SEM provides an effective way to analyze these particles and their surfaces, and a medium to measure the particle size and diameter as well [81] (Figure 5).

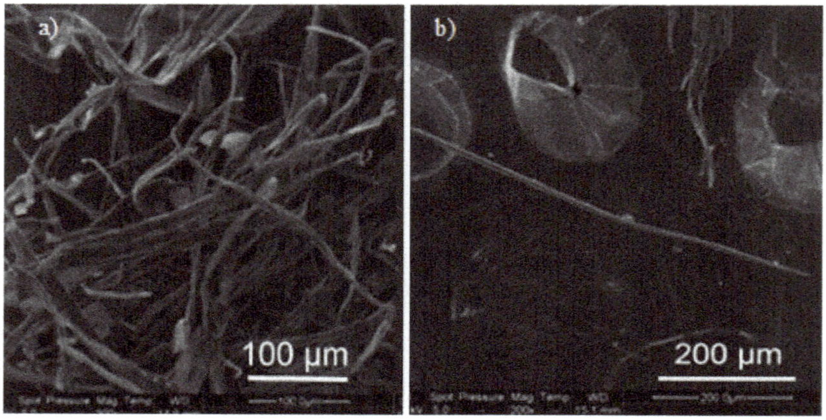

Figure 5. Scanning electron microscopy (SEM) images of different cellulose fibers; (**a**) bundles of unbleached fibers; (**b**) individual bleached fibers [81]. (Reprinted with permission from Nagalakshmaiah et al., 2016.)

SEM imaging is widely used to determine the surface topography, homogeneity, and any phase separation between different components in a biopolymer and composite film [116]. These properties have a direct correlation with the mechanical properties of the film, which in turn determines its end-use applications. SEM studies on electrospun nanofibers from natural polymers were helpful in determining the diameter and length of the fibers [19] and could distinguish the change in surface topology arising from the plasma treatment; untreated nanofibers had a smooth surface while after plasma treatment SEM micrographs showed significant surface roughness (Figure 6) [19]. In the case of porous structures like scaffolds or membranes, SEM micrographs assist in the determination of pore size, structure, and density [131–133].

In nanocomposite characterization, SEM is primarily used to determine the dispersion and distribution of the fillers in the polymer matrix. In addition, SEM images also show the presence of any agglomeration of the fillers or additives within the polymer matrix [134,135]. Intercalation and exfoliation of layered nanoparticles, such as montmorillonite within the soy protein isolate nanocomposites, have been predicted by analyzing the SEM images [136,137]. In addition to the surface morphology, SEM has been successful in observing the alignment of the crystals in cellulose nanocrystal (CNC) films [70]; high magnification SEM image of a fractured cross-section of the CNC films revealed that the direction of the chiral nematic axis changes with the location in the film. Figure 7a shows a high magnification image of the layered structure of the film while the fan-like structures seen in Figure 7b correspond to the cross-section of a left-handed helicoidal arrangement of the nanocrystals, wherein the helicoid axis is from top to bottom of the image. The results obtained from the SEM images can help in determining the failure mechanisms of the chiral nematic model [70].

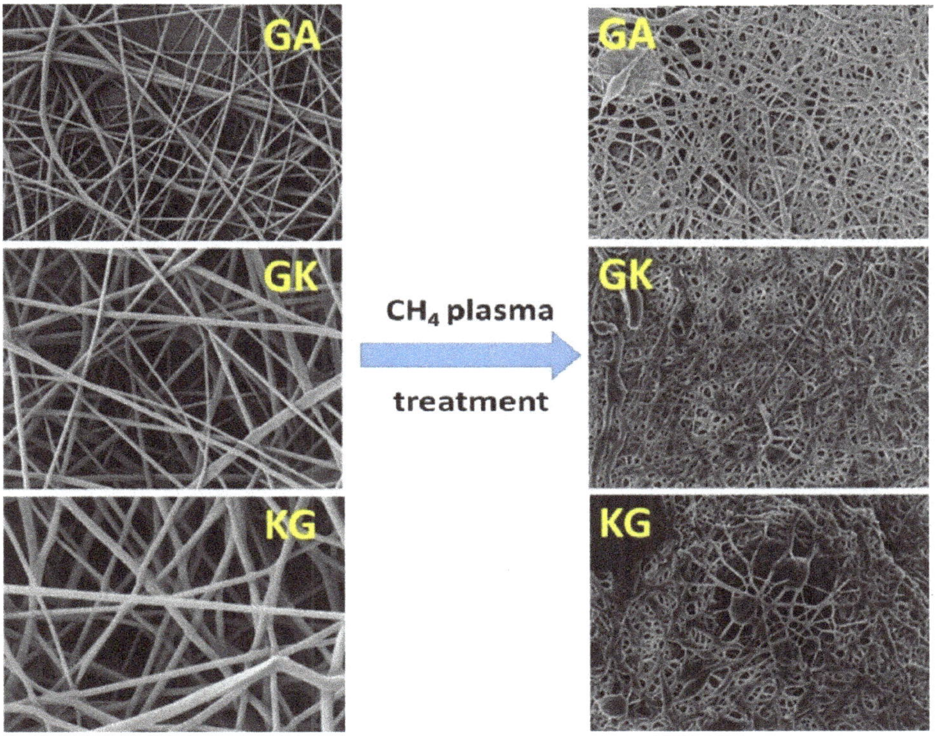

Figure 6. SEM images of polyvinyl alcohol (PVA)/gum Arabic (GA), PVA/gum Karaya (GK), and PVA/Kondagogu gum (KG) before and after plasma treatment [19]. (Reprinted with permission from Padil et al., 2016.)

Field emission scanning electron microscopy (FESEM) is a variant of conventional SEM, which provides higher resolution images and a greater energy range. One of the main differences between SEM and FESEM is that the latter uses a field emission gun as an electron generation system. This provides highly focused electron beams, which improves spatial resolution and enables the analysis of samples at low potentials; FESEM has already been used efficiently instead of SEM to obtain high-resolution images of biopolymer systems [18,138–141]. An environmental scanning electron microscope (ESEM) is another variant of SEM, which unlike conventional SEM does not require any special sample preparation like coating, and can examine the specimen at various temperature regimes. ESEM can operate in gaseous atmospheres like in air, nitrogen, argon, oxygen, and even water vapor, thus making it possible to carry out 'wet imaging' of samples. ESEM has the ability to analyze dynamic phenomena such as crystallization, wetting, swelling, drying, melting, freezing, as well as material deformation [71,142–145]. Rizzieri et al. studied the strain deformation and failure behavior of biopolymer gel mixture through in situ ESEM [71]. From the images acquired by ESEM, they carried out the dynamic observations of the microstructure response to mechanical deformations in biopolymer gel mixture systems that were kept hydrated throughout the whole experiment.

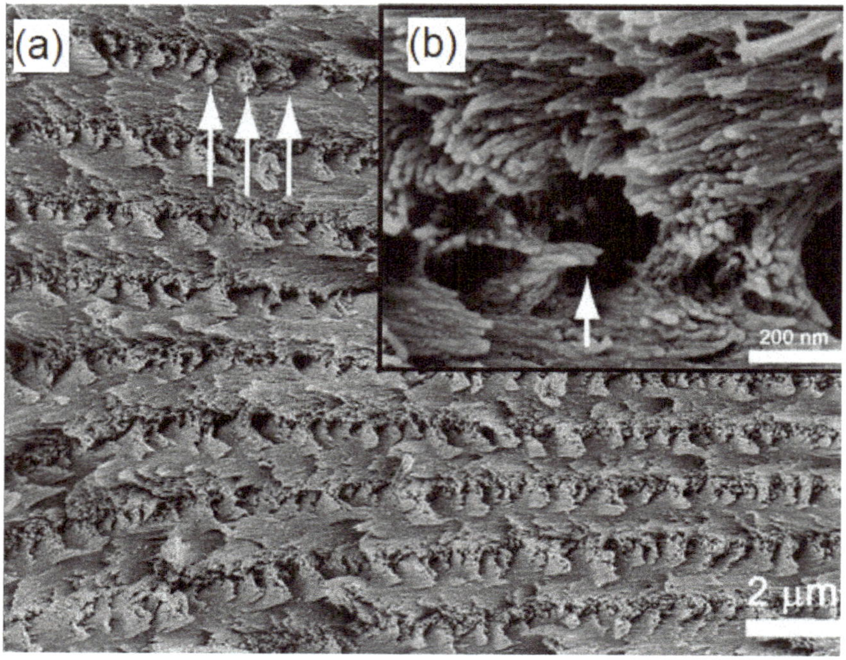

Figure 7. SEM images of the fractured surface across the cellulose nanocrystal (CNC) film (**a**) High magnification image of the layered structure of the film, (**b**) Cross-sectional image showing (inset Figure b) the helicoidal arrangement of nanocrystals. [70]. (Reprinted with permission from Majoinen et al., 2012.)

In this study, phase-separated gelatin and maltodextrin composite gels comprising of spherical maltodextrin phases inside a continuous gelatin matrix were subjected to increasing applied strain. ESEM micrographs were taken at a frame refresh rate of 0.5 fps and the applied strain rate was in the order of 10 µm s^{-1}. Figure 8 shows that a crack was initiated from an existing defect on the surface, which acts as a stress concentration point. This crack propagates revealing the underlying maltodextrin particles and the fibrillation of gelatin matrix around the particles until the complete failure of the sample is seen from the micrographs.

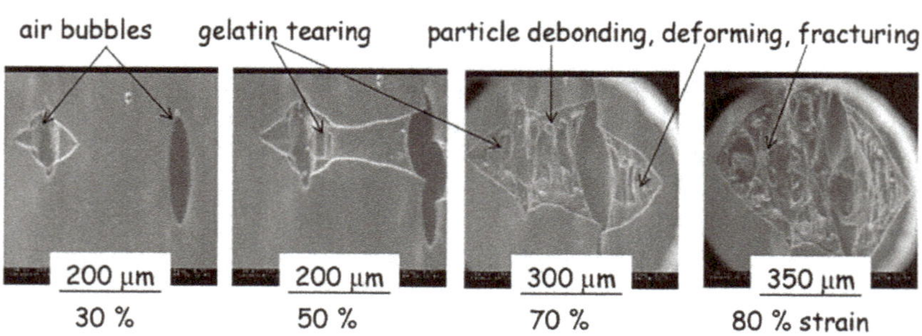

Figure 8. Environmental scanning electron microscope (ESEM) micrographs following the continuous crack propagation on the surface gelatin/maltodextrin sample [71]. (Reprinted with permission from Rizzieri et al., 2003.)

Energy-dispersive X-ray spectroscopy (EDX) is an additional accessory included in SEM which functions as an analytical technique to determine the elemental composition or chemical characterization of a sample [146]. EDX works on the basis that every element shows a unique set of peaks on its X-ray spectrum corresponding to its unique atomic structure. A high-energy incident beam excites an electron in an inner shell, and it will be ejected from the shell creating a hole in its place. Another electron from a higher energy outer shell will fill this hole and the energy difference between the shells will be released in the form of X-rays which will be characteristic of the atomic structure of the emitting element [146]. EDX studies on biopolymer systems help in determining the elemental composition, impurities, chemical modifications, and functionalization of the samples [147–150]. EDX analysis can be very helpful in analyzing biopolymers with chemical modification as this enables researchers to determine the presence of heteroatoms (chlorine or sulphur) used in the modification [151]. EDX assists by providing the elemental composition of the analyzed material and EDX mapping can provide elemental mapping in the analyzed sample. In addition to providing the elemental composition and content, it helps in the analysis of dispersion and distribution of compounds within the sample. For example, EDX mapping can be helpful in determining the distribution of carboxymethyl chitosan and calcium alginate in a composite by analyzing the N element content of chitosan and Ca element of calcium alginate (Figure 9) [152].

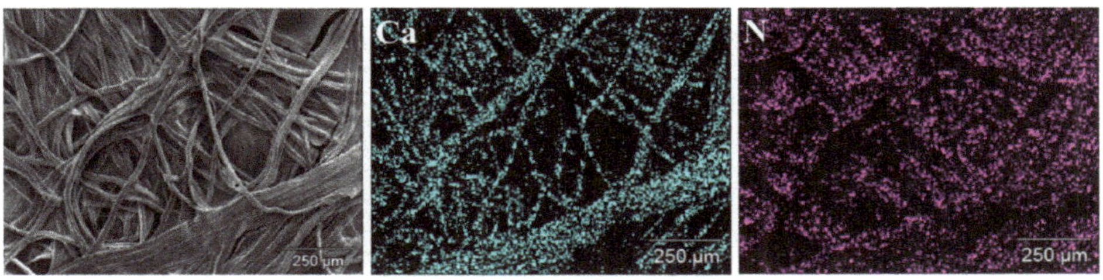

Figure 9. Energy-dispersive X-ray (EDX) mapping images of carboxymethyl chitosan and calcium alginate composite dressings [152]. (Reprinted with permission from Gao et al., 2019.)

3.3. Transmission Electron Microscopy

A transmission electron microscope (TEM) is one of the most powerful microscopes available today which is used as an analytical tool to analyze and visualize the samples in the realm of a nanoscale [153,154]. TEM and light microscopes both operate on the same basic principles; however, the major difference between them is that TEM uses electrons instead of light. Since the wavelengths of electrons are much smaller when compared to that of light, images obtained from TEM have much higher optical resolution than those of a light microscope. This enables TEM to reveal the tiniest details of the sample, sometimes as small as individual atoms. The name 'transmission' meaning, 'to pass through' is essentially how the TEM operates, by passing a beam of high-energy electrons through very thin samples [155]. The electron beam may not pass through some sections of the sample and can be deflected by some sections. The electrons transmitted through the samples are collected from below through a camera or onto a phosphorescent screen to obtain the images. The dark part of the image represents the areas in the sample through which very few or no electrons are transmitted, the brighter part represents the areas through which more electrons are transmitted, and a range of grey color patterns are obtained depending on the way the electrons interact with the sample [153,154]. The most tedious part of TEM analysis is sample preparation, which requires a lot of time as the samples should be very thin in order to transmit sufficient electrons through with minimum energy loss [155]. TEM can be used to obtain information related to particle size, shape, distribution, crystallinity, crystallite size structure, and orientation [156–158].

TEM is a well-established analytical technique in many fields including but not limited to nanotechnology, material science, medical research, biomedical, biological, and semiconductor research.

TEM has found extensive use in biopolymer research over the past decades and has helped reveal information at the atomic scale. TEMs have been known to provide information on the topological, morphological, compositional, and crystalline information of biopolymer systems. In biopolymer nanocomposites, TEM is highly recommended when there is a need for in-depth study to understand the quality of the internal structure, spatial distribution of different phases, and defect structure of nanocomposites. It has been used in studying the dispersion of nanoparticles within the matrix. As can be seen from Figure 10, the TEM images clearly show the nanoparticle dispersion on cellulose nanofiber. Further magnification allowed for the confirmation of the metal nanoparticle attachment onto fiber surface and assisted in calculating the average particle size [74].

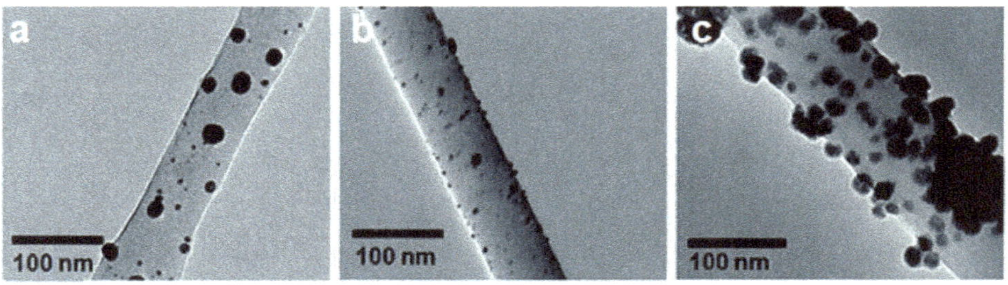

Figure 10. Transmission electron microscope (TEM) image of (**a**) Ag, (**b**) Au, and (**c**) Ni nanoparticles on cellulose nanofiber surface [74]. (Reprinted with permission from Gopiraman et al., 2018.)

TEM has been an invaluable asset in characterizing core-shell and hollow nanoparticles and nanofibers by observing the contrast created by the electron beam diffraction, which represents the distinctive phases present in the analyzed sample. TEM images of core-shell structured polyethylene oxide (PEO)-chitosan electrospun fibers depicted in Figure 11 reveal two distinct phases [75], where the dark region represents the core while the bright region represents the shell structure. Further, TEM assists in the identification of different concentric and eccentric structures as well as in the measurement of diameters of the core and the shell components of the fiber. Additionally, the fibers once washed with water to remove the PEO core and the TEM analysis of the ensuing hollow structure can confirm the complete removal of the PEO core while the chitosan shell maintains its physical structure [75]. High-resolution transmission electron microscopy (HRTEM) is a variant of TEM with resolutions below 0.5 Å [159] which facilitates the imaging of specimens at an atomic scale and enables the analysis of the atomic structure of the samples [159]. HRTEM has been an invaluable asset in the analysis of biopolymer-assisted formation of nanoparticles including the study of their crystal planes [160–162], the crystal structure of cellulose [163], nanocomposites [164,165], and even in the molecular orientations of biopolymers [166].

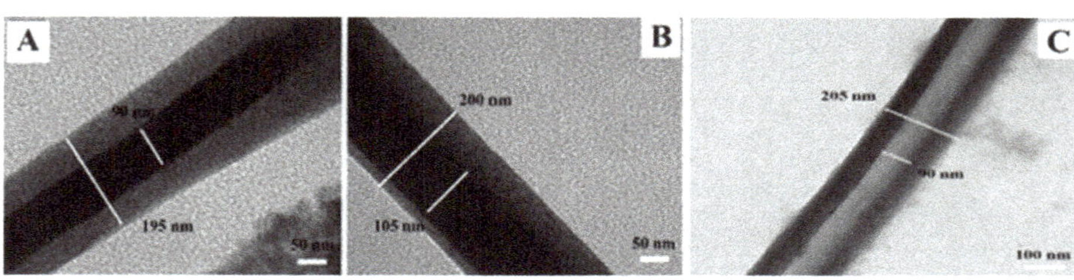

Figure 11. TEM images of PEO-chitosan biocomponent electrospun fibers showing (**A**) concentric core-shell structure, (**B**) eccentric core-shell structure, and (**C**) hollow chitosan shell structure [75]. (Reprinted with permission from Pakravan et al., 2012.)

Conventional TEM utilizes electron beams in high vacuum conditions, but certain materials are not compatible with these high vacuum conditions. In such cases, a specialized variant of TEM called cryo-TEM is employed for the analysis of the samples [167,168] where samples are often frozen by rapidly plunging into a liquid ethane bath for preservation purposes. The solvent around the specimen will be frozen upon dipping in the cold medium thus ensuring the cryogenic preservation of the specimen. Water or salt solutions are commonly used as solvents to ensure the stability of the samples. However, care must be taken to expedite the process of freezing the sample very quickly, as this will prevent the formation of cubic ice, which interferes by absorbing the electron beam and obscuring the sample. There are many advantages associated with cryo-TEM including the analysis of the sample in vitreous ice, which actually preserves the structural information largely and reflects the state of the sample prior to freezing. In addition, since the sample is not exposed to any adhering surface, the shape observed is the true shape of the sample, which is not affected by any kind of attachments that might result in flattening. Cryo-TEM is also useful to analyze nanoparticle suspensions, which prevents any changes to the sample induced by drying or staining such as agglomeration, degradation, deformation. Several researchers have utilized cryo-TEM to study cellulose nanocrystals in its native state, as they tend to agglomerate upon drying [169,170]. Scanning transmission electron microscopy (STEM) is a technique which combines the operational modes of both SEM and TEM [171] and operates by focusing a convergent electron beam onto a small area of the sample. To obtain an image, the electron probe raster scans thereafter propagate through the sample [172,173]. STEM imaging could be done by using secondary or backscattered electrons, however, superior spatial resolution and better signal levels are obtained by recording the transmitted electrons [172,173]. Several detectors are employed in the STEM and each of them provides a distinct complementary view of the sample. Electrons transmitted through the sample in a path close to the optical axis are collected by a bright field detector so the holes appear bright, whereas an annular dark field or a high-angle annular dark field detector (HAADF) collect the scattered electrons excluding the transmitted beam so the holes appear dark [172,173]. This ensures the recording of maximum possible details about the sample from each scan. STEM has been used in the analysis of biopolymers and biopolymer nanocomposites. The bright field and dark field imaging of the STEM has assisted in the determination of particle size, shape, filler dispersion and agglomeration, and biopolymer coatings as well [174–183]. The HAADF-STEM images obtained for metal oxide nanoparticles stabilized in carboxymethyl-cellulose (CMC)/cetyltrimethyl-ammonium-bromide (CTAB) templates have assisted authors to distinguish between the two phases [179]. The metal oxide nanoparticles, seen as brighter zones, are found to be embedded in the darker zones of CMC/CTAB templates. Furthermore, the bright field and HAADF images have assisted in determining experimental parameters controlling the size and morphology of the nanostructures [179].

Selected area electron diffraction (SAED) is a crystallographic analysis technique often accompanied by TEM. The majority of the TEM equipment, having a parallel electron beam source, are capable of carrying out the SAED analysis, which provides diffraction patterns as a result of the electron beam scattering by the sample lattice. The electrons elastically scattered by the lattice obey Bragg's law and hence it is possible to index the diffraction spots in the pattern and in turn identify the phases and analyze the sample structure. SAED aperture has the capability to select and analyze 0.5–1 μm length of the sample; therefore, it is possible to selectively analyze the sample. This is helpful while analyzing polycrystalline samples because when more than one crystal contributes to the diffraction pattern it can be difficult to analyze the diffractogram. One can obtain information regarding the structure of the sample including the crystalline symmetry, crystal defects, unit cell parameter, and texture of the sample. SAED has been used in the analysis of biopolymers to obtain information regarding their crystalline structure [184–187]. However, SAED is proven to be a more prominent technique in the analysis of the biopolymer-assisted synthesis of metallic nanoparticles [22,23,188–191]; SAED analysis of biopolymer gum guar capped silver nanoparticles revealed information about the crystalline structure of the nanoparticles. The diffraction pattern (Figure 12) consisted of concentric rings with bright dots, which suggests that the synthesized particles are highly crystalline in nature; the rings were further assigned to (111), (200), (220), and (311) planes of the face-centered cubic structure of the particles [192].

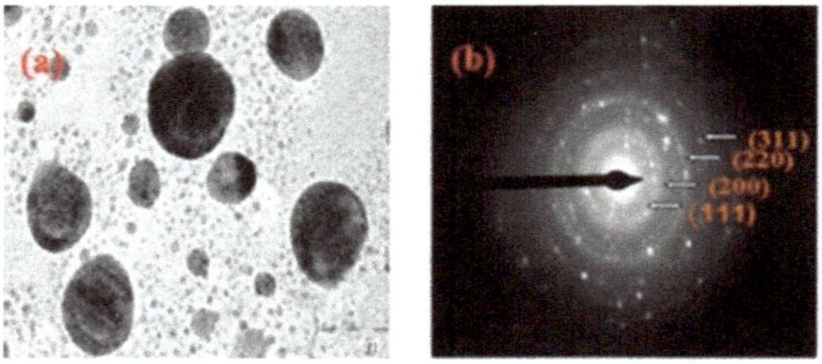

Figure 12. TEM micrograph of (**a**) guar gum capped silver nanoparticles; (**b**) selected area electron diffraction (SAED) pattern [192]. (Reprinted with permission from Vanamudan et al., 2016.)

3.4. Scanning Probe Microscopy

Scanning probe microscopy (SPM) encompasses a family of several unique techniques, which are capable of imaging not only the nanoscale structures and surfaces but atoms as well [193]. The main distinguishing feature that sets apart the scanning probe microscopy from other techniques is that it uses no lenses but a probe to interact with the sample [193]. The probe moves over the sample surface and a computer gathers and analyses this data to create an image of the surface. The probe, which is as sharp as an atom, is usually mounted at the end of a cantilever and is moved over the surface precisely. During the scanning process, different forces, including electrostatic forces, magnetic forces, mechanical contact, Van der Waals interactions, chemical bonding, and capillary forces, can deflect the cantilever tip and the SPMs are capable of measuring these deflections. A laser, which is focused onto the cantilever, is reflected off its top onto a detector and measures the vertical deflection caused by any of the above-mentioned forces [194,195]. Multifarious interactions are studied and analyzed depending on the probe sensors being used and the interactions being measured between the probe and the sample surface determines the variant of SPM being used [194,195]. For example, atomic force microscopy (AFM) measures electrostatic forces, magnetic force microscopy (MFM) measures magnetic forces, chemical force microscopy (CFM) measures chemical interactions, while scanning tunneling microscopy (STM) measures electric current flowing between the probe and sample surface [196]. A wide variety of techniques fall under the family of SPMs, herein, the major techniques used in biopolymer field are discussed.

(a) Atomic Force Microscopy

Atomic force microscopy is a distinguished analytical tool with a capacity to determine surface structures with high spatial resolution [197,198]. An AFM operates on the basic principles of SPMs wherein it analyzes the sample surface by means of a very sharp tip, which is often a 3–6-micron long pyramid with a diameter of less than 100 Å. The tip is present at the free end of a 100–200 μm long cantilever, which undergoes deflection/bending due to the forces between the probe and the sample [199–201]. AFMs are capable of measuring these lateral or vertical deflections by means of an optical lever that reflects a laser beam off the cantilever [197]. A position-sensitive photodetector detects these deflections and allows a computer to generate an image of the surface topology of the sample being analyzed. Over the years, like any other technique, many variations have been developed for AFM to suit modern needs and they are applicable to all the samples, however, not all variants yield the desired quality results. Proper use of these variants enables one to study and analyze the samples at fundamental, even at the atomic level. The three most popular modes employed in AFM are contact mode, non-contact mode, and tapping mode.

In contact mode, as the name suggests, the tip is in constant contact with the sample. Contact mode further has two subclasses, namely constant height mode and constant force mode. In constant height mode, the height of the scanner is fixed during the scanning process and the spatial variations of the cantilever tip deflection are used to generate the topographic image. This technique is used when measuring samples with atomically flat surfaces to obtain atomic-scale images and is capable of quick scans [202]. While in the constant force mode, the cantilever deflection is kept constant and hence the force on the sample is constant. A feedback circuit is used to move the sample or the tip up and down to keep the deflection constant, and measuring the z-movement provides the changes in heights in the sample. As the tip will be in constant contact with the sample, there is a chance of destruction of samples due to frictional forces [203]. In the non-contact mode, the cantilever tip will not be in contact with the sample surface but will be hovering above it at a very close distance of 5–10 nm. In this mode, the cantilevers used will have high spring constants and have the advantage of having negligible frictional forces. This is very helpful in analyzing the soft samples without altering their surface characteristics [204]. A major drawback of this mode is that when compared to the contact mode it has very low lateral and z-resolution, which is circumvented by employing the tapping mode or intermittent contact mode, wherein, instead of hovering, the cantilever tip vibrates and scans above the surface and momentarily makes contact with the sample surface. The amplitude of vibrations decreases, and a phase shift occurs because of the interactions of the cantilever tip with the surface of the sample. A feedback circuit can be used to move the sample or the tip in the z-direction and to keep the amplitude or the phase shift constant. This mode requires stiffer cantilevers with very small damping factors. Due to its high-resolution imaging and close to non-destructive nature, as well as its applicability in liquid and air conditions, amplitude modulation mode is the most popular mode of AFM [205].

AFM has been deployed in the field of biopolymers for quite some time where it has been used in the analysis of complex structures formed by the association of biopolymers. AFM has been a powerful technique in the analysis and understanding of molecular interactions, nature, and strength of surface forces in the biopolymer systems. Previously, when AFM was first introduced in the biopolymer field, the imaging was carried out in air by depositing biopolymers on mica substrates. This technique works well for stiffer samples [206,207]; however, for soft samples, the results obtained by this technique were unreliable due to the interference of residual water present on the tip and the sample surface [208]. This residual water resulted in adhesive forces causing damage or displacement of molecules and was overcome by employing contact mode imaging in liquids [209], tapping mode [210], and non-contact imaging [211] modes. AFM has been used in the analysis of single molecule [76], conformations [78], local structure and elasticity of gels [212], gelation mechanisms [213], and nanoparticle topography [79,80]. Quantitative information including the contour length and the molecular heights has been measured accurately using high-resolution AFM images. Studies on the single chitosan strands deposited on mica surface revealed the contour lengths of chitosan strands around 94–178 nm with an average molecular height of 0.45 ± 0.04 nm [214]. Similar studies have been conducted to determine the size of a single cellulose nanocrystal; AFM has provided an efficient way to evaluate the length, width, and aspect ratios of these individual crystals (Figure 13) [81].

The difference in the molecular structure of the biopolymers has also been determined by the AFM technique; the molecular structure of xanthan biopolymer produced by several different strains of Xanthomonas campestris was studied by AFM [77].

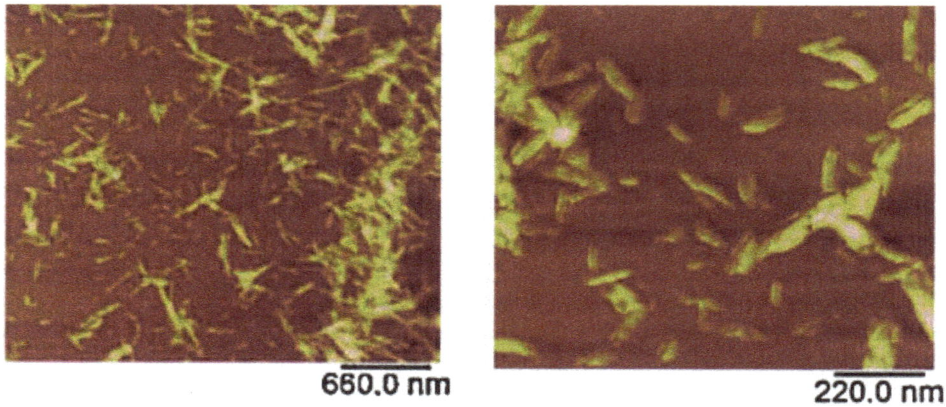

Figure 13. Atomic force microscopy (AFM) images of cellulose nanocrystals at **a**) low and **b**) high magnification. The length, width, and aspect ratio were 90–180 nm, 4–6 nm, and 26, respectively [81]. (Reprinted with permission from Nagalakshmaiah et al., 2016.)

The AFM images depicted in Figure 14 revealed different structural features for different strains. Xanthan from the wild-type strain Xanthomonas campestris B100 showed branching and overlapping sections while the one from the strain JBL007 showed no branching. The acetate-free xanthan shows no branching with single and double-stranded areas and the pyruvate-free xanthan show a homogenous and branched structure with single and double strands (Figure 14).

(b) Chemical Force Microscopy

Chemical force microscopy (CFM) is a form of AFM, which relies on the modification of probe tips by functionalization to introduce chemical specificity into the measurements [215]. It is a powerful technique for the quantitative analysis of intermolecular interactions between distinct chemical functionalities [215,216]. CFM detects chemical bonding forces between functional groups on the surface of a specimen by attaching ligands on to the probe tip [215,216]. The principle involves bringing a chemically modified tip in contact with the sample surface and measuring the resulting attractive or repulsive forces as the tip is approached or withdrawn from the sample [217,218]. The measured forces are then mapped and compared to the surface structures detected by the topological imaging for further analysis. These measurements are proven to be helpful in identification purposes, to determine compatibility between two materials, and to predict interactions between materials [219]. CFM has proven its use in the analysis of biomaterial and biopolymer fields, and has been used in the characterization of cellulose [220], chitosan [82], and DNA [221]. Lee et al. analyzed the cellulose in biomass samples through CFM by functionalizing the silicon nitride cantilever probes with triethoxysilyl *N*-propyl gluconamide [222]. AFM images obtained by tapping mode, topographic force-volume mode and their corresponding adhesion force measurements, along with the false color scheme for measuring adhesion force strengths are depicted in Figure 15. AFM images of switch-grass cellulose obtained by tapping mode showed micro-fibrils over the entire sample and the images obtained by topographic force-volume mode with lower spatial resolution showed parallel microfibril bundles. The corresponding adhesion force-mapping image revealed a uniform adhesion force throughout the sample. In the case of extractives-free switch-grass cellulose, the tapping mode images revealed a heterogeneous structure with aligned fibrillary structure along with less defined globular masses embedded in the hemicellulose and lignin matrix. The topographic force-volume mode images showed only general contours; however, the adhesion force image shows areas with adhesion force similar to that observed for cellulose microfibrils bound by regions of low adhesion probably the lignin and hemicellulose as observed in tapping mode [222]. All this information obtained from CFM suggests that it is an invaluable asset in the microscopical analysis of biopolymers.

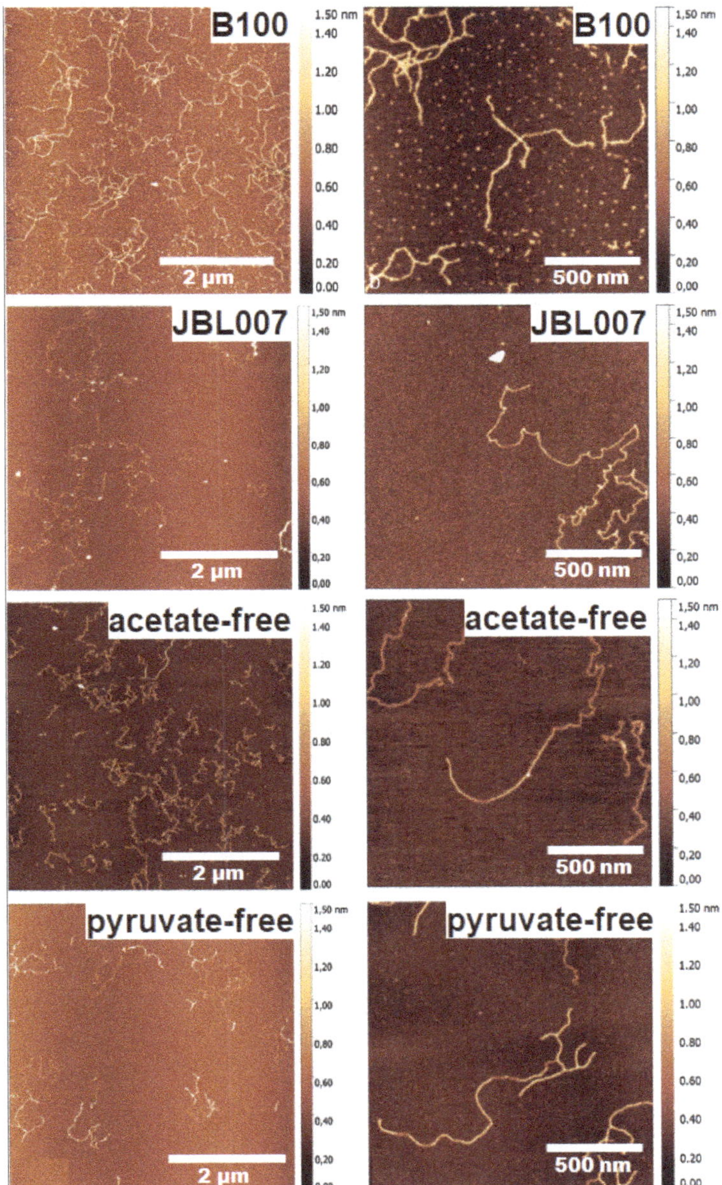

Figure 14. AFM topographical images of xanthan from different sources. Images left provide the overview while right show the detailed structure of the samples [77] (Reprinted with permission from Teckentrup et al., 2017.)

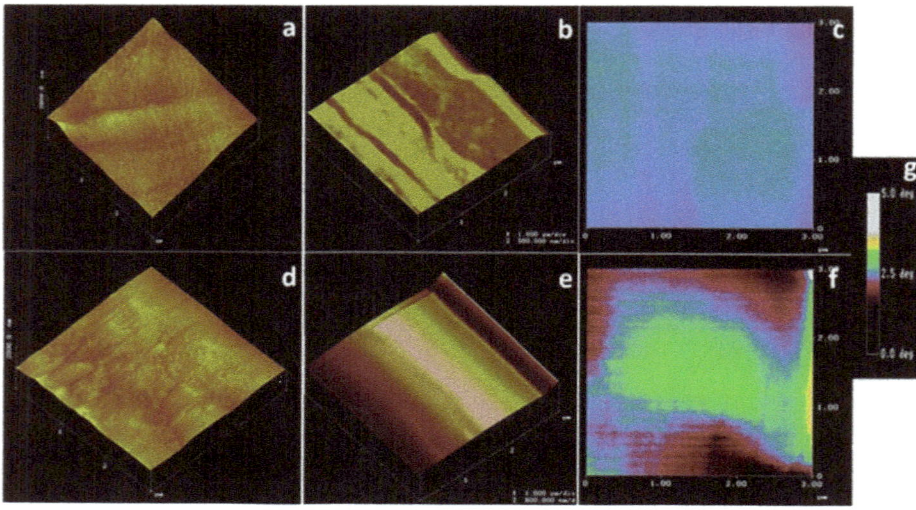

Figure 15. (**a**) Tapping mode image of cellulose isolated from switch-grass; (**b**) topographic image of switch-grass cellulose obtained in force-volume mode; (**c**) adhesion force image of switch-grass cellulose obtained in force-volume mode; (**d**) tapping mode image of extractives-free switch-grass; (**e**) topographic image of extractives-free switch-grass obtained in force-volume mode; (**f**) adhesion force image of extractives-free switch-grass obtained in force-volume mode; (**g**) a false color scheme based on the calibrations used for the depiction of the adhesion forces [222]. (Reprinted with permission from Lee et al., 2015.)

(c) Magnetic Force Microscopy

Magnetic force microscopy (MFM) is another variant of AFM, which is capable of analyzing surfaces with magnetic properties at the nanoscale [223,224]. In MFM, the surface of the sample is scanned by means of a standard tip coated with a ferromagnetic film of a few nanometers thickness [223,224]. The operating mechanism of MFM is very similar to that of AFM with both static and dynamic modes of analysis available, but the dynamic mode is more popular as it offers higher sensitivity. While scanning the specimen with the tip of the cantilever, the magnetic forces between the sample and the tip cause the cantilever to bend and the oscillations are recorded. However, while scanning the specimen, along with the magnetic forces the Van der Waals forces will also be present. This can be controlled by regulating the distance of the tip from the sample as the effect of Van der Waals forces wear off at longer tip-to-sample distances while the magnetic forces persist [225]. Van der Waals forces are used to obtain the topographical image of the surface as the forces vary with the tip-to-sample distance. In general, for successful MFM imaging, the sample is first scanned in the close range where the Van der Waals forces are dominant to acquire a topographical image and then the tip is lifted to a region where the magnetic forces are dominant and scanned for MFM image. This technique is advantageous as it minimizes the effects caused by non-magnetic forces and ensures only the record of magnetic forces [226,227].

The use of MFM in biopolymer systems is marginally explored and is limited to the analysis of magnetic biocomposites [83,228–231]. The magnetic properties of the magnetic nanoparticles loaded biopolymeric systems can be evaluated by means of MFM. Since the biopolymers have no response to the external magnetic field (Figure 16) [228], MFM provides information regarding the dispersion and encapsulation of magnetic nanoparticles inside the biopolymer matrix [83]. The aggregations of magnetic nanoparticles can also be determined by measuring the variation in frequency shifts vs. The position of the tip [228].

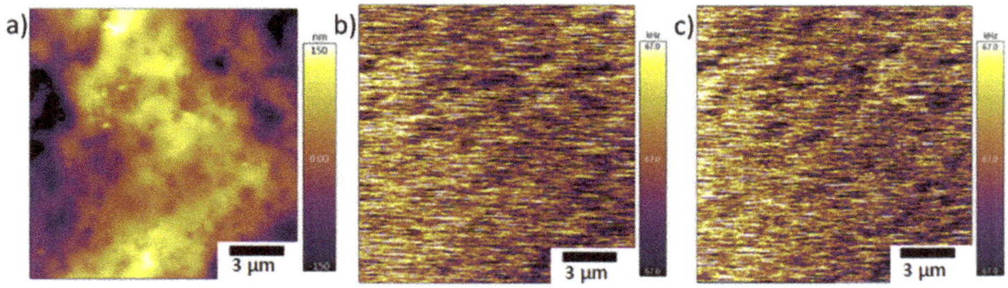

Figure 16. Magnetic force microscopy (MFM) images of pure gelatin films (**a**) topographical image; (**b**) frequency image at 0 Tl (**c**) frequency image at 0.47 T; both frequency images are identical indicating no response to external magnetic field [228]. (Reprinted with permission from Marín et al., 2018.)

3.5. Scanning Tunneling Microscopy

A scanning tunneling microscope (STM) is a non-optical microscope, which consists of an electric probe tip that is used to scan over a sample surface at a constant spacing. STM operates on the principle of quantum mechanical phenomenon called tunneling, which occurs when the wave-like properties of electrons allow them to pass through a barrier that, in general, they should not be able to pass through. The effect of tunneling reduces as the gap between the two surfaces increases. To analyze a sample by STM, it must be capable of conducting electricity. The extreme end of the tip is very sharp, down to a single atom, and a voltage is applied between the tip and the sample surface, resulting in the tunneling of electrons. As the electrons begin tunneling, a current starts flowing and this current can be measured. When the probe is moved over the surface of the sample, variations in the tunneling current corresponding to the surface structural changes are observed. A feedback circuit is employed to monitor and make necessary changes to maintain a constant tunneling current, which is recorded and processed by a computer to provide a topological image of the sample surface. There are two modes in STM—constant current mode and constant height mode. In the constant current mode, a feedback loop adjusts the height to keep the current constant and the image of the sample surface is generated by recording the vertical position of the tip. In the constant height mode, the vertical position of the tip remains unchanged and the change in current as a function of the position is recorded to obtain the topographical image.

STM has a high resolution of 1 Å in the sample plane and as high as 0.1 Å in the vertical plane, and is capable of imaging at ambient pressures and in liquids with minimum specimen damage. Despite these advantages, the use of STM in the analysis of the biopolymer field is limited by poor conductivity, instability, flexible elasticity of biopolymers. However, these drawbacks have been overcome by imaging carbon cast replicas [232], STM related AFM [233], and by conductive coating [234,235]. The use of STM in the biopolymeric field is very limited and has been used in the analysis of the molecular structure of bacterial polysaccharides like gellan gum and xanthan gum, deposited on highly oriented pyrolytic graphite (HOPG). The STM images of xanthan gum deposited from a high concentration aqueous solution onto HOPG revealed stiff aligned rod-like molecules [84]. In the case of gellan gum, the double-helical structure forming into cation-mediated aggregates was observed in the STM images [85], suggesting that STM has the ability to analyze intricate details related to the molecular structure of biopolymers. STM has been further used in the studies of cellulose crystalline fibrils [86], surface modification of methylcellulose [87], biocomposites [236], collagen [237], and other polysaccharides [238]. Abdullah et al. studied the surface modification of methylcellulose/cobalt nitrate polymer electrolytes by H_2S, and the STM analysis of samples revealed an increase in surface roughness of the samples post H_2S treatment (Figure 17) [87]; maximum height and root mean square deviation of the roughness profile were increased to 4.296 and 0.752 pA from 3.734 and 0.705 pA, respectively. This increase in surface roughness was attributed to the formation of CoS nanoparticles, which were dispersed homogenously throughout the film as indicated by the relatively low value of root mean square of the gas treated samples.

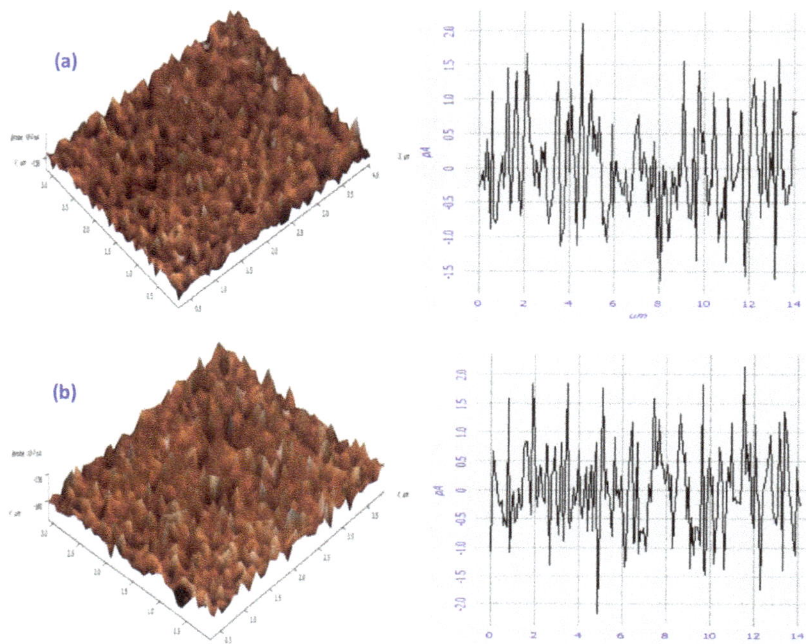

Figure 17. Scanning tunneling microscopy (STM) images and height profiles of methylcellulose-based polymer electrolyte (**a**) before H_2S treatment, (**b**) after H_2S treatment [87]. (Reprinted with permission from Abdullah et al., 2018.)

4. Conclusions

The emergence of new biopolymer-derived materials creates a need to gain insights into their complex structure and morphology. The influence of the micro and nanostructures of biopolymers on the macroscopic scale dictates the need for their study as these structural and morphological variations influence the physical and mechanical properties of the final material. In order to understand new material, it is crucial to correlate the structural observations with the intrinsic material properties as microscopic analysis can obtain images revealing the intricate details of the microstructures present within the samples. These microscopic techniques are of paramount importance in the analysis of fundamental structures, morphology, surface properties, molecular structure, microstructure, chemical composition, topography, and interfaces. They also provide information related to the dispersion, distribution, intercalation, exfoliation, and aggregation of nanoparticles in the composites. However, care should be taken while interpreting these images. Materials' intrinsic nature, analysis technique, sample preparation methods can have a significant influence on the final structures revealed in the images. Biopolymers are often sensitive to the electron beams employed in the electron microscopes and can undergo changes or impairments. If these changes are not considered while interpreting the images obtained, the results may not be reliable. Despite providing a plethora of information, sometimes the microscopic techniques can yield ambiguous results. Hence, more often different microscopic techniques are used in conjunction with each other and with other characterization techniques to obtain 'in-depth' knowledge and understanding of the material. The advent and development of microscopic techniques such as aberration-corrected electron microscopy and 3D electron microscopy have opened new horizons in the material understanding. These microscopic techniques have served as an indispensable tool in the development and commercialization of biopolymers.

Author Contributions: Conceptualization, A.V. M.N.; writing—original draft preparation, A.V.; writing—review and editing, M.N. S.W. V.V.T.P., R.S.V; supervision, V.V.T.P., R.S.V.; project administration, S.W; V.V.T.P; M.Č.; funding acquisition, V.V.T; M.Č. All authors have read and agreed to the published version of the manuscript.

Funding: The authors acknowledge the support rendered by the Ministry of Education, Youth and Sports in the framework of the targeted support of the OPR&DI project 'Extension of CxI facilities' (CZ.1.05/2.1.00/19.0386) and the Research Infrastructure NanoEnviCz (Project No. LM2015073). These authors also acknowledge the assistance provided by the projects Pro-NanoEnviCz (Reg. No. CZ.02.1.01/0.0/0.0/16_013/0001821) and Project Hybrid Materials for Hierarchical Structures (HyHi), Reg. No. CZ.02.1.01/0.0/0.0/16_019/0000843) supported by the Ministry of Education, Youth and Sports of the Czech Republic and the European Union–European Structural and Investment Funds in the frames of Operational Program Research, Development and Education. This work was also supported by the project "Tree Gum Polymers and their Modified Bioplastics for Food Packaging Application" granted by Bavarian-Czech-Academic-Agency (BTHA) (registration numbers LTAB19007 and BTHA-JC-2019-26) and the Ministry of Education, Youth and Sports in the Czech Republic under the "Inter Excellence–Action programme" within the framework of project "Bio-based Porous 2D Membranes and 3D Sponges Based on Functionalized Tree Gum Polysaccharides and their Environmental Application" (registration number LTAUSA19091)-TUL internal No.: 18309/136.

Conflicts of Interest: The authors declare no conflict of interest.

References

1. Bassas-Galià, M. Rediscovering Biopolymers. In *Consequences of Microbial Interactions with Hydrocarbons, Oils, and Lipids: Production of Fuels and Chemicals*; Springer International Publishing: Cham, Switzerland, 2017; pp. 529–550.
2. Hernández, N.; Williams, R.C.; Cochran, E.W. The battle for the "green" polymer. Different approaches for biopolymer synthesis: Bioadvantaged vs. bioreplacement. *Org. Biomol. Chem.* **2014**, *12*, 2834–2849. [CrossRef] [PubMed]
3. Cacciotti, I.; Mori, S.; Cherubini, V.; Nanni, F. Eco-sustainable systems based on poly(lactic acid), diatomite and coffee grounds extract for food packaging. *Int. J. Biol. Macromol.* **2018**, *112*, 567–575. [CrossRef] [PubMed]
4. Garavand, F.; Rouhi, M.; Razavi, S.H.; Cacciotti, I.; Mohammadi, R. Improving the integrity of natural biopolymer films used in food packaging by crosslinking approach: A review. *Int. J. Biol. Macromol.* **2017**, *104*, 687–707. [CrossRef]
5. Venkateshaiah, A.; Cheong, J.Y.; Habel, C.; Wacławek, S.; Lederer, T.; Černík, M.; Kim, I.-D.; Padil, V.V.T.; Agarwal, S. Tree Gum–Graphene Oxide Nanocomposite Films as Gas Barriers. *ACS Appl. Nano Mater.* **2020**, *3*, 633–640. [CrossRef]
6. Babu, R.P.; O'Connor, K.; Seeram, R. Current progress on bio-based polymers and their future trends. *Prog. Biomater.* **2013**, *2*, 8. [CrossRef]
7. He, M.; Lu, A.; Zhang, L. *Advances in Cellulose Hydrophobicity Improvement*; American Chemical Society: Washington, DC, USA, 2014.
8. Sajid, M.A.; Shahzad, S.A.; Hussain, F.; Skene, W.G.; Khan, Z.A.; Yar, M. Synthetic modifications of chitin and chitosan as multipurpose biopolymers: A review. *Synth. Commun.* **2018**, *48*, 1893–1908. [CrossRef]
9. Abraham, A.; Soloman, P.A.; Rejini, V.O. Preparation of Chitosan-Polyvinyl Alcohol Blends and Studies on Thermal and Mechanical Properties. *Procedia Technol.* **2016**, *24*, 741–748. [CrossRef]
10. Raj, A.; Prashantha, K.; Samuel, C. Compatibility in biobased poly(L-lactide)/polyamide binary blends: From melt-state interfacial tensions to (thermo)mechanical properties. *J. Appl. Polym. Sci.* **2020**, *137*, 48440. [CrossRef]
11. Resano-Goizueta, I.; Ashokan, B.K.; Trezza, T.A.; Padua, G.W. Effect of Nano-Fillers on Tensile Properties of Biopolymer Films. *J. Polym. Environ.* **2018**, *26*, 3817–3823. [CrossRef]
12. Agusnar, H.; Wirjosentono, B.; Rihayat, T. Improving the quality of biopolymer (poly lactic acid) with the addition of bentonite as filler. *IOP Conf. Ser. Mater. Sci. Eng.* **2017**, *222*, 012008.
13. Mekonnen, T.; Mussone, P.; Khalil, H.; Bressler, D. Progress in bio-based plastics and plasticizing modifications. *J. Mater. Chem. A* **2013**, *1*, 13379–13398. [CrossRef]

14. Dufresne, A.; Thomas, S.; Pothen, L.A. *Biopolymer Nanocomposites*; John Wiley & Sons, Inc.: Hoboken, NJ, USA, 2013.
15. Okamoto, M.; John, B. Synthetic biopolymer nanocomposites for tissue engineering scaffolds. *Prog. Polym. Sci.* **2013**, *38*, 1487–1503. [CrossRef]
16. Jacob, J.; Haponiuk, J.T.; Thomas, S.; Gopi, S. Biopolymer based nanomaterials in drug delivery systems: A review. *Mater. Today Chem.* **2018**, *9*, 43–55. [CrossRef]
17. Joye, I.J.; McClements, D.J. Biopolymer-based nanoparticles and microparticles: Fabrication, characterization, and application. *Curr. Opin. Colloid Interface Sci.* **2014**, *19*, 417–427. [CrossRef]
18. Gupta, B.; Tummalapalli, M.; Deopura, B.L.; Alam, M.S. Preparation and characterization of in-situ crosslinked pectin–gelatin hydrogels. *Carbohydr. Polym.* **2014**, *106*, 312–318. [CrossRef] [PubMed]
19. Padil, V.V.T.; Senan, C.; Wacławek, S.; Černík, M. Electrospun fibers based on Arabic, karaya and kondagogu gums. *Int. J. Biol. Macromol.* **2016**, *91*, 299–309. [CrossRef]
20. Silvestri, D.; Mikšíček, J.; Wacławek, S.; Torres-Mendieta, R.; Padil, V.V.; Černík, M. Production of electrospun nanofibers based on graphene oxide/gum Arabic. *Int. J. Biol. Macromol.* **2019**, *124*, 396–402. [CrossRef]
21. Vinod, V.T.P.; Saravanan, P.; Sreedhar, B.; Devi, D.K.; Sashidhar, R.B. A facile synthesis and characterization of Ag, Au and Pt nanoparticles using a natural hydrocolloid gum kondagogu (*Cochlospermum gossypium*). *Colloids Surf. B Biointerfaces* **2011**, *83*, 291–298. [CrossRef]
22. Thekkae Padil, V.V.; Černík, M. Green synthesis of copper oxide nanoparticles using gum karaya as a biotemplate and their antibacterial application. *Int. J. Nanomed.* **2013**, *8*, 889–898.
23. Silvestri, D.; Wacławek, S.; Sobel, B.; Torres-Mendieta, R.; Novotný, V.; Nguyen, N.H.; Ševců, A.; Padil, V.V.; Müllerová, J.; Stuchlík, M.; et al. A poly(3-hydroxybutyrate)–chitosan polymer conjugate for the synthesis of safer gold nanoparticles and their applications. *Green Chem.* **2018**, *20*, 4975–4982. [CrossRef]
24. Venkateshaiah, A.; Silvestri, D.; Ramakrishnan, R.K.; Wacławek, S.; Padil, V.V.; Černík, M.; Varma, R.S. Gum kondagoagu/reduced graphene oxide framed platinum nanoparticles and their catalytic role. *Molecules* **2019**, *24*, 3643. [CrossRef] [PubMed]
25. Padil, V.V.T.; Wacławek, S.; Černík, M.; Varma, R.S. Tree gum-based renewable materials: Sustainable applications in nanotechnology, biomedical and environmental fields. *Biotechnol. Adv.* **2018**, *36*, 1984–2016. [CrossRef] [PubMed]
26. Mülhaupt, R. Green Polymer Chemistry and Bio-based Plastics: Dreams and Reality. *Macromol. Chem. Phys.* **2013**, *214*, 159–174. [CrossRef]
27. Hu, B. Biopolymer-Based Lightweight Materials for Packaging Applications. *ACS Symposium Series, Am. Chem. Soc.* **2014**, *1175*, 239–255.
28. Yadav, P.; Yadav, H.; Shah, V.G.; Shah, G.; Dhaka, G. Biomedical biopolymers, their origin and evolution in biomedical sciences: A systematic review. *J. Clin. Diagn. Res.* **2015**, *9*, 21–25. [CrossRef]
29. Cao, Y.; Wu, J.; Zhang, J.; Li, H.; Zhang, Y.; He, J. Room temperature ionic liquids (RTILs): A new and versatile platform for cellulose processing and derivatization. *Chem. Eng. J.* **2009**, *147*, 13–21. [CrossRef]
30. Rajinipriya, M.; Nagalakshmaiah, M.; Astruc, J.; Robert, M.; Elkoun, S. Single stage purification of flax, hemp, and milkweed stem and their physical and morphological properties. *Int. J. Polym. Anal. Charact.* **2018**, *23*, 78–88. [CrossRef]
31. Heinze, T.; El Seoud, O.A.; Koschella, A. *Production and Characteristics of Cellulose from Different Sources*; Springer: Cham, Switzerland, 2018.
32. Gellerstedt, G. Softwood kraft lignin: Raw material for the future. *Ind. Crops Prod.* **2015**, *77*, 845–854. [CrossRef]
33. Iravani, S.; Varma, R.S. Greener synthesis of lignin nanoparticles and their applications. *Green Chem.* **2020**, *22*, 612–636.
34. Agrawal, A.; Kaushik, N.; Biswas, S. Derivatives and Applications of Lignin: An Insight. *Scitech J.* **2014**, *1*, 30–36.
35. Bertoft, E. Understanding Starch Structure: Recent Progress. *Agronomy* **2017**, *7*, 56. [CrossRef]
36. Ogunsona, E.; Ojogbo, E.; Mekonnen, T. Advanced material applications of starch and its derivatives. *Eur. Polym. J.* **2018**, *108*, 570–581. [CrossRef]
37. Stamov, D.R.; Pompe, T. Structure and function of ECM-inspired composite collagen type i scaffolds. *Soft Matter* **2012**, *8*, 10200–10212. [CrossRef]

38. Sionkowska, A.; Skrzyński, S.; Śmiechowski, K.; Kołodziejczak, A. The review of versatile application of collagen. *Polym. Adv. Technol.* **2017**, *28*, 4–9. [CrossRef]
39. Silvestri, D.; Wacławek, S.; KRamakrishnan, R.; Venkateshaiah, A.; Krawczyk, K.; Padil, V.V.; Sobel, B.; Černík, M. The Use of a Biopolymer Conjugate for an Eco-Friendly One-Pot Synthesis of Palladium-Platinum Alloys. *Polymers Basel* **2019**, *11*, 1948. [CrossRef]
40. Crini, G. Historical review on chitin and chitosan biopolymers. *Environ. Chem. Lett.* **2019**, *17*, 1623–1643. [CrossRef]
41. Morin-Crini, N.; Lichtfouse, E.; Torri, G.; Crini, G. Fundamentals and Applications of Chitosan. In *Sustainable Agriculture Reviews*; Springer: Cham, Switzerland, 2019.
42. Zhao, D.; Yu, S.; Sun, B.; Gao, S.; Guo, S.; Zhao, K. Biomedical applications of chitosan and its derivative nanoparticles. *Polymers Basel* **2018**, *10*, 462. [CrossRef]
43. Shariatinia, Z. Pharmaceutical applications of chitosan. *Adv. Colloid Interface Sci.* **2019**, *263*, 131–194. [CrossRef]
44. Vahedikia, N.; Garavand, F.; Tajeddin, B.; Cacciotti, I.; Jafari, S.M.; Omidi, T.; Zahedi, Z. Biodegradable zein film composites reinforced with chitosan nanoparticles and cinnamon essential oil: Physical, mechanical, structural and antimicrobial attributes. *Colloids Surf. B Biointerfaces* **2019**, *177*, 25–32. [CrossRef]
45. Cacciotti, I.; Lombardelli, C.; Benucci, I.; Esti, M. Clay/chitosan biocomposite systems as novel green carriers for covalent immobilization of food enzymes. *J. Mater. Res. Technol.* **2019**, *8*, 3644–3652. [CrossRef]
46. Li, Z.; Yang, J.; Loh, X.J. Polyhydroxyalkanoates: Opening doors for a sustainable future. *NPG Asia Mater.* **2016**, *8*, e265. [CrossRef]
47. Winnacker, M. Polyhydroxyalkanoates: Recent Advances in Their Synthesis and Applications. *Eur. J. Lipid Sci. Technol.* **2019**, *121*, 1900101. [CrossRef]
48. Możejko-Ciesielska, J.; Kiewisz, R. Bacterial polyhydroxyalkanoates: Still fabulous? *Microbiol. Res.* **2016**, *192*, 271–282. [CrossRef] [PubMed]
49. Balaji, S.; Gopi, K.; Muthuvelan, B. A review on production of poly β hydroxybutyrates from cyanobacteria for the production of bio plastics. *Algal Res.* **2013**, *2*, 278–285. [CrossRef]
50. Ray, S.; Kalia, V.C. Biomedical Applications of Polyhydroxyalkanoates. *Indian J. Microbiol.* **2017**, *57*, 261–269. [CrossRef]
51. Yeo, J.C.C.; Muiruri, J.K.; Thitsartarn, W.; Li, Z.; He, C. Recent advances in the development of biodegradable PHB-based toughening materials: Approaches, advantages and applications. *Mater. Sci. Eng. C* **2018**, *92*, 1092–1116. [CrossRef]
52. Bianco, A.; Calderone, M.; Cacciotti, I. Electrospun PHBV/PEO co-solution blends: Microstructure, thermal and mechanical properties. *Mater. Sci. Eng. C* **2013**, *33*, 1067–1077. [CrossRef]
53. Cacciotti, I.; Calderone, M.; Bianco, A. Tailoring the properties of electrospun PHBV mats: Co-solution blending and selective removal of PEO. *Eur. Polym. J.* **2013**, *49*, 3210–3222. [CrossRef]
54. Mehta, R.; Kumar, V.; Bhunia, H.; Upadhyay, S.N. Synthesis of Poly(Lactic Acid): A Review. *J. Macromol. Sci. Part C Polym. Rev.* **2005**, *45*, 325–349. [CrossRef]
55. Singhvi, M.S.; Zinjarde, S.S.; Gokhale, D.V. Polylactic acid: Synthesis and biomedical applications. *J. Appl. Microbiol.* **2019**, *127*, 1612–1626. [CrossRef]
56. Hu, Y.; Daoud, W.; Cheuk, K.; Lin, C. Newly Developed Techniques on Polycondensation, Ring-Opening Polymerization and Polymer Modification: Focus on Poly(Lactic Acid). *Materials Basel* **2016**, *9*, 133. [CrossRef] [PubMed]
57. Chen, Y.; Geever, L.M.; Killion, J.A.; Lyons, J.G.; Higginbotham, C.L.; Devine, D.M. Review of Multifarious Applications of Poly (Lactic Acid). *Polym. Plast. Technol. Eng.* **2016**, *55*, 1057–1075. [CrossRef]
58. Farah, S.; Anderson, D.G.; Langer, R. Physical and mechanical properties of PLA, and their functions in widespread applications—A comprehensive review. *Adv. Drug Deliv. Rev.* **2016**, *107*, 367–392. [CrossRef] [PubMed]
59. Sisson, A.L.; Ekinci, D.; Lendlein, A. The contemporary role of ε-caprolactone chemistry to create advanced polymer architectures. *Polymer* **2013**, *54*, 4333–4350. [CrossRef]
60. Fuoco, T.; Finne-Wistrand, A. Enhancing the Properties of Poly(ε-caprolactone) by Simple and Effective Random Copolymerization of ε-Caprolactone with *p* -Dioxanone. *Biomacromolecules* **2019**, *20*, 3171–3180. [CrossRef]

61. Malikmammadov, E.; Tanir, T.E.; Kiziltay, A.; Hasirci, V.; Hasirci, N. PCL and PCL-based materials in biomedical applications. *J. Biomater. Sci. Polym. Ed.* **2018**, *29*, 863–893. [CrossRef]
62. Espinoza, S.M.; Patil, H.I.; San Martin Martinez, E.; Casañas Pimentel, R.; Ige, P.P. Poly-ε-caprolactone (PCL), a promising polymer for pharmaceutical and biomedical applications: Focus on nanomedicine in cancer. *Int. J. Polym. Mater. Polym. Biomater.* **2020**, *69*, 85–126. [CrossRef]
63. Bianco, A.; Di Federico, E.; Cacciotti, I. Electrospun poly(ε-caprolactone)-based composites using synthesized β-tricalcium phosphate. *Polym. Adv. Technol.* **2011**, *22*, 1832–1841. [CrossRef]
64. Michler, G.H. *Overview of Techniques*; Springer: Berlin/Heidelberg, Germany, 2008.
65. Malheiro, V.N.; Caridade, S.G.; Alves, N.M.; Mano, J.F. New poly(ε-caprolactone)/chitosan blend fibers for tissue engineering applications. *Acta Biomater.* **2010**, *6*, 418–428. [CrossRef]
66. Khalid, S.; Yu, L.; Meng, L.; Liu, H.; Ali, A.; Chen, L. Poly(lactic acid)/starch composites: Effect of microstructure and morphology of starch granules on performance. *J. Appl. Polym. Sci.* **2017**, *134*, 45504. [CrossRef]
67. Govindaraju, I.; Pallen, S.; Umashankar, S.; Mal, S.S.; Kaniyala Melanthota, S.; Mahato, D.R.; Zhuo, G.Y.; Mahato, K.K.; Mazumder, N. Microscopic and spectroscopic characterization of rice and corn starch. *Microsc. Res. Tech.* **2020**, 1–9. [CrossRef]
68. Vigneshwaran, N.; Ammayappan, L.; Huang, Q. Effect of Gum arabic on distribution behavior of nanocellulose fillers in starch film. *Appl. Nanosci.* **2011**, *1*, 137–142. [CrossRef]
69. Zhang, Y.; Xu, Y.; Xi, X.; Shrestha, S.; Jiang, P.; Zhang, W.; Gao, C. Amino acid-modified chitosan nanoparticles for Cu2+ chelation to suppress CuO nanoparticle cytotoxicity. *J. Mater. Chem. B* **2017**, *5*, 3521–3530. [CrossRef]
70. Majoinen, J.; Kontturi, E.; Ikkala, O.; Gray, D.G. SEM imaging of chiral nematic films cast from cellulose nanocrystal suspensions. *Cellulose* **2012**, *19*, 1599–1605. [CrossRef]
71. Rizzieri, R.; Baker, F.S.; Donald, A.M. A study of the large strain deformation and failure behaviour of mixed biopolymer gels via in situ ESEM. *Polymer* **2003**, *44*, 5927–5935. [CrossRef]
72. Karhale, S.; Bhenki, C.; Rashinkar, G.; Helavi, V. Covalently anchored sulfamic acid on cellulose as heterogeneous solid acid catalyst for the synthesis of structurally symmetrical and unsymmetrical 1,4-dihydropyridine derivatives. *New J. Chem.* **2017**, *41*, 5133–5141. [CrossRef]
73. Shankar, S.; Oun, A.A.; Rhim, J.W. Preparation of antimicrobial hybrid nano-materials using regenerated cellulose and metallic nanoparticles. *Int. J. Biol. Macromol.* **2018**, *107*, 17–27. [CrossRef]
74. Gopiraman, M.; Deng, D.; Saravanamoorthy, S.; Chung, I.M.; Kim, I.S. Gold, silver and nickel nanoparticle anchored cellulose nanofiber composites as highly active catalysts for the rapid and selective reduction of nitrophenols in water. *RSC Adv.* **2018**, *8*, 3014–3023. [CrossRef]
75. Pakravan, M.; Heuzey, M.-C.; Ajji, A. Core–Shell Structured PEO-Chitosan Nanofibers by Coaxial Electrospinning. *Biomacromolecules* **2012**, *13*, 412–421. [CrossRef]
76. Camesano, T.A.; Wilkinson, K.J. Single Molecule Study of Xanthan Conformation Using Atomic Force Microscopy. *Biomacromolecules* **2001**, *2*, 1184–1191.
77. Teckentrup, J.; Al-Hammood, O.; Steffens, T.; Bednarz, H.; Walhorn, V.; Niehaus, K.; Anselmetti, D. Comparative analysis of different xanthan samples by atomic force microscopy. *J. Biotechnol.* **2017**, *257*, 2–8. [CrossRef] [PubMed]
78. Moffat, J.; Morris, V.J.; Al-Assaf, S.; Gunning, A.P. Visualisation of xanthan conformation by atomic force microscopy. *Carbohydr. Polym.* **2016**, *148*, 380–389. [CrossRef] [PubMed]
79. Lahiji, R.R.; Xu, X.; Reifenberger, R.; Raman, A.; Rudie, A.; Moon, R.J. Atomic Force Microscopy Characterization of Cellulose Nanocrystals. *Langmuir* **2010**, *26*, 4480–4488. [CrossRef] [PubMed]
80. Robert, R.W.; Raman, J.M.A. Mechanical properties of cellulose nanomaterials studied by contact resonance atomic force microscopy. *Cellulose* **2016**, 23.
81. Nagalakshmaiah, M.; kissi NEl Mortha, G.; Dufresne, A. Structural investigation of cellulose nanocrystals extracted from chili leftover and their reinforcement in cariflex-IR rubber latex. *Carbohydr. Polym.* **2016**, *136*, 945–954. [CrossRef] [PubMed]
82. Haugstad, K.E.; Håti, A.G.; Nordgård, C.T.; Adl, P.S.; Maurstad, G.; Sletmoen, M.; Draget, K.I.; Dias, R.S.; Stokke, B.T. Direct Determination of Chitosan–Mucin Interactions Using a Single-Molecule Strategy: Comparison to Alginate–Mucin Interactions. *Polymers Basel* **2015**, *7*, 161–185. [CrossRef]

83. Jaiswal, M.K.; Banerjee, R.; Pradhan, P.; Bahadur, D. Thermal behavior of magnetically modalized poly(N-isopropylacrylamide)-chitosan based nanohydrogel. *Colloids Surf. B Biointerfaces* **2010**, *81*, 185–194. [CrossRef]
84. Gunning, A.P.; McMaster, T.J.; Morris, V.J. Scanning tunnelling microscopy of xanthan gum. *Carbohydr. Polym.* **1993**, *21*, 47–51. [CrossRef]
85. Nakajima, K.; Ikehara, T.; Nishi, T. Observation of gellan gum by scanning tunneling microscopy. *Carbohydr. Polym.* **1996**, *30*, 77–81. [CrossRef]
86. Zhang, Y.Z. Size and arrangement of elementary fibrils in crystalline cellulose studied with scanning tunneling microscopy. *J. Vac. Sci. Technol. B Microelectron. Nanometer Struct.* **1997**, *15*, 1502. [CrossRef]
87. Abdullah, O.G.; Aziz, B.K.; Aziz, S.B.; Suhail, M.H. Surfaces modification of methylcellulose: Cobalt nitrate polymer electrolyte by sulfurated hydrogen gas treatment. *J. Appl. Polym. Sci.* **2018**, *135*, 46676. [CrossRef]
88. Weisenburger, S.; Sandoghdar, V. Light microscopy: An ongoing contemporary revolution. *Contemp. Phys.* **2015**, *56*, 123–143. [CrossRef]
89. Chen, N.; Rehman, S.; Sheppard, C.J.R. Recent advances in optical microscopy methods for subcellular imaging of thick biological tissues. *Crit. Rev. Biomed. Eng.* **2013**, *41*, 393–403. [CrossRef] [PubMed]
90. Agocs, E.; Attota, R.K. Enhancing optical microscopy illumination to enable quantitative imaging. *Sci. Rep.* **2018**, *8*, 1–9. [CrossRef] [PubMed]
91. Coceancigh, H.; Higgins, D.A.; Ito, T. Optical Microscopic Techniques for Synthetic Polymer Characterization. *Anal. Chem.* **2019**, *91*, 405–424. [CrossRef] [PubMed]
92. Ma, Y.; Wang, X.; Liu, H.; Wei, L.; Xiao, L. Recent advances in optical microscopic methods for single-particle tracking in biological samples. *Anal. Bioanal. Chem.* **2019**, *411*, 4445–4463. [CrossRef]
93. Murphy, D.B.; Davidson, M.W. *Fundamentals of Light Microscopy and Electronic Imaging*, 2nd ed.; John Wiley and Sons: Hoboken, NJ, USA, 2012.
94. Vielreicher, M.; Schürmann, S.; Detsch, R.; Schmidt, M.A.; Buttgereit, A.; Boccaccini, A.; Friedrich, O. Taking a deep look: Modern microscopy technologies to optimize the design and functionality of biocompatible scaffolds for tissue engineering in regenerative medicine. *J. R. Soc. Interface* **2013**, *10*, 20130263. [CrossRef]
95. Hubbe, M.A.; Chandra, R.P.; Dogu, D.; Van Velzen, S.T.J. Analytical Staining of Cellulosic Materials: A Review. *BioRes.* **2019**, *14*, 7387–7464.
96. Daemen, S.; van Zandvoort, M.A.M.J.; Parekh, S.H.; Hesselink, M.K.C. Microscopy tools for the investigation of intracellular lipid storage and dynamics. *Mol. Metab.* **2016**, *5*, 153–163. [CrossRef]
97. Westphal, V.; Hell, S.W. Nanoscale resolution in the focal plane of an optical microscope. *Phys. Rev. Lett.* **2005**, *94*, 143903. [CrossRef]
98. Hayes, B.S.; Gammon, L.M. *Optical Microscopy of Fiber Reinforced Composites*; ASM International: Materials Park, OH, USA, 2010.
99. Vanderghem, C.; Jacquet, N.; Danthine, S.; Blecker, C.; Paquot, M. Effect of Physicochemical Characteristics of Cellulosic Substrates on Enzymatic Hydrolysis by Means of a Multi-Stage Process for Cellobiose Production. *Appl. Biochem. Biotechnol.* **2012**, *166*, 1423–1432. [CrossRef] [PubMed]
100. Gutiérrez, T.J.; Pérez, E.; Guzmán, R.; Tapia, M.S.; Famá, L. Physicochemical and Functional Properties of Native and Modified by Crosslinking, Dark-Cush-Cush Yam (*Dioscorea Trifida*) and Cassava (*Manihot Esculenta*) Starch. *J. Polym. Biopolym. Phys. Chem.* **2014**, *2*, 1–5.
101. Gao, C.; Bao, X.; Yu, L.; Liu, H.; Simon, G.P.; Chen, L.; Liu, X. Thermal properties and miscibility of semi-crystalline and amorphous PLA blends. *J. Appl. Polym. Sci.* **2014**, *131*, 41205. [CrossRef]
102. Gao, M.; Ren, Z.; Yan, S.; Sun, J.; Chen, X. An optical microscopy study on the phase structure of poly(L-lactide acid)/poly(propylene carbonate) blends. *J. Phys. Chem. B* **2012**, *116*, 9832–9837. [CrossRef]
103. Sokhal, K.S.; Dasaroju, G.; Bulasara, V.K. Formation, stability and comparison of water/oil emulsion using gum arabic and guar gum and effect of aging of polymers on drag reduction percentage in water/oil flow. *Vacuum* **2018**, *159*, 247–253. [CrossRef]
104. Rousi, Z.; Malhiac, C.; Fatouros, D.G.; Paraskevopoulou, A. Complex coacervates formation between gelatin and gum Arabic with different arabinogalactan protein fraction content and their characterization. *Food Hydrocoll.* **2019**, *96*, 577–588. [CrossRef]
105. Lippincott-Schwartz, J.; Manley, S. Putting super-resolution fluorescence microscopy to work. *Nat. Methods* **2009**, *6*, 21–23. [CrossRef]

106. Thorley, J.A.; Pike, J.; Rappoport, J.Z. *Super-Resolution Microscopy: A Comparison of Commercially Available Options*; Elsevier Inc.: Philadelphia, PA, USA, 2014.
107. Jia, S.; Han, B.; Kutz, J.N. Example-Based Super-Resolution Fluorescence Microscopy. *Sci. Rep.* **2018**, *8*, 1–8. [CrossRef]
108. Abitbol, T.; Palermo, A.; Moran-Mirabal, J.M.; Cranston, E.D. Fluorescent Labeling and Characterization of Cellulose Nanocrystals with Varying Charge Contents. *Biomacromolecules* **2013**, *14*, 3278–3284. [CrossRef]
109. Shazali, N.A.H.; Zaidi, N.E.; Ariffin, H.; Abdullah, L.C.; Ghaemi, F.; Abdullah, J.M.; Takashima, I.; Rahman, N.A.; Afizan, N.M. Characterization and cellular internalization of Spherical Cellulose Nanocrystals (CNC) into normal and cancerous fibroblasts. *Materials Basel* **2019**, *12*, 3251. [CrossRef]
110. Ur-Rehman, A.; Khan, N.M.; Ali, F.; Khan, H.; Khan, Z.U.; Jan, A.K.; Khan, G.S.; Ahmad, S. Kinetics Study of Biopolymers Mixture with the Help of Confocal Laser Scanning Microscopy. *J. Food Process Eng.* **2016**, *39*, 533–541. [CrossRef]
111. Zammarano, M.; Maupin, P.H.; Sung, L.P.; Gilman, J.W.; McCarthy, E.D.; Kim, Y.S.; Fox, D.M. Revealing the Interface in Polymer Nanocomposites. *ACS Nano* **2011**, *5*, 3391–3399. [CrossRef] [PubMed]
112. Spence, J.C.H. *High-Resolution Electron Microscopy*; Oxford University Press: New York, NY, USA, 2013.
113. Yu, X.; Arey, B.; Chatterjee, S.; Chun, J. Improving in situ liquid SEM imaging of particles. *Surf. Interface Anal.* **2019**, *51*, 1325–1331. [CrossRef]
114. Nguyen, J.N.T.; Harbison, A.M. *Scanning Electron Microscopy Sample Preparation and Imaging*; Humana Press Inc.: Totowa, NJ, USA, 2017; Volume 1606.
115. Shekarforoush, E.; Mirhosseini, H.; Amid, B.T.; Ghazali, H.; Muhammad, K.; Sarker, M.Z.I.; Paykary, M. Rheological Properties and Emulsifying Activity of Gum Karaya (*Sterculia Urens*) in Aqueous System and Oil in Water Emulsion: Heat Treatment and Microwave Modification. *Int. J. Food Prop.* **2016**, *19*, 662–679. [CrossRef]
116. Pang, J.; Liu, X.; Zhang, X.; Wu, Y.; Sun, R. Fabrication of Cellulose Film with Enhanced Mechanical Properties in Ionic Liquid 1-Allyl-3-methylimidaxolium Chloride (AmimCl). *Materials Basel* **2013**, *6*, 1270–1284. [CrossRef] [PubMed]
117. Phinichka, N.; Kaenthong, S. Regenerated cellulose from high alpha cellulose pulp of steam-exploded sugarcane bagasse. *J. Mater. Res. Technol.* **2018**, *7*, 55–65. [CrossRef]
118. İlgü, H.; Turan, T.; Şanli-Mohamed, G. Preparation, Characterization and Optimization of Chitosan Nanoparticles as Carrier for Immobilization of Thermophilic Recombinant Esterase. *J. Macromol. Sci. Part A* **2011**, *48*, 713–721.
119. Saari, H.; Fuentes, C.; Sjöö, M.; Rayner, M.; Wahlgren, M. Production of starch nanoparticles by dissolution and non-solvent precipitation for use in food-grade Pickering emulsions. *Carbohydr. Polym.* **2017**, *157*, 558–566. [CrossRef]
120. Mun, S.; Kim, H.C.; Yadave, M.; Kim, J. Graphene oxide–gellan gum–sodium alginate nanocomposites: Synthesis, characterization, and mechanical behavior. *Compos. Interfaces* **2015**, *22*, 249–263. [CrossRef]
121. Wittmar, A.; Vorat, D.; Ulbricht, M. Two step and one step preparation of porous nanocomposite cellulose membranes doped with TiO 2. *RSC Adv.* **2015**, *5*, 88070–88078. [CrossRef]
122. Wang, Q.; Chen, D. Synthesis and characterization of a chitosan based nanocomposite injectable hydrogel. *Carbohydr. Polym.* **2016**, *136*, 1228–1237. [CrossRef] [PubMed]
123. Bagheri-Khoulenjani, S.; Mirzadeh, H.; Etrati-Khosroshahi, M.; Ali Shokrgozar, M. Particle size modeling and morphology study of chitosan/gelatin/nanohydroxyapatite nanocomposite microspheres for bone tissue engineering. *J. Biomed. Mater. Res. Part A* **2013**, *101*, 1758–1767. [CrossRef] [PubMed]
124. Rodriguez, S.A.; Weese, E.; Nakamatsu, J.; Torres, F. Development of Biopolymer Nanocomposites Based on Polysaccharides Obtained from Red Algae Chondracanthus chamissoi Reinforced with Chitin Whiskers and Montmorillonite. *Polym.-Plast. Technol. Eng.* **2016**, *55*, 1557–1564. [CrossRef]
125. Tarus, B.; Fadel, N.; Al-Oufy, A.; El-Messiry, M. Effect of polymer concentration on the morphology and mechanical characteristics of electrospun cellulose acetate and poly (vinyl chloride) nanofiber mats. *Alexandria Eng. J.* **2016**, *55*, 2975–2984. [CrossRef]
126. Diantoro, M.; Kusumaatmaja, A.; Triyana, K. Preparation of PVA/Chitosan/TiO 2 nanofibers using electrospinning method. *AIP Conf. Proc.* **2016**, *1755*, 150002.

127. Qasim, S.B.; Najeeb, S.; Delaine-Smith, R.M.; Rawlinson, A.; Ur Rehman, I. Potential of electrospun chitosan fibers as a surface layer in functionally graded GTR membrane for periodontal regeneration. *Dent. Mater.* **2017**, *33*, 71–83. [CrossRef] [PubMed]
128. Padil, V.V.; Senan, C.; Wacławek, S.; Cerník, M.; Agarwal, S.; Varma, R.S. Bioplastic Fibers from Gum Arabic for Greener Food Wrapping Applications. *ACS Sustain. Chem. Eng.* **2019**, *7*, 5900–5911. [CrossRef]
129. Fazeli, M.; Florez, J.P.; Simão, R.A. Improvement in adhesion of cellulose fibers to the thermoplastic starch matrix by plasma treatment modification. *Compos. Part B Eng.* **2019**, *163*, 207–216. [CrossRef]
130. Anžlovar, A.; Kunaver, M.; Krajnc, A.; Žagar, E. Nanocomposites of LLDPE and Surface-Modified Cellulose Nanocrystals Prepared by Melt Processing. *Molecules* **2018**, *23*, 1782. [CrossRef]
131. Naranda, J.; Sušec, M.; Maver, U.; Gradišnik, L.; Gorenjak, M.; Vukasović, A.; Ivković, A.; Rupnik, M.S.; Vogrin, M.; Krajnc, P. Polyester type polyHIPE scaffolds with an interconnected porous structure for cartilage regeneration. *Sci. Rep.* **2016**, *6*, 1–11. [CrossRef]
132. Zhou, F.L.; Parker, G.J.M.; Eichhorn, S.J.; Hubbard Cristinacce, P.L. Production and cross-sectional characterization of aligned co-electrospun hollow microfibrous bulk assemblies. *Mater. Charact.* **2015**, *109*, 25–35. [CrossRef] [PubMed]
133. Yin, H.M.; Qian, J.; Zhang, J.; Lin, Z.F.; Li, J.S.; Xu, J.Z.; Li, Z.M. Engineering porous poly(lactic acid) scaffolds with high mechanical performance via a solid state extrusion/porogen leaching approach. *Polymers Basel* **2016**, *8*, 213. [CrossRef]
134. Mafirad, S.; Mehrnia, M.R.; Zahedi, P.; Hosseini, S.-N. Chitosan-based nanocomposite membranes with improved properties: Effect of cellulose acetate blending and TiO_2 nanoparticles incorporation. *Polym. Compos.* **2017**, *39*, 4452–4466. [CrossRef]
135. Zhang, Z.-H.; Han, Z.; Zeng, X.-A.; Xiong, X.-Y.; Liu, Y.-J. Enhancing mechanical properties of chitosan films via modification with vanillin. *Int. J. Biol. Macromol.* **2015**, *81*, 638–643. [CrossRef] [PubMed]
136. Nanjunda, R.B.; Venkata, L.V.; Vishnu, M.K.; Mylarappa, M.; Raghavendra, N.; Venkatesh, T. Preparation of chitosan/different organomodified clay polymer nanocomposites: Studies on morphological, swelling, thermal stability and anti-bacterial properties. *Nanosyst. Phys. Chem. Math.* **2016**, *7*, 667–674.
137. Kumar, P.; Sandeep, K.P.; Alavi, S.; Truong, V.D.; Gorga, R.E. Effect of Type and Content of Modified Montmorillonite on the Structure and Properties of Bio-Nanocomposite Films Based on Soy Protein Isolate and Montmorillonite. *J. Food Sci.* **2010**, *75*, N46–N56. [CrossRef]
138. Jain, R.; Mahto, V.; Mahto, T.K. Study of the Effect of Xanthan Gum Based Graft Copolymer on Water Based Drilling Fluid. *J. Macromol. Sci. Part A* **2014**, *51*, 976–982. [CrossRef]
139. Abdullah, N.H.; Shameli, K.; Nia, P.M.; Etesami, M.; Abdullah, E.C.; Abdullah, L.C. Electrocatalytic activity of starch/Fe3O4/zeolite bionanocomposite for oxygen reduction reaction. *Arab. J. Chem.* **2017**, *13*, 1297–1308. [CrossRef]
140. Yusof, Y.M.; Shukur, M.F.; Illias, H.A.; Kadir, M.F.Z. Conductivity and electrical properties of corn starch-chitosan blend biopolymer electrolyte incorporated with ammonium iodide. *R. Swedish Acad. Sci. Phys. Scr. Phys. Scr* **2014**, *89*, 10. [CrossRef]
141. Rajisha, K.R.; Maria, H.J.; Pothan, L.A.; Ahmad, Z.; Thomas, S. Preparation and characterization of potato starch nanocrystal reinforced natural rubber nanocomposites. *Int. J. Biol. Macromol.* **2014**, *67*, 147–153. [CrossRef]
142. Daio, T.; Bayer, T.; Ikuta, T.; Nishiyama, T.; Takahashi, K.; Takata, Y.; Sasaki, K.; Lyth, S.M. In-Situ ESEM and EELS Observation of Water Uptake and Ice Formation in Multilayer Graphene Oxide. *Sci. Rep.* **2015**, *5*, 11807. [CrossRef]
143. Jenkins, L.M.; Donald, A.M. Use of the environmental scanning electron microscope for the observation of the swelling behaviour of cellulosic fibres. *Scanning* **2006**, *19*, 92–97. [CrossRef]
144. Podor, R.; Ravaux, J.; Brau, H.-P. *In Situ Experiments in the Scanning Electron Microscope Chamber, Scanning electron Microscopy*; IntechOpen: London, UK, 2012.
145. Jansson, A.; Nafari, A.; Sanz-Velasco, A.; Svensson, K.; Gustafsson, S.; Hermansson, A.M.; Olsson, E. Novel Method for Controlled Wetting of Materials in the Environmental Scanning Electron Microscope. *Microsc. Microanal.* **2013**, *19*, 30–37. [CrossRef] [PubMed]
146. Girão, A.V.; Caputo, G.; Ferro, M.C. Application of Scanning Electron Microscopy–Energy Dispersive X-Ray Spectroscopy (SEM-EDS). *Compr. Anal. Chem.* **2017**, *75*, 153–168.

147. Chauhan, K.; Priya, V.; Singh, P.; Chauhan, G.S.; Kumari, S.; Singhal, R.K. A green and highly efficient sulfur functionalization of starch. *RSC Adv.* **2015**, *5*, 51762–51772. [CrossRef]
148. Anjum, F.; Bukhari, S.A.; Siddique, M.; Shahid, M.; Potgieter, J.H.; Jaafar, H.Z.; Ercisli, S.; Zia-Ul-Haq, M. Microwave Irradiated Copolymerization of Xanthan Gum with Acrylamide for Colonic Drug Delivery. *BioResources* **2015**, *10*, 1434–1451. [CrossRef]
149. Ghannam, H.E.; STalab, A.; VDolgano, N.; MSHusse, A.; Abdelmagui, N.M. Characterization of Chitosan Extracted from Different Crustacean Shell Wastes. *J. Appl. Sci.* **2016**, *16*, 454–461.
150. Sofla, M.R.K.; Brown, R.J.; Tsuzuki, T.; Rainey, T.J. A comparison of cellulose nanocrystals and cellulose nanofibres extracted from bagasse using acid and ball milling methods. *Adv. Nat. Sci. Nanosci. Nanotechnol.* **2016**, *7*, 035004. [CrossRef]
151. Singh, P.; Chauhan, K.; Priya, V.; Singhal, R.K. A greener approach for impressive removal of As(iii)/As(v) from an ultra-low concentration using a highly efficient chitosan thiomer as a new adsorbent. *RSC Adv.* **2016**, *6*, 64946–64961. [CrossRef]
152. Gao, Y.; Zhang, X.; Jin, X. Preparation and Properties of Minocycline-Loaded Carboxymethyl Chitosan Gel/Alginate Nonwovens Composite Wound Dressings. *Mar. Drugs* **2019**, *17*, 575. [CrossRef]
153. Kuei, B.; Aplan, M.P.; Litofsky, J.H.; Gomez, E.D. New opportunities in transmission electron microscopy of polymers. *Mater. Sci. Eng. R Rep.* **2020**, *139*, 100516. [CrossRef]
154. Libera, M.R.; Egerton, R.F. Advances in the Transmission Electron Microscopy of Polymers. *Polym. Rev.* **2010**, *50*, 321–339. [CrossRef]
155. Winey, M.; Meehl, J.B.; O'Toole, E.T.; Giddings, T.H. Conventional transmission electron microscopy. *Mol. Biol. Cell* **2014**, *25*, 319–323. [CrossRef] [PubMed]
156. Rice, S.B.; Chan, C.; Brown, S.C.; Eschbach, P.; Han, L.; Ensor, D.S.; Stefaniak, A.B.; Bonevich, J.; Vladár, A.E.; Walker, A.R.H. Particle size distributions by transmission electron microscopy: an interlaboratory comparison case study. *Metrologia* **2013**, *50*, 663. [CrossRef] [PubMed]
157. Mielańczyk, Ł.; Matysiak, N.; Klymenko, O.; Wojnicz, R. *Transmission Electron Microscopy of Biological Samples*; IntechOpen: Vienna, Austria, 2015.
158. Tang, C.Y.; Yang, Z. *Transmission Electron Microscopy (TEM)*; Elsevier Inc.: Philadelphia, PA, USA, 2017.
159. Kirkland, A.I.; Chang, S.L.Y.; Hutchison, J.L. *Springer Handbooks*; Springer: Berlin/Heidelberg, Germany, 2019; pp. 3–47.
160. Alnarabiji, M.S.; Yahya, N.; Hamed, Y.; Ardakani, S.E.M.; Azizi, K.; Klemeš, J.J.; Abdullaha, B.; Tasfyd, S.F.H.; Hamide, S.B.A.; Nashed, O. Scalable bio-friendly method for production of homogeneous metal oxide nanoparticles using green bovine skin gelatin. *J. Clean. Prod.* **2017**, *162*, 186–194. [CrossRef]
161. Santos, E.D.B.; Lima, E.C.N.L.; Oliveira, C.S.; De Sigoli, F.A.; Mazali, I.O. Fast detection of paracetamol on a gold nanoparticle-chitosan substrate by SERS. *Anal. Methods* **2014**, *6*, 3564–3568. [CrossRef]
162. Bhagyaraj, S.; Krupa, I. Alginate-Mediated Synthesis of Hetero-Shaped Silver Nanoparticles and Their Hydrogen Peroxide Sensing Ability. *Molecules* **2020**, *25*, 435. [CrossRef]
163. Josefsson, G.; Tanem, B.S.; Li, Y.; Vullum, P.E.; Gamstedt, E.K. Prediction of elastic properties of nanofibrillated cellulose from micromechanical modeling and nano-structure characterization by transmission electron microscopy. *Cellulose* **2013**, *20*, 761–770. [CrossRef]
164. Ramesh, S.; Sivasamy, A.; Kim, H.S.; Kim, J.H. High-performance N-doped MWCNT/GO/cellulose hybrid composites for supercapacitor electrodes. *RSC Adv.* **2017**, *7*, 49799–49809. [CrossRef]
165. Qin, X.; Zhang, H.; Wang, Z.; Jin, Y. Magnetic chitosan/graphene oxide composite loaded with novel photosensitizer for enhanced photodynamic therapy. *RSC Adv.* **2018**, *8*, 10376–10388. [CrossRef]
166. Kondo, T.; Kasai, W.; Brown, R.M. Formation of nematic ordered cellulose and chitin. *Cellulose* **2004**, *11*, 463–474. [CrossRef]
167. Danev, R.; Yanagisawa, H.; Kikkawa, M. Cryo-Electron Microscopy Methodology: Current Aspects and Future Directions. *Trends Biochem. Sci.* **2019**, *44*, 837–848. [CrossRef] [PubMed]
168. Frank, J. Story in a sample-the potential (and limitations) of cryo-electron microscopy applied to molecular machines. *Biopolymers* **2013**, *99*, 832–836. [CrossRef]
169. Majoinen, J.; Haataja, J.S.; Appelhans, D.; Lederer, A.; Olszewska, A.; Seitsonen, J.; Aseyev, V.; Kontturi, E.; Rosilo, H.; Österberg, M.; et al. Supracolloidal Multivalent Interactions and Wrapping of Dendronized Glycopolymers on Native Cellulose Nanocrystals. *J. Am. Chem. Soc.* **2014**, *136*, 866–869. [CrossRef] [PubMed]

170. Kaushik, M.; Basu, K.; Benoit, C.; Cirtiu, C.M.; Vali, H.; Moores, A. Cellulose Nanocrystals as Chiral Inducers: Enantioselective Catalysis and Transmission Electron Microscopy 3D Characterization. *J. Am. Chem. Soc.* **2015**, *137*, 6124–6127. [CrossRef] [PubMed]
171. Pennycook, S.J.; Lupini, A.R.; Varela, M.; Borisevich, A.; Peng, Y.; Oxley, M.P.; Van Benthem, K.; Chisholm, M.F. Scanning transmission electron microscopy for nanostructure characterization. In *Scanning Microscopy for Nanotechnology: Techniques and Applications*; Springer New York: New York, NY, USA, 2007.
172. Nellist, P.D. *Springer Handbooks*; Springer: Berlin/Heidelberg, Germany, 2019; pp. 49–99.
173. Ponce, A.; Mejía-Rosales, S.; José-Yacamán, M. Scanning transmission electron microscopy methods for the analysis of nanoparticles. *Methods Mol. Biol.* **2012**, *906*, 453–471.
174. Rodríguez-Argüelles, M.C.; Sieiro, C.; Cao, R.; Nasi, L. Chitosan and silver nanoparticles as pudding with raisins with antimicrobial properties. *J. Colloid Interface Sci.* **2011**, *364*, 80–84. [CrossRef]
175. Teodoro, K.B.R.; Migliorini, F.L.; Facure, M.H.M.; Correa, D.S. Conductive electrospun nanofibers containing cellulose nanowhiskers and reduced graphene oxide for the electrochemical detection of mercury(II). *Carbohydr. Polym.* **2019**, *207*, 747–754. [CrossRef]
176. Liu, K.; Nasrallah, J.; Chen, L.; Huang, L.; Ni, Y. Preparation of CNC-dispersed Fe3O4 nanoparticles and their application in conductive paper. *Carbohydr. Polym.* **2015**, *126*, 175–178. [CrossRef]
177. Karakus, S.; Ilgar, M.; Kahyaoglu, I.M.; Kilislioglu, A. Influence of ultrasound irradiation on the intrinsic viscosity of guar gum–PEG/rosin glycerol ester nanoparticles. *Int. J. Biol. Macromol.* **2019**, *141*, 1118–1127. [CrossRef]
178. Motahharifar, N.; Nasrollahzadeh, M.; Taheri-Kafrani, A.; Varma, R.S.; Shokouhimehr, M. Magnetic chitosan-copper nanocomposite: A plant assembled catalyst for the synthesis of amino- and N-sulfonyl tetrazoles in eco-friendly media. *Carbohydr. Polym.* **2020**, *232*, 115819. [CrossRef] [PubMed]
179. Torres-Martínez, N.E.; Garza-Navarro, M.A.; García-Gutiérrez, D.; González-González, V.A.; Torres-Castro, A.; Ortiz-Méndez, U. Hybrid nanostructured materials with tunable magnetic characteristics. *J. Nanoparticle Res.* **2014**, *16*, 1–12. [CrossRef]
180. Teixeira ED, M.; Lotti, C.; Corrêa, A.C.; Teodoro, K.B.; Marconcini, J.M.; Mattoso, L.H. Thermoplastic corn starch reinforced with cotton cellulose nanofibers. *J. Appl. Polym. Sci.* **2011**, *120*, 2428–2433. [CrossRef]
181. Omerzu, A.; Saric, I.; Piltaver, I.K.; Petravic, M.; Kapun, T.; Zule, J.; Stifter, S.; Salamon, K. Prevention of spontaneous combustion of cellulose with a thin protective Al2O3 coating formed by atomic layer deposition. *Surf. Coat. Technol.* **2018**, *333*, 81–86. [CrossRef]
182. Song, S.; Zhao, Y.; Yuan, X.; Zhang, J. β-Chitin nanofiber hydrogel as a scaffold to in situ fabricate monodispersed ultra-small silver nanoparticles. *Colloids Surf. A Physicochem. Eng. Asp.* **2019**, *574*, 36–43. [CrossRef]
183. Liu, K.; Liang, H.; Nasrallah, J.; Chen, L.; Huang, L.; Ni, Y. Preparation of the CNC/Ag/beeswax composites for enhancing antibacterial and water resistance properties of paper. *Carbohydr. Polym.* **2016**, *142*, 183–188. [CrossRef]
184. Sosiati, H.; Wijayanti, D.A.; Triyana, K.; Kamiel, B. Morphology and crystallinity of sisal nanocellulose after sonication. *AIP Conf. Proc.* **2017**, *1755*, 20029.
185. Gupta, K.; Kaushik, A.; Tikoo, K.B.; Kumar, V.; Singhal, S. Enhanced catalytic activity of composites of NiFe2O4 and nano cellulose derived from waste biomass for the mitigation of organic pollutants. *Arab. J. Chem.* **2017**, *13*, 783–798. [CrossRef]
186. Celebi, H.; Kurt, A. Effects of processing on the properties of chitosan/cellulose nanocrystal films. *Carbohydr. Polym.* **2015**, *133*, 284–293. [CrossRef]
187. Guirguis, O.; Abdelzaher, N.; El-Bassyouni, G.; Moselhey, M. Structural, Thermal and Optical Modifications of Chitosan due to UV-Ozone Irradiation. *Egypt. J. Chem.* **2018**, *61*, 350–360. [CrossRef]
188. Anandan, M.; Gurumallesh Prabu, H. Dodonaea viscosa Leaf Extract Assisted Synthesis of Gold Nanoparticles: Characterization and Cytotoxicity against A549 NSCLC Cancer Cells. *J. Inorg. Organomet. Polym. Mater.* **2018**, *28*, 932–941. [CrossRef]
189. Yulizar, Y.; Utari, T.; Ariyanta, H.A.; Maulina, D. Green Method for Synthesis of Gold Nanoparticles Using *Polyscias scutellaria* Leaf Extract under UV Light and Their Catalytic Activity to Reduce Methylene Blue. *J. Nanomater.* **2017**, *2017*, 1–6. [CrossRef]

190. Kiruba Daniel, S.C.G.; Vinothini, G.; Subramanian, N.; Nehru, K.; Sivakumar, M. Biosynthesis of Cu, ZVI, and Ag nanoparticles using Dodonaea viscosa extract for antibacterial activity against human pathogens. *J. Nanoparticle Res.* **2013**, *15*, 1319. [CrossRef]
191. Guidelli, E.J.; Ramos, A.P.; Zaniquelli, M.E.D.; Baffa, O. Green synthesis of colloidal silver nanoparticles using natural rubber latex extracted from Hevea brasiliensis. *Spectrochim. Acta Part A Mol. Biomol. Spectrosc.* **2011**, *82*, 140–145. [CrossRef]
192. Vanamudan, A.; Sudhakar, P.P. Biopolymer capped silver nanoparticles with potential for multifaceted applications. *Int. J. Biol. Macromol.* **2016**, *86*, 262–268. [CrossRef]
193. Karoutsos, V. Scanning probe microscopy: Instrumentation and applications on thin films and magnetic multilayers. *J. Nanosci. Nanotechnol.* **2009**, *9*, 6783–6798. [CrossRef]
194. Huey, B.D.; Luria, J.; Bonnell, D.A. *Springer Handbooks*; Springer: Berlin/Heidelberg, Germany, 2019; pp. 1239–1277.
195. Wallace, A.F. *Scanning Probe Microscopy, Analytical Geomicrobiology*; Cambridge University Press: Cambridge, UK, 2019.
196. Raigoza, A.F.; Dugger, J.W.; Webb, L.J. Review: Recent Advances and Current Challenges in Scanning Probe Microscopy of Biomolecular Surfaces and Interfaces. *ACS Appl. Mater. Interfaces* **2013**, *5*, 9249–9261. [CrossRef]
197. Cohen, S.H.; Bray, M.T.; Lightbody, M.L. *Atomic Force Microscopy/Scanning Tunneling Microscopy*; Springer: Boston, MA, USA, 1994.
198. Binnig, G.; Quate, C.F.; Gerber, C. Atomic Force Microscope. *Phys. Rev. Lett.* **1986**, *56*, 930–933. [CrossRef]
199. Hansma, P.K.; Elings, V.B.; Marti, O.; Bracker, C.E. Scanning tunneling microscopy and atomic force microscopy: Application to biology and technology. *Science* **1988**, *242*, 209–216. [CrossRef]
200. Santos, N.C.; Castanho, M.A.R.B. An overview of the biophysical applications of atomic force microscopy. *Biophys. Chem.* **2004**, *107*, 133–149. [CrossRef]
201. Giessibl, F.J. Advances in atomic force microscopy. *Rev. Mod. Phys.* **2003**, *75*, 949–983. [CrossRef]
202. Ando, T.; Uchihashi, T.; Kodera, N.; Yamamoto, D.; Miyagi, A.; Taniguchi, M.; Yamashita, H. High-speed AFM and nano-visualization of biomolecular processes. *Pflüg. Arch. Eur. J. Physiol.* **2008**, *456*, 211–225. [CrossRef] [PubMed]
203. Alsteens, D.; Dupres, V.; Dague, E.; Verbelen, C.; André, G.; Francius, G.; Dufrêne, Y.F. *Imaging Chemical Groups and Molecular Recognition Sites on Live Cells Using AFM*; Springer: Berlin/Heidelberg, Germany, 2009.
204. Barish, J.A.; Goddard, J.M. Topographical and chemical characterization of polymer surfaces modified by physical and chemical processes. *J. Appl. Polym. Sci.* **2011**, *120*, 2863–2871. [CrossRef]
205. Nnebe, I.; Schneider, J.W. Characterization of distance-dependent damping in tapping-mode atomic force microscopy force measurements in liquid. *Langmuir* **2004**, *20*, 3195–3201. [CrossRef]
206. Gunning, A.P.; Kirby, A.R.; Ridout, M.J.; Brownsey, G.J.; Morris, V.J. Investigation of Gellan Networks and Gels by Atomic Force Microscopy. *Macromolecules* **1996**, *29*, 6791–6796. [CrossRef]
207. Kirby, A.R.; Gunning, A.P.; Waldron, K.W.; Morris, V.J.; Ng, A. Visualization of plant cell walls by atomic force microscopy. *Biophys. J.* **1996**, *70*, 1138–1143. [CrossRef]
208. Hansma, H.G.; Hoh, J.H. Biomolecular Imaging with the Atomic Force Microscope. *Ann. Rev. Biophys. Biomol. Struct.* **1994**, *23*, 115–140. [CrossRef]
209. Kirby, A.R.; Gunning, A.P.; Morris, V.J. Imaging polysaccharides by atomic force microscopy. *Biopolymers* **1998**, *38*, 355–366. [CrossRef]
210. Patrick Gunning, A.; Kirby, A.R.; Morris, V.J. Imaging xanthan gum in air by ac "tapping" mode atomic force microscopy. *Ultramicroscopy* **1996**, *63*, 1–3. [CrossRef]
211. McIntire, T.M.; Penner, R.M.; Brant, D.A. Observations of a circular, triple-helical polysaccharide using noncontact atomic force microscopy. *Macromolecules* **1995**, *28*, 6375–6377. [CrossRef]
212. Uricanu, V.I.; Duits, M.H.G.; Nelissen, R.M.F.; MLBennink, A.; Mellema, J. Local Structure and Elasticity of Soft Gelatin Gels Studied with Atomic Force Microscopy. *Langmuir* **2003**, *19*, 8182–8194. [CrossRef]
213. Iijima, M.; Shinozaki, M.; Hatakeyama, T.; Takahashi, M.; Hatakeyama, H. AFM studies on gelation mechanism of xanthan gum hydrogels. *Carbohydr. Polym.* **2007**, *68*, 701–707. [CrossRef]
214. Kocun, M.; Grandbois, M.; Cuccia, L.A. Single molecule atomic force microscopy and force spectroscopy of chitosan. *Colloids Surf. B Biointerfaces* **2011**, *82*, 470–476. [CrossRef] [PubMed]

215. Maver, U.; Maver, T.; Persin, Z.; Mozetic, M.; Vesel, A.; Gaberšček, M.; Stana-Kleinschek, K. Polymer Characterization with the Atomic Force Microscope. In *Polymer Science*; IntechOpen: London, UK, 2013.
216. Vezenov, D.V.; Noy, A.; Ashby, P. Chemical force microscopy: Probing chemical origin of interfacial forces and adhesion. *J. Adhes. Sci. Technol.* **2005**, *19*, 313–364. [CrossRef]
217. Steffens, C.; Leite, F.L.; Bueno, C.C.; Manzoli, A.; Herrmann, P.S.D.P. Atomic Force Microscopy as a Tool Applied to Nano/Biosensors. *Sensors* **2012**, *12*, 8278–8300. [CrossRef]
218. Ito, T.; Ibrahim, S.; Grabowska, I. Chemical-force microscopy for materials characterization. *TrAC Trends Anal. Chem.* **2010**, *29*, 225–233. [CrossRef]
219. Arslan, B.; Ju, X.; Zhang, X.; Abu-Lail, N.I. Heterogeneity and Specificity of Nanoscale Adhesion Forces Measured between Self-Assembled Monolayers and Lignocellulosic Substrates: A Chemical Force Microscopy Study. *Langmuir* **2015**, *31*, 10233–10245. [CrossRef]
220. Le Troëdec, M.; Rachini, A.; Peyratout, C.; Rossignol, S.; Max, E.; Kaftan, O.; Fery, A.; Smith, A. Influence of chemical treatments on adhesion properties of hemp fibres. *J. Colloid Interface Sci.* **2011**, *356*, 303–310. [CrossRef]
221. Boland, T.; Ratner, B.D. Direct measurement of hydrogen bonding in DNA nucleotide bases by atomic force microscopy. *Proc. Natl. Acad. Sci. USA.* **1995**, *92*, 5297–5301. [CrossRef]
222. Lee, I.; Evans, B.R.; Foston, M.; Ragauskas, A.J. Silicon cantilever functionalization for cellulose-specific chemical force imaging of switchgrass. *Anal. Methods* **2015**, *7*, 4541–4545. [CrossRef]
223. Passeri, D.; Angeloni, L.; Reggente, M.; Rossi, M. *Magnetic force microscopy, Magnetic Characterization Techniques for Nanomaterials*; Springer: Berlin/Heidelberg, Germany, 2017; pp. 209–259.
224. Kazakova, O.; Puttock, R.; Barton, C.; Corte-León, H.; Jaafar, M.; Neu, V.; Asenjo, A. Frontiers of magnetic force microscopy. *J. Appl. Phys.* **2019**, *125*, 060901. [CrossRef]
225. Torre, B.; Bertoni, G.; Fragouli, D.; Falqui, A.; Salerno, M.; Diaspro, A.; Cingolani, R.; Athanassiou, A. Magnetic force microscopy and energy loss imaging of superparamagnetic iron oxide nanoparticles. *Sci. Rep.* **2011**, *1*, 202. [CrossRef] [PubMed]
226. Passeri, D.; Dong, C.; Reggente, M.; Angeloni, L.; Barteri, M.; Scaramuzzo, F.A.; Angelis, F.D.; Marinelli, F.; Antonelli, F.; Rinaldi, F.; et al. Magnetic force microscopy: Quantitative issues in biomaterials. *Biomatter* **2014**, *4*, e29507. [CrossRef] [PubMed]
227. Nisticò, R. Magnetic materials and water treatments for a sustainable future. *Res. Chem. Intermed.* **2017**, *43*, 6911–6949. [CrossRef]
228. Marín, T.; Montoya, P.; Arnache, O.; Pinal, R.; Calderón, J. Development of magnetite nanoparticles/gelatin composite films for triggering drug release by an external magnetic field. *Mater. Des.* **2018**, *152*, 78–87. [CrossRef]
229. Nisticò, R.; Franzoso, F.; Cesano, F.; Scarano, D.; Magnacca, G.; Parolo, M.E.; Carlos, L. Chitosan-Derived Iron Oxide Systems for Magnetically Guided and Efficient Water Purification Processes from Polycyclic Aromatic Hydrocarbons. *ACS Sustain. Chem. Eng.* **2017**, *5*, 793–801.
230. Cesano, F.; Fenoglio, G.; Carlos, L.; Nisticò, R. One-step synthesis of magnetic chitosan polymer composite films. *Appl. Surf. Sci.* **2015**, *345*, 175–181. [CrossRef]
231. Lewandowska-Łańcucka, J.; Staszewska, M.; Szuwarzyński, M.; Kępczyński, M.; Romek, M.; Tokarz, W.; Szpak, A.; Kania, G.; Nowakowska, M. Synthesis and characterization of the superparamagnetic iron oxide nanoparticles modified with cationic chitosan and coated with silica shell. *J. Alloys Compd.* **2014**, *586*, 45–51.
232. Zasadzinski, J.A.; Schneir, J.; Gurley, J.; Elings, V.; Hansma, P.K. Scanning tunneling microscopy of freeze-fracture replicas of biomembranes. *Science* **1988**, *239*, 1013–1015. [CrossRef]
233. Marti, O.; Ribi, H.O.; Drake, B.; Albrecht, T.R.; Quate, C.F.; Hansma, P.K. Atomic force microscopy of an organic monolayer. *Science* **1988**, *239*, 50–52. [CrossRef]
234. Travaglini, G.; Rohrer, H.; Amrein, M.; Gross, H. Scanning tunneling microscopy on biological matter. *Surf. Sci.* **1987**, *181*, 380–390. [CrossRef]
235. Sonnenfeld, R.; Hansma, P.K.; Gross, H.; Stoll, E.; Travaglini, G. Atomic-resolution microscopy in water. *Science* **1986**, *232*, 211–213. [CrossRef] [PubMed]
236. Abdel-Kareem, O.; Abdel-Rahim, H.; Ezzat, I.; Essa, D.M. Evaluating the use of chitosan coated Ag nano-SeO 2 composite in consolidation of Funeral Shroud from the Egyptian Museum of Cairo. *J. Cult. Herit.* **2015**, *16*, 486–495. [CrossRef]

237. Hameroff, S.R.; Simić-Krstić, J.; Kelley, M.F.; Voelker, M.A.; He, J.D.; Dereniak, E.L.; McCuskey, R.S.; Schneiker, C.W. Scanning tunneling microscopy of biopolymers: Conditions for microtubule stabilization. *J. Vac. Sci. Technol. A Vac. Surf. Film.* **1989**, *7*, 2890–2894. [CrossRef]
238. Golovnya, R.V.; Terenina, M.B.; Krikunova, N.I.; Yuryev, V.P.; Misharina, T.A. Formation of Supramolecular Structures of Aroma Compounds with Polysaccharides of Corn Starch Cryotextures. *Starch Stärke* **2001**, *53*, 269–277. [CrossRef]

© 2020 by the authors. Licensee MDPI, Basel, Switzerland. This article is an open access article distributed under the terms and conditions of the Creative Commons Attribution (CC BY) license (http://creativecommons.org/licenses/by/4.0/).

polymers

MDPI

Review

Electrospun Environment Remediation Nanofibers Using Unspinnable Liquids as the Sheath Fluids: A Review

Menglong Wang, Ke Wang *, Yaoyao Yang, Yanan Liu and Deng-Guang Yu *

School of Materials Science & Engineering, University of Shanghai for Science and Technology, Shanghai 200093, China; 172442574@st.usst.edu.cn (M.W.); yyyang@usst.edu.cn (Y.Y.); yananliu@usst.edu.cn (Y.L.)

* Correspondence: wangk2017@usst.edu.cn (K.W.); ydg017@usst.edu.cn (D.-G.Y.)

Received: 19 November 2019; Accepted: 16 December 2019; Published: 4 January 2020

Abstract: Electrospinning, as a promising platform in multidisciplinary engineering over the past two decades, has overcome major challenges and has achieved remarkable breakthroughs in a wide variety of fields such as energy, environmental, and pharmaceutics. However, as a facile and cost-effective approach, its capability of creating nanofibers is still strongly limited by the numbers of treatable fluids. Most recently, more and more efforts have been spent on the treatments of liquids without electrospinnability using multifluid working processes. These unspinnable liquids, although have no electrospinnability themselves, can be converted into nanofibers when they are electrospun with an electrospinnable fluid. Among all sorts of multifluid electrospinning methods, coaxial electrospinning is the most fundamental one. In this review, the principle of modified coaxial electrospinning, in which unspinnable liquids are explored as the sheath working fluids, is introduced. Meanwhile, several typical examples are summarized, in which electrospun nanofibers aimed for the environment remediation were prepared using the modified coaxial electrospinning. Based on the exploration of unspinnable liquids, the present review opens a way for generating complex functional nanostructures from other kinds of multifluid electrospinning methods.

Keywords: coaxial electrospinning; core-sheath nanofibers; environmental remediation; unspinnable liquid; nanocoating

---

1. Introduction

Electrospinning, a simple and straightforward method for preparing nanofibers [1–4], is quickly moving forward along two important directions. One is the production on a large scale for potential commercial applications [5–7]. The other is the simultaneous treatment of multiple-fluid working fluids for generating nanofibers with complicated nanostructures [8–11].

Shown in Figure 1, the traditional single-fluid blending electrospinning is differentiating into different sorts of double-fluid and three-fluid processes. Based on the spatial positions, there are two kind of double-fluid electrospinning, i.e., the coaxial electrospinning and the side-by-side electrospinning. As for the three-fluid processes, there are tri-axial process (an abbreviation of tri-layer working fluids organized in a coaxial manner), coaxial process with a side-by-side core, side-by-side process with one coaxial side, and tri-layer parallel side-by-side one. These advanced techniques should greatly expand the capability of electrospinning in generating a wide variety of complex nanostructures.

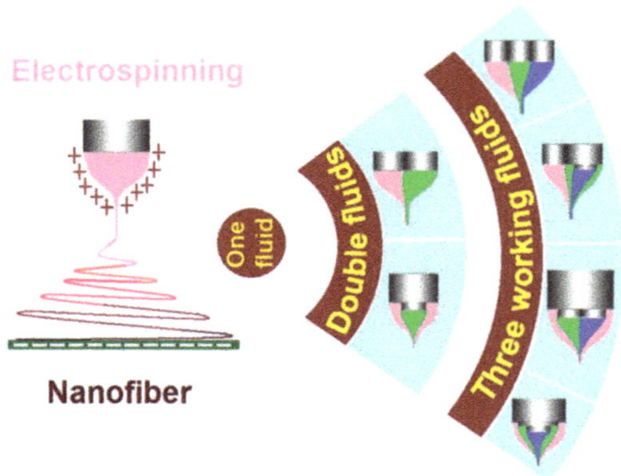

Figure 1. The development of electrospinning from the traditional one-fluid blending process to the double-fluid coaxial/side-by-side and three-fluid processes.

However, among all the complex nanostructures, the core-shell should be the most fundamental and the most important one [12]. It is not only because that many other complicated nanostructures can be viewed as a derivative of core-shell structure, but also this fundamental nanostructure has been broadly demonstrated to be extremely useful in designing and developing a wide variety of functional nanomaterials for applications in almost all the scientific fields [13,14]. Correspondingly, the methods for creating core-sheath structures are always highly desired in scientific fields. This should be one of the most important reasons that coaxial electrospinning [15,16], and also coaxial electrospraying [17–21], are receiving the increasing attention nearly from all the applied functional material fields.

2. The Modified Coaxial Electrospinning with an Unspinnable Sheath Fluid

In 2002, the first publication about core-shell structure was published based on electrospraying, or electrohydrodynamic atomization, a sister technique of electrospinning [22]. Later, in 2004, coaxial electrospinning was regarded as one of the most important three breakthroughs in this field [23]. In 2008, Moghe & Gupta gave the first review on coaxial electrospinning. Based on the previous publications before this time point, they concluded that the shell working fluids must be electrospinnable for successfully carrying out a coaxial process and for creating a core-shell nanostructure [24]. However, Yu, et al. broke this concept to develop a modified coaxial electrospinning, in which the sheath working fluid could be a pure solvent [12]. The modified process was successfully conducted to generate fibers from a concentrated polymer solution, which was impossible for a single-fluid process owing to the frequent clogging of the spinneret's nozzle.

Shown in Figure 2, as the differentiation of traditional processes, it can be anticipated that the modified coaxial process can also differentiated into modified tri-axial process and even tri-layer side-by-side process. Most recently, modified tri-axial electrospinning with only one of the three working fluids having electrospinnability was demonstrated to be very useful in generating high quality core-shell nanofibers [25,26]. What is more, the functional sheath's thickness can be easily manipulated through adjusting the fluid flow rate ratio in a suitable range.

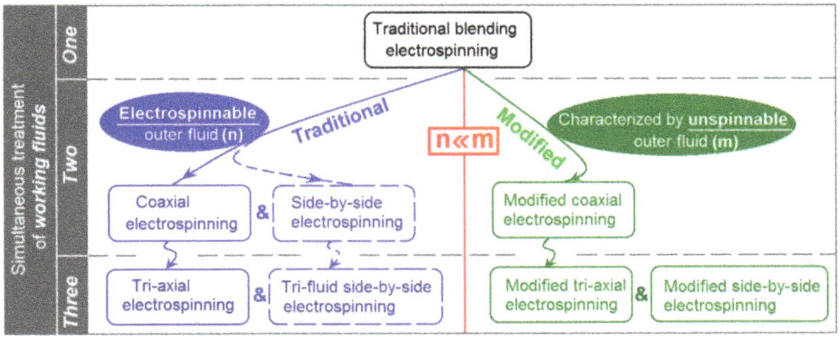

Figure 2. The differentiation of electrospinning from the traditional blending electrospinning to a series of double-fluid and tri-fluid electrospinning methods.

Apparently, the involvement of unspinnable sheath fluid can greatly promote the capability of creating complex nanostructures for functional applications. This is because the spinnable filament-forming polymers are very limited, slightly over 100 kinds by estimation [27], but the un-spinnable liquids are numerous. Thus, the modified coaxial electrospinning provides those materials without filament-forming properties the previous opportunity to taking advantage of the large surface, high porosity and 3D web structure of electrospun nanofiber mats to exert their functional performances. Shown in Figure 3, those unspinnable liquids include solvents, surfactants or salt solutions, and liquids containing functional components. All the liquids are possible to be converted into electrospun nanofibers, regardless of they are solutions, emulsions, nano suspensions or even a slurry.

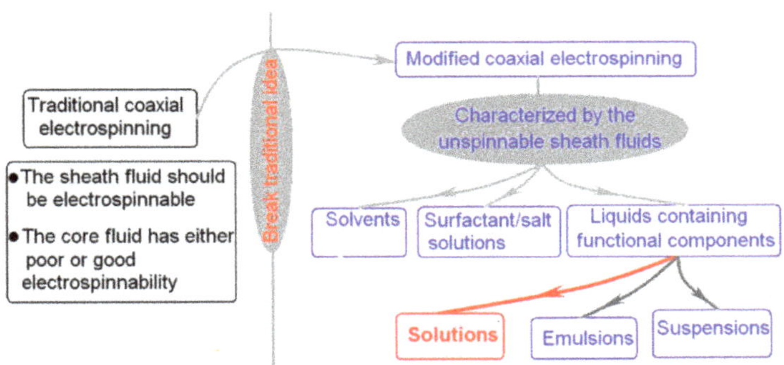

Figure 3. The differences between the traditional coaxial electrospinning and the modified coaxial electrospinning, and many kinds of unspinnable fluids can greatly enrich the electrospun nanoproducts.

3. The Nanostructures Created by the Modified Coaxial Electrospinning and the Key Parameters for Carrying out the Working Processes

The advantages of modified coaxial electrospinning over the traditional one lies on not only the great treatment capability about unspinnable fluids prepared from numerous raw materials, but also in creating different kinds of nanostructures. Shown in Figure 4 is a comparison between the traditional coaxial and modified coaxial process in generating nanostructures [28]. The traditional coaxial electrospinning can only directly create solid core-shell structures, although which can be utilized to further prepare nanotubes and other derivatives [29]. In sharp contrast, the modified coaxial electrospinning can be exploited for producing core-shell structures, just as the traditional ones [30], but also can be utilized to prepare nanocoating on the core polymeric fibers [31]. What is more, when solvents are explored as the sheath working fluids, modified coaxial electrospinning can be utilized to stabilize the working process [32], keep the working processes from clogging, manipulate the nanofibers' diameter without the additions of salt or other additives, and systematically improve the nanofibers' quality.

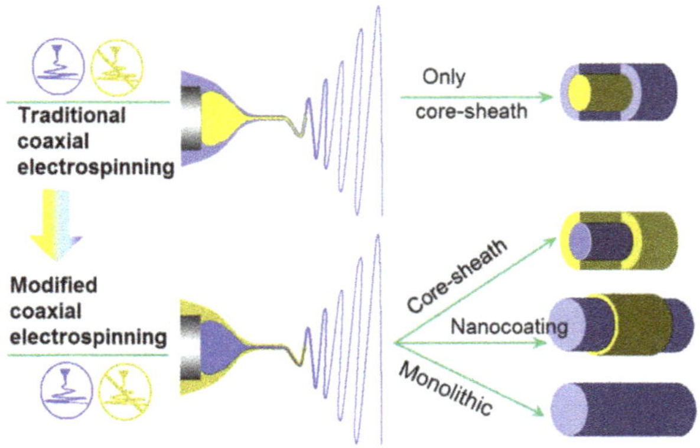

Figure 4. The nanostructures created by the traditional coaxial electrospinning and the modified one [28].

In the traditional blending electrospinning, the working process is regulated by a series of variables, including the properties of the solution, experimental parameters and surrounding conditions [33–39]. The properties of solution further include the viscosity, conductivity, surface tension, molecular weight of polymer and dielectric constant; the operational variables include flow rate, electric field force, distance between needle and receiving screen, diameter and shape of needle, material composition and surface morphology of receiving screen, and so on; the surrounding parameters include temperature, humidity, wind speed, and maybe the vacuum state. But the most important thing is that these different parameters do not exert their impacts on the electrospinning process solely, they are not independent of each other, but interrelated. Thus, although electrospinning is a simple one-step processes, but it is very difficult to be optimized for robust and repetitive preparation.

No matter how many fluids are simultaneously electrospun using the multiple-fluid electrospinning processes, the above-mentioned parameters will similarly influence the working processes and the final products' quality. The additional key parameter for the modified coaxial electrospinning should be the flow rates of sheath/core working fluids or the sheath-to-core fluid flow rate ratio. This is particularly vital for the modified coaxial electrospinning, in which solvent is exploited as a sheath fluid. Shown in Figure 5 is a diagram about the comparison between a single-fluid electrospinning and a modified process with a sheath solvent [40]. Compared with the former, the latter has a layer of solvent around the core spinnable polymeric solution during the previous stage of the whole electrospinning process, i.e., the Taylor cone, the straight fluid jet, and partial the unstable region. On one hand, the solvent can lubricate the spinneret to prevent clogging, can decrease the negative drawing forces from the capillary's surface and the capillary forces, and can keep the core fluid from disturbance by the environmental changes. On the other hand, the sheath solvent may result in a failure preparation of linear nanofibers or even a failure of solid products [41].

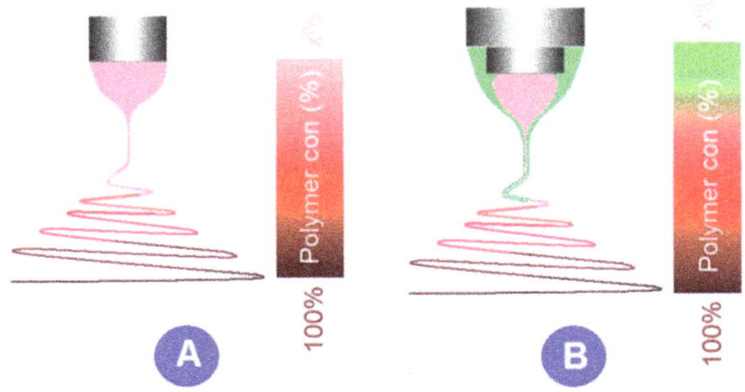

Figure 5. A diagram showing the solidification processes during (**A**) the traditional and (**B**) modified coaxial electrospinning using a solvent as a sheath working fluid [40].

The sheath-to-core fluid flow rate ratio R can be expressed as $R = F_S/F_C$, where Fs and F_c are flow rates of sheath and core working fluids, respectively. When R is manipulated in a suitable range, the sheath solvent will make positive influences on the whole coaxial working process and benefit the quality of final solid nanofibers. However, when R value is beyond the limitation, i.e., the sheath solvent flow rate is too large, the beads-on-a-string or spindles-on-a-string may be generated. This is because the sheath solvent can't be completely exhausted in time and in turn influence the solidification of the core fluid [42].

4. The Nanofibers Prepared from Unspinnable Liquids for Environment Remediation

Based on coaxial electrospinning, the applications of electrospun nanofibers from unspinnable fluids for pharmaceutical applications have been broadly reported, which can provide different kinds of drug-controlled release profiles such as immediate release, sustained release and multiple-phase release [43–46]. However, the publications about their environmental applications are still very limited, the published jobs have demonstrated the usefulness of this advanced electrospinning method in creating novel functional nanostructures for removing heavy metal ions from the polluted water and for antibacterial applications to keep a healthy environment.

4.1. Zein Nanoribbons from Sheath Surfactant Solution Electrospinning for Environmental Remediation

It is well-known that biomacromolecules have the potential to become fine adsorbents for the removal of heavy metals [47,48]. Particularly, protein molecules contain numerous groups with lone pairs of electrons and partial negative charges, and these groups allow protein molecules to absorb metal ions through electrostatic interaction and/or chelation. Zein is a typical plant protein, which is naturally abundant and eco-friendly. Shown in Figure 6, the interactions between the metal ions and the protein molecules comprise both the mechanism for modified coaxial electrospinning with unspinnable sheath fluid and mechanism for the ion removement from polluted water. On one hand, during the modified coaxial process, the sheath solution can eliminate the Mn^{+}-protein interactions for avoiding the frequent clogging, and thus ensure a continuous and robust preparation process. On the other hand, during the applications, this interaction is exploited to absorb lead ions from the polluted water for environmental remediation.

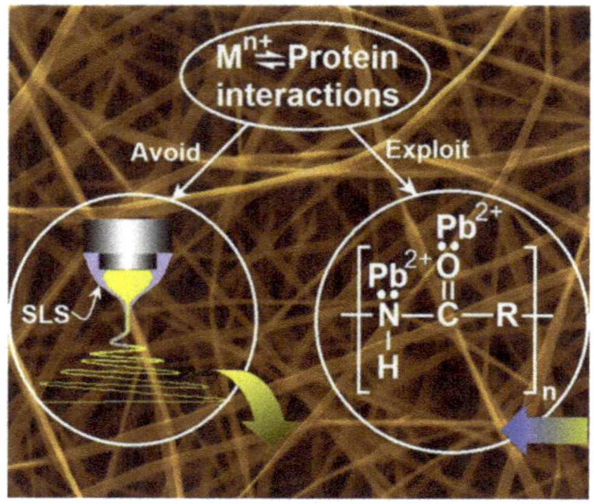

Figure 6. The mechanisms for the preparation of electrospun zein nanoribbons using unspinnable surfactant solution as a sheath fluid, and their applications in removing the lead ions from polluted water. Reprinted with permission from [49].

The difference between the traditional single-fluid blending process and the multiple-fluid process mainly lie in how to guide the fluids in an organized manner. Shown in Figure 7a,b are digital photos about the modified coaxial methods for creating zein nanoribbons. The upper-right inset shows a core-shell droplet, in which the core spinnable zein solution consisting of 30 g zein in 100 mL of a 75%/25% (*v/v*) ethanol/water mixture was surrounded by a transparent and unspinnable sheath fluid composed of 0.5% (*w/v*) sodium lauryl sulfate in 75%/25% (*v/v*) ethanol/water. An enlarged image shown in Figure 7b clearly tells the typical three steps of electrospinning, i.e., the compound Taylor cone (the upper-left inset), the straight fluid jet, and the bending and whipping regions with enlarged loops in the instable region.

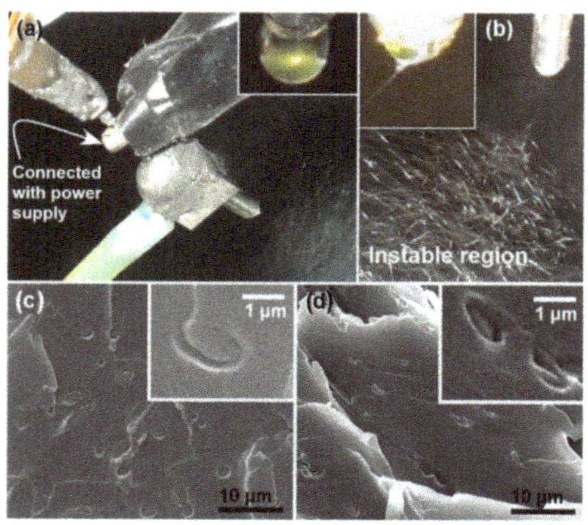

Figure 7. The working processes of modified coaxial electrospinning (**a**,**b**) using unspinnable surfactant solution as a sheath working fluid and the prepared zein nanoribbons (**c**,**d**). Reprinted with permission from [49].

Cross-sections of different zein nanoribbons were prepared, which are compared in Figure 7c,d. The former image is zein ribbons prepared using the single-fluid electrospinning. The latter image is zein ribbons created using a modified coaxial electrospinning, in which unspinnable SDS ethanol aqueous solution was exploited as a sheath working fluid. By estimation, the surface area of zein ribbons from the coaxial process was enlarged 2.29 times than those from the single-fluid process, which was favorite for lead ion adsorption. By exploiting the favorable interactions between metal and protein, the zein nanoribbon mats were used to treat Pb^{2+}-polluted water. Adsorption results indicated that the adsorption process can be described using the pseudo-second-order model. Isotherm data fitted well to the Langmuir isotherm model, with a maximum adsorption of 89.37 mg/g for the nanoribbons prepared from the modified coaxial process. Desorption results showed that the adsorption capacity can remain up to 82.3% even after 5 cycles of re-use. These positive results apparently demonstrated that the unspinnable working fluids utilized in the modified coaxial electrospinning were able to stabilize the preparation process, to further downsize the nanoproducts' width, and to improve their functional performances in environmental remediation.

4.2. Pure Solvent as a Sheath Fluid to Create Oligomer-Loaded Functional Nanofibers for Removing Mercury Ions

For the traditional single-fluid electrospinning, the filament-forming polymeric matrices must have enough molecular weight to support enough physical entanglements for forming electrospun fibers within their electrospinnable windows [50]. Few reports can be found on the directly electrospun nanofibers of oligomers because they haven't enough molecular weight and in turn no enough physical entanglements in their working fluids. These oligomers solutions are often un-electrospinnable and can't be converted into nanofibers.

However, in a recent investigation, an oligomer Poly(2-aminothiazole) (PAT) was synthesized, and then this functional oligomer solution (0.1 g of PAT was dissolved in a mixed solvent of 2 mL DMAC and 2 mL ethanol) was mixed with a cellulose acetate (CA, 1.44 g of CA was dissolved in 8 mL of acetone) solution to create composite nanofibers. The sheath fluid was pure solvent acetone, which was pumped quantitatively through another booster to another inlet of a concentric spinneret for a smooth working process.

Figure 8 shows two representative high-magnification TEM images of the PAT/CA fibers, which clearly present their core-sheath nanostructures. The sheath thickness ranged from 35 to 205 nm, and the core thickness ranged from 105 to 480 nm. The sheath section was mainly composed of small PAT nanoparticles with only few nanoparticles appeared in the core section. This phenomenon is beyond anticipation because there was no any solutes in the sheath working fluid. The resultant core-shell structures should be a reason of solid phase separation during the modified coaxial processes. The concrete forming mechanism is still waited to be further disclosed.

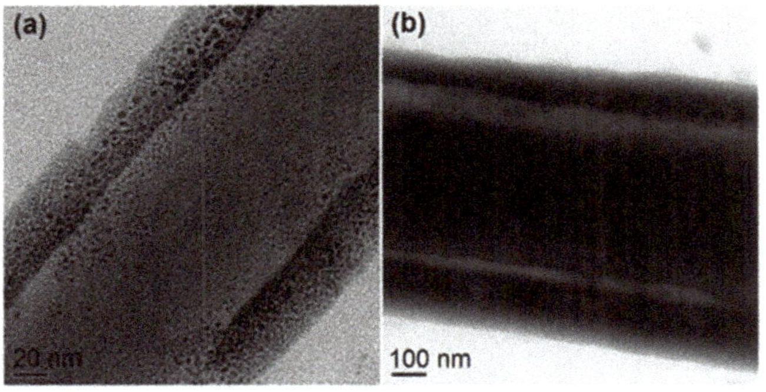

Figure 8. Both (**a**) and (**b**) are representative TEM images of 0.1 g PAT/1.44 g CA nanofibers with unexpected core-shell nanostructures. Reprinted with permission from [50].

The Hg(II) adsorption properties of the PAT/CA composite nanofiber mat from polluted water was tested. The adsorption capability was highly dependent on the pH value of solution, with a maximum at pH 6.5. The isotherm data fitted well to the Langmuir isothermal model. The adsorption kinetics data fitted well to the pseudo-second-order model. At a temperature of 298 K, the maximum adsorption capacity of 177 mg/g for PAT/CA fiber membrane with a very low PAT percentage (6.5 wt.%) was calculated. Through conversion, the adsorption capacity of the PAT in the composite fiber membrane is much larger than that of the nanoparticle-type PAT. The desorption experiments revealed that the desorption percentage remained at 80.2% after being used for three times. Here the unspinnable sheath solvent smoothed the electrospinning processes to prevent the spinneret clogging of CA, resulted in solid phase separation to lead to a surface distribution of functional ingredient, i.e., PAT particles, and thus provided an improved functional performance of removing Hg(II) from water. Certainly, the PAT-loaded nanofibers, as a non-woven mat, were more convenient than the PAT powders themselves for environmental remediations and less possibility of secondary pollutions due to nanoparticles themselves.

4.3. Nanocoating of Inorganic Materials on Polymeric Fibers for Antibacterial Applications Using Unspinnable Salt Solution as a Sheath Fluid

Through the modified coaxial electrospinning, it is possible to directly distribute the inorganic particles on the surface of polymeric nanofibers, in which nanosuspension can be explored as the sheath working fluid. Unfortunately, the inorganic nanoparticles are very easy to precipitate during the working processes, which may stop the continuous preparation. However, a combined strategy based on the modified coaxial electrospinning using salt solution as a sheath working fluid can overcome this issue.

With a solution of 15% (*w/v*) polyacrylonitrile (PAN) in N, N-dimethylacetamide (DMAc) as the electrospinnable core working fluid, a 10% (*w/v*) $AgNO_3$ DMAc solution was employed as a sheath fluid, a modified coaxial process was successfully carried out to create the salt coating PAN nanofibers. The electrospun nanofibers were then exposed to 254 nm ultraviolet light for 24 h to reduce the Ag^+ to Ag nanoparticles (NPs). The prepared products are shown in Figure 9a,b. Both nanofibers have kind linear morphology with the Ag NPs uniformly distributed on the PAN surface. This is very useful for the Ag NPs to exert their functional performances.

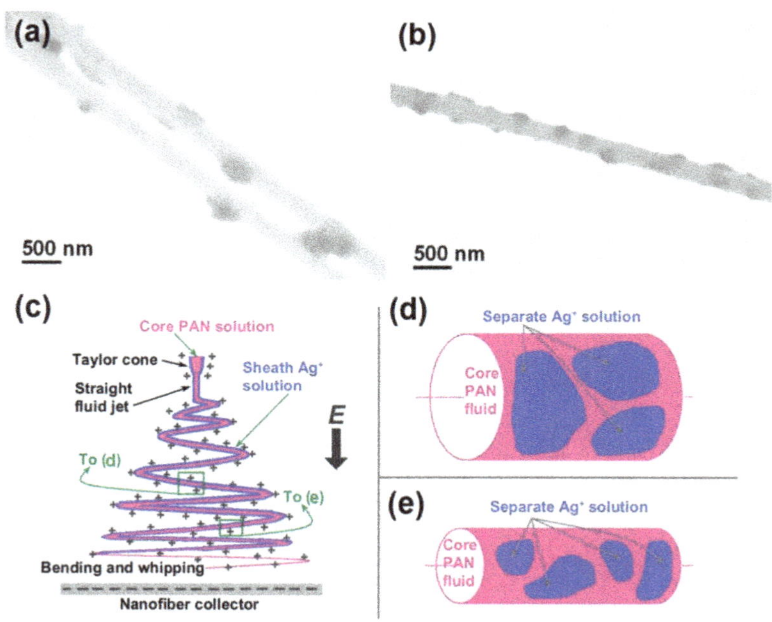

Figure 9. The Ag nanoparticles coating nanofibers (**a**,**b**) and the mechanisms for fabricating inorganic nanoparticles distributed PAN hybrids for antibacterial applications (**c**–**e**). Reprinted with permission from [51].

The influence of the key parameter, i.e., the sheath-to-core fluid flow rate ratio was investigated. The results demonstrated that the smaller flow rate of $AgNO_3$ solution of 0.1 was not able to follow the core PAN fluid jet during the fast bending and whipping process. This results in a separation of the sheath $AgNO_3$ solution and a non-uniform deposition of $AgNO_3$ on the surface of nanofibers (Figure 9c,d). Thus, a higher sheath-to-core fluid rate ratio of 0.2 was selected to carry out the experiments, which resulted in a more homogeneous distribution of $AgNO_3$ on the surface of nanofibers. When the Ag^+ on the surface of PAN nanofibers was reduced to Ag NPs, the nanofibers from a larger sheath fluid flow rate ratio had a more uniform but smaller size, and a tighter distribution than those from a smaller ratio (Figure 9c,e). It is anticipated that reduction of Ag^+ to Ag NP on the surface of nanofibers should be beneficial to long-term stability of the system and prevent Ag leaching from the fibers. These Ag NPs-loaded nanofibers showed fine antibacterial performances and should be candidates for friendly ecological environmental applications.

The popularity of electrospun nanofibers has a close relationship with their small diameters, huge surface area and large porosity [52–54]. However, for a single-fluid electrospinning process, the treated fluids must be electrospinnable to keep the formation of solid linear nanofibers, or they will be degraded into particles or even wet films [40,55]. The modified coaxial electrospinning, with an unspinnable sheath fluid on a spinnable core fluid, can't be utilized to prepare some advanced nanostructures presently such as hollow fibers from a traditional coaxial processes [56,57], which can be viewed as a special core-shell structure with an empty core. However, all the liquids, containing or not the filament-forming polymers, can be converted as a thin film on the solid core fibers to take advantages of the nanofibers' unique properties. This give a hint on many new possibilities in future. There are many publications about the environmental applications of electrospun composite nanofibers, in which the guest functional ingredients scattered within host polymer matrices [58–61]. With the valuable hints from these investigations, new types of nanofibers can be fabricated using the similar raw materials for an improved functional performance. The functional ingredients can be prepared as their owns' working fluids in the sheath section to superficialize them on the solid hybrid nanofibers. Meanwhile, the pure polymers core may increase the mechanical properties of the core-sheath structures.

5. Conclusions and Perspectives

As the fast development of electrospinning, modified coaxial electrospinning is upgraded from the traditional one and shows strong vitality. Its usefulness lies in both stabilizing working process for creating high quality monolithic nanofibers, but also in generating solid core-shell nanostructure and nanocoating. What is more, almost any kinds of functional raw materials, regardless of filament-forming property, can be treated into nanofibers to exert their functional performances by taking advantages of the unique properties of electrospun non-woven mats. The above-mentioned several examples demonstrated that the modified coaxial processes can be utilized to manipulate the nano products' shape and size facilely, to incorporate oligomer into the nanofibers, and to coat inorganic salt on the surface of polymeric nanofibers. These preparations are impossible using the traditional processes. Based on the combination of modified coaxial electrospinning and the tremendous active ingredients for treating pollutants in our environment, a new platform can be built to develop more and more functional nanostructures for environmental remediation.

The new nanomaterials and the corresponding applications resulted from the modified coaxial electrospinning using unspinnable liquid as a sheath working fluid is still at its infant time period. It is anticipated there are more and more possibilities in future, an unspinnable fluid can be treated with one or more spinnable fluids in many ways such as as tri-axial [62,63] and side-by-side [64]. Electrospinning and the electrospun nanofibers may find their applications in almost all the environmental issues such as gas pollution, water pollution and solid waste. They can also find applications for a better ecological environment, such as producing antibacterial materials, anti-UV materials, noise reduction and electromagnetic shielding [65]. Similarly, modified coaxial electrospinning and the corresponding nanofibers and nanostructures can upgrade these applications and generate new materials from an even wide selections of raw materials such as inorganic NPs, carbon nanotube, graphene and small molecules. In traditional monolithic nanofibers, the active ingredients are homogeneously distributed all over the polymeric matrices, means that most of them are buried into the nanofibers and can't play their roles for removing the pollutants [66]. With the unspinnable solution as a sheath working fluid, a wide variety of functional ingredients (such as cyclodextrin) can be effectively distributed on the surface of polymeric nanofibers [67,68], and thus their functional performances on environmental remediation can be easily enhanced. Certainly, the modified coaxial electrospinning can further expand its capability of creating novel functional material through a combination with other spinning methods or numerous traditional physical and chemical manners [69,70].

Just as the traditional electrospinning, one of the most important challenges for the modified coaxial electrospinning is to create nanofibers on a large scale [71–73]. Another important challenge is get more and more knowledge about the special working processes [74], here the key point is about the unspinnable sheath fluid's role during the transmission of solvents through air-fluid-fluid interfaces and the related solidification mechanism of electrospun nanostructures. Certainly, the structure-performance relationships for environmental and also other applications based on the nanoproducts prepared using modified coaxial electrospinning should be a very interesting topic deservin.

Author Contributions: Conceptualization, D.-G.Y. and K.W.; funding acquisition, K.W. and D.-G.Y.; project administration, D.-G.Y.; writing—original draft, M.W. and Y.Y.; writing—review and editing, D.-G.Y. and Y.L. All authors have read and agreed to the published version of the manuscript.

Funding: This research and the APC were funded by the National Natural Science Foundation of China (No. 51,803,121), Natural Science Foundation of University of Shanghai for Science and Technology (No. ZR18PY05), Program of Shanghai Pu-Jiang Program (No. 18PJ1409100), and Shanghai young Oriental scholar project.

Conflicts of Interest: The authors declare no conflicts of interest.

References

1. Cao, R.; Wang, J.N.; Zhao, S.Y.; Yang, W.; Yuan, Z.Q.; Yin, Y.Y.; Du, X.Y.; Li, N.W.; Zhang, X.L.; Li, X.Y.; et al. Self-powered nanofiber-based screen-print triboelectric sensors for respiratory monitoring. *Nano Res.* **2018**, *11*, 3771–3779. [CrossRef]
2. Lu, X.F.; Wang, C.; Wei, Y. One-dimensional composite nanomaterials: Synthesis by electrospinning and their applications. *Small* **2009**, *5*, 2349–2370. [CrossRef] [PubMed]
3. Wang, B.; Sun, Z.M.; Sun, Q.; Wang, J.; Du, Z.X.; Li, C.J.; Li, X.Y. The preparation of bifunctional electrospun air filtration membranes by introducing attapulgite for the efficient capturing of ultrafine PMs and hazardous heavy metal ions. *Environ. Pollut.* **2019**, *249*, 851–859. [CrossRef] [PubMed]
4. Wu, F.; Li, C.J.; Cao, R.; Du, X.Y. High-performance electronic cloth for facilitating the rehabilitation of human joints. *ACS Appl. Mater. Interfaces* **2019**, *11*, 22722–22729. [CrossRef] [PubMed]
5. Farkas, B.; Balogh, A.; Cselkó, R.; Molnár, K.; Farkas, A.; Borbás, E.; Marosi, G.; Nagy, Z.K. Corona alternating current electrospinning: A combined approach for increasing the productivity of electrospinning. *Int. J. Pharm.* **2019**, *561*, 219–227. [CrossRef] [PubMed]
6. Liu, W.C.; Zhang, J.W.; Liu, H. Conductive bicomponent fibers containing polyaniline produced via side-by-side electrospinning. *Polymers* **2019**, *11*, 945. [CrossRef]
7. Vass, P.; Hirsch, E.; Kóczián, R.; Démuth, B.; Farkas, A.; Fehér, C.; Szabó, E.; Németh, Á.; Andersen, K.S.; Vigh, T.; et al. Scaled-up production and tableting of grindable electrospun fibers containing a protein-type drug. *Pharmaceutics* **2019**, *11*, 329. [CrossRef]
8. Yu, D.G.; Li, J.J.; Zhang, M.; Williams, G.R. High-quality Janus nanofibers prepared using three-fluid electrospinning. *Chem. Commun.* **2017**, *53*, 4542–4545. [CrossRef]
9. Zhao, Y.; Cao, X.Y.; Jiang, L. Bio-mimic multichannel microtubes by a facile method. *J. Am. Chem. Soc.* **2007**, *129*, 764–765. [CrossRef]
10. Zhao, Y.; Jiang, L. Hollow micro/nanomaterials with multilevel interior structures. *Adv. Mater.* **2009**, *21*, 3621–3638. [CrossRef]

11. Zhao, K.; Wang, W.; Yang, Y.Y.; Wang, K.; Yu, D.G. From Taylor cone to solid nanofiber in tri-axial electrospinning: Size relationships. *Results Phys.* **2019**, *15*, 102770. [CrossRef]
12. Yu, D.G.; Li, J.J.; Williams, G.R.; Zhao, M. Electrospun amorphous solid dispersions of poorly water-soluble drugs: A review. *J. Control. Release* **2018**, *292*, 91–110. [CrossRef] [PubMed]
13. Liu, Z.; Qin, D.T.; Zhao, J.H.; Feng, Q.; Li, Z.T.; Bai, H.W.; Sun, D.D. Efficient oil/water separation membrane derived from super-flexible and superhydrophilic core-shell organic/inorganic nanofibrous architectures. *Polymers* **2019**, *11*, 974. [CrossRef] [PubMed]
14. Wang, X.X.; Pei, Y.X.; Hou, Y.; Pei, Z.C. Fabrication of core-shell magnetic molecularly imprinted nanospheres towards hypericin via click polymerization. *Polymers* **2019**, *11*, 313. [CrossRef] [PubMed]
15. Rascón-Chu, A.; Díaz-Baca, A.J.; Carvajal-Millan, E.; Pérez-López, E.; Hotchkiss, T.A.; González-Ríos, H.; Balandrán-Quintana, R.; Campa-Mada, C.A. Electrosprayed core-shell composite microbeads based on pectin-arabinoxylans for insulin carrying: Aggregation and size dispersion control. *Polymers* **2018**, *10*, 108. [CrossRef]
16. Caimi, S.; Timmerer, E.; Banfi, M.; Storti, G.; Morbidelli, M. Core-shell morphology of redispersible powders in polymer-cement waterproof mortars. *Polymers* **2018**, *10*, 1122. [CrossRef]
17. Huang, W.D.; Hou, Y.; Lu, X.Y.; Gong, Z.Y.; Yang, Y.Y.; Lu, X.J.; Liu, X.L.; Yu, D.G. The process–property–performance relationship of medicated nanoparticles prepared by modified coaxial electrospraying. *Pharmaceutics* **2019**, *11*, 226. [CrossRef]
18. Wang, K.; Wen, H.F.; Yu, D.G.; Yang, Y.Y.; Zhang, D.F. Electrosprayed hydrophilic nanocomposites coated with shellac for colon-specific delayed drug delivery. *Mater. Des.* **2018**, *143*, 248–255. [CrossRef]
19. Liu, Z.P.; Zhang, L.L.; Yang, Y.Y.; Wu, D.; Jiang, G.; Yu, D.G. Preparing composite nanoparticles for immediate drug release by modifying electrohydrodynamic interfaces during electrospraying. *Powder Technol.* **2018**, *327*, 179–187. [CrossRef]
20. Yu, D.G.; Zheng, X.L.; Yang, Y.Y.; Li, X.Y.; Williams, G.R.; Zhao, M. Immediate release of helicid from nanoparticles produced by modified coaxial electrospraying. *Appl. Surf. Sci.* **2019**, *473*, 148–155. [CrossRef]
21. Yang, Y.Y.; Zhang, M.; Liu, Z.P.; Wang, K.; Yu, D.G. Meletin sustained-release gliadin nanoparticles prepared via solvent surface modification on blending electrospraying. *Appl. Surf. Sci.* **2018**, *434*, 1040–1047. [CrossRef]
22. Loscertales, I.G.; Barrero, A.; Guerrero, I.; Cortijo, R.; Marquez, M.; Gañán-Calvo, A.M. Micro/nano encapsulation via electrified coaxial liquid jets. *Science* **2002**, *295*, 1695–1698. [CrossRef] [PubMed]
23. Dzenis, Y. Spinning continuous fibers for nanotechnology. *Science* **2004**, *304*, 1917–1919. [CrossRef]
24. Moghe, A.K.; Gupta, B.S. Co-axial electrospinning for nanofiber structures: Preparation and applications. *Polym. Rev.* **2008**, *48*, 353–377. [CrossRef]
25. Liu, X.K.; Yang, Y.Y.; Yu, D.G.; Zhu, M.J.; Zhao, M.; Williams, G.R. Tunable zero-order drug delivery systems created by modified triaxial electrospinning. *Chem. Eng. J.* **2019**, *356*, 886–894. [CrossRef]
26. Yu, D.G.; Wang, M.L.; Li, X.Y.; Liu, X.K.; Zhu, L.M.; Annie Bligh, S.W. Multifluid electrospinning for the generation of complex nanostructures. *Wiley Interdiscip. Rev. Nanomed. Nanobiotechnol.* **2019**. [CrossRef] [PubMed]
27. Yoon, J.; Yang, H.S.; Lee, B.S.; Yu, W.R. Recent progress in coaxial electrospinning: New parameters, various structures, and wide applications. *Adv. Mater.* **2018**, *30*, 1704765. [CrossRef]
28. Li, X.Y.; Hai, T.; Wang, Y.T.; Chen, X.; He, Y.Y.; Yu, D.G. Innovation training-from analogy to deepen comprehension and to innovation with modified coaxial electrospinning as an example. *Adv. Soc. Sci. Edu. Humanit. Res.* **2018**, *176*, 628–632.
29. Panthi, G.; Park, M.; Kim, H.Y.; Lee, S.Y.; Park, S.J. Electrospun ZnO hybrid nanofibers for photodegradation of wastewater containing organic dyes: A review. *J. Ind. Eng. Chem.* **2015**, *21*, 26–35. [CrossRef]
30. Wang, Q.; Yu, D.G.; Zhang, L.L.; Liu, X.K.; Deng, Y.C.; Zhao, M. Electrospun hypromellose-based hydrophilic composites for rapid dissolution of poorly water-soluble drug. *Carbohydr. Polym.* **2017**, *174*, 617–625. [CrossRef]
31. Zhou, H.; Shi, Z.R.; Wan, X.; Fang, H.L.; Yu, D.G.; Chen, X.H.; Liu, P. The relationships between process parameters and polymeric nanofibers fabricated using a modified coaxial electrospinning. *Nanomaterials* **2019**, *9*, 843. [CrossRef] [PubMed]
32. Wang, M.L.; Hai, T.; Feng, Z.B.; Yu, D.G.; Yang, Y.Y.; Annie Bligh, S.W. The relationships between the working fluids, process characteristics and products from the modified coaxial electrospinning of zein. *Polymers* **2019**, *11*, 1287. [CrossRef] [PubMed]
33. Chen, B.Y.; Lung, Y.C.; Kuo, C.C.; Liang, F.C.; Tsai, T.L.; Jiang, D.H.; Satoh, T.; Jeng, R.J. Novel multifunctional luminescent electrospun fluorescent nanofiber chemosensor-filters and their versatile sensing of pH, temperature, and metal ions. *Polymers* **2018**, *10*, 1259. [CrossRef] [PubMed]

34. Yew, H.C.; Azari, P.; Choi, R.J.; Muhamad, F.; Pingguan-Murphy, B. Electrospun polycaprolactone nanofibers as a reaction membrane for lateral flow assay. *Polymers* **2018**, *10*, 1387. [CrossRef] [PubMed]
35. Wu, Y.K.; Wang, L.; Fan, J.; Shou, W.; Zhou, B.M.; Liu, Y. Multi-jet electrospinning with auxiliary electrode: The influence of solution properties. *Polymers* **2018**, *10*, 572. [CrossRef] [PubMed]
36. Kim, Y.I.; An, S.; Kim, M.W.; Jo, H.S.; Kim, T.G.; Swihart, M.T.; Yarin, A.L.; Yoon, S.S. Highly transparent, conducting, body-attachable metallized fibers as a flexible and stretchable film. *J. Alloys Compd.* **2019**, *790*, 1127–1136. [CrossRef]
37. Kim, M.W.; An, S.; Seok, H.; Yoon, S.S.; Yarin, A.L. Electrostatic transparent air filter membranes composed of metallized microfibers for particulate removal. *ACS Appl. Mater. Interfaces* **2019**, *11*, 26323–26332. [CrossRef]
38. Yang, Y.Y.; Zhu, T.H.; Liu, Z.P.; Luo, M.Y.; Yu, D.G.; Annie Bligh, S.W. The key role of straight fluid jet in predicting the drug dissolution from electrospun nanofibers. *Int. J. Pharm.* **2019**, *569*, 118634. [CrossRef]
39. Li, X.; Zhang, Y.C.; Li, H.; Chen, H.B.; Ding, Y.M.; Yang, W.M. Effect of oriented fiber membrane fabricated via needleless melt electrospinning on water filtration efficiency. *Desalination* **2014**, *344*, 266–273. [CrossRef]
40. Liu, Z.; Feng, Z.; Zhao, G.; He, C.; Yu, D. Lessons from a key parameter of coaxial electrospinning for engineering education to college students. *Adv. Soc. Sci. Edu. Humanit. Res.* **2018**, *176*, 590–593.
41. Yang, G.Z.; Li, J.J.; Yu, D.G.; He, M.F.; Yang, J.H.; Williams, G.R. Nanosized sustained-release drug depots fabricated using modified tri-axial electrospinning. *Acta Biomater.* **2017**, *53*, 233–241. [CrossRef] [PubMed]
42. Huang, W.D.; Yang, Y.Y.; Zhao, B.W.; Liang, G.Q.; Liu, S.W.; Liu, X.L.; Yu, D.G. Fast dissolving of ferulic acid via electrospun ternary amorphous composites produced by a coaxial process. *Pharmaceutics* **2018**, *10*, 115. [CrossRef] [PubMed]
43. Nair, R.P. Delivering combination chemotherapies and targeting oncogenic pathways via polymeric drug delivery systems. *Polymers* **2019**, *11*, 630. [CrossRef] [PubMed]
44. Liu, X.K.; Shao, W.Y.; Luo, M.Y.; Bian, J.Y.; Yu, D.G. Electrospun blank nanocoating for improved sustained release profiles from medicated gliadin nanofibers. *Nanomaterials* **2018**, *8*, 184.
45. Wang, Q.; Li, H.P.; Yang, C.; Li, J.J.; Yu, D.G. Beads-on-a-string amorphous solid dispersion fabricated using a modified coaxial electrospinning. *J. Control. Release* **2017**, *259*, e111–e112. [CrossRef]
46. Li, J.J.; Yang, C.; Li, H.P.; Wang, Q.; Yu, D.G. Oral controlled release in accordance with drug adsorption biological rhythm provided by an electrospun structural amorphous solid dispersion. *J. Control. Release* **2017**, *259*, e61–e62. [CrossRef]
47. Hong, J.H.; An, S.; Song, K.Y.; Kim, Y.I.; Yarin, A.L.; Kim, J.J.; Yoon, S.S. Eco-friendly lignin nanofiber mat for protection of wood against attacks by environmentally hazardous fungi. *Polym. Test.* **2019**, *74*, 113–118. [CrossRef]
48. Amin, M.T.; Alazba, A.A. A review of nanomaterials-based membranes for removal of contaminants from polluted waters. *Membr. Water Treat.* **2014**, *5*, 123–146. [CrossRef]
49. Wen, H.F.; Yang, C.; Yu, D.G.; Li, X.Y.; Zhang, D.F. Electrospun zein nanoribbons for treatment of lead-contained wastewater. *Chem. Eng. J.* **2016**, *290*, 263–272. [CrossRef]
50. Zou, H.; Lv, P.F.; Wang, X.; Wu, D.; Yu, D.G. Electrospun poly(2-aminothiazole)/cellulose acetate fiber membrane for removing Hg(II) from water. *J. Appl. Polym. Sci.* **2017**, *134*. [CrossRef]
51. Yu, D.G.; Zhou, J.; Chatterton, N.P.; Li, Y.; Huang, J.; Wang, X. Polyacrylonitrile nanofibers coated with silver nanoparticles using a modified coaxial electrospinning process. *Int. J. Nanomed.* **2012**, *7*, 5725–5732. [CrossRef] [PubMed]
52. Jiang, S.; Uch, B.; Agarwal, S.; Greiner, A. Ultralight, thermally insulating, compressible polyimide fiber assembled sponges. *ACS Appl. Mater. Interfaces* **2017**, *9*, 32308–32315. [CrossRef] [PubMed]
53. Sharma, P.R.; Joshi, R.; Sharma, S.K.; Hsiao, B.S. A simple approach to prepare carboxycellulose nanofibers from untreated biomass. *Biomacromolecules* **2017**, *18*, 2333–2342. [CrossRef] [PubMed]
54. Sharma, P.R.; Chattopadhyay, A.; Sharma, S.K.; Geng, L.; Amiralian, N.; Martin, D.; Hsiao, B.S. Nanocellulose from spinifex as an effective adsorbent to remove cadmium(II) from water. *ACS Sustain. Chem. Eng.* **2018**, *6*, 3279–3290. [CrossRef]
55. Geng, L.; Peng, X.; Zhan, C.; Naderi, A.; Sharma, P.R.; Mao, Y.; Hsiao, B.S. Structure characterization of cellulose nanofiber hydrogel as functions of concentration and ionic strength. *Cellulose* **2017**, *24*, 5417–5429. [CrossRef]
56. Duan, G.; Greiner, A. Air-blowing-assisted coaxial electrospinning toward high productivity of core/sheath and hollow fibers. *Macromol. Mater. Eng.* **2019**, *304*, 1800669. [CrossRef]
57. Ouyang, W.; Liu, S.; Yao, K.; Zhao, L.; Cao, L.; Jiang, S.; Hou, H. Ultrafine hollow TiO_2 nanofibers from core-shell composite fibers and their photocatalytic properties. *Compos. Commun.* **2018**, *9*, 76–80. [CrossRef]

58. Sharma, P.R.; Sharma, S.K.; Antoine, R.; Hsiao, B.S. Efficient removal of arsenic using zinc oxide nanocrystal-decorated regenerated microfibrillated cellulose scaffolds. *ACS Sustain. Chem. Eng.* **2019**, *7*, 6140–6151. [CrossRef]
59. Sharma, P.R.; Varma, A.J. Functional nanoparticles obtained from cellulose: Engineering the shape and size of 6-carboxycellulose. *Chem. Commun.* **2013**, *49*, 8818–8820. [CrossRef]
60. Rajinipriya, M.; Nagalakshmaiah, M.; Robert, M.; Elkoun, S. Importance of agricultural and industrial waste in the field of nanocellulose and recent industrial developments of wood based nanocellulose: A review. *ACS Sustain. Chem. Eng.* **2018**, *6*, 2807–2828. [CrossRef]
61. Yu, H.Y.; Zhang, D.Z.; Lu, F.F.; Yao, J. New approach for single-step extraction of carboxylated cellulose nanocrystals for their use as adsorbents and flocculants. *ACS Sustain. Chem. Eng.* **2016**, *4*, 2632–2643. [CrossRef]
62. Yu, D.G.; Li, H.P.; Yang, C.; Li, J.J.; Wang, Q.; Williams, G. Double-pulsatile release core-shell fibers fabricated using modified tri-axial electrospinning. *J. Control. Release* **2017**, *259*, e24–e25. [CrossRef]
63. Yang, C.; Yu, D.G.; Pan, D.; Liu, X.K.; Wang, X.; Bligh, S.A.; Williams, G.R. Electrospun pH-sensitive core–shell polymer nanocomposites fabricated using a tri-axial process. *Acta Biomater.* **2016**, *35*, 77–86. [CrossRef] [PubMed]
64. Yu, D.G.; Yang, C.; Jin, M.; Williams, G.R.; Zou, H.; Wang, X.; Bligh, S.A. Medicated Janus fibers fabricated using a Teflon-coated side-by-side spinneret. *Colloids Surf. B Biointerfaces* **2016**, *138*, 110–116. [CrossRef] [PubMed]
65. Omura, T.; Chan, C.H.; Wakisaka, M.; Nishida, H. Organic thin paper of cellulose nanofiber/polyaniline doped with (±)-10-camphorsulfonic acid nanohybrid and its application to electromagnetic shielding. *ACS Omega* **2019**, *4*, 9446–9452. [CrossRef] [PubMed]
66. Dhandayuthapani, B.; Mallampati, R.; Sriramulu, D.; Dsouza, R.F.; Valiyaveettil, S. PVA/gluten hybrid nanofibers for removal of nanoparticles from water. *ACS Sustain. Chem. Eng.* **2014**, *2*, 1014–1021. [CrossRef]
67. Forouharshad, M.; Putti, M.; Basso, A.; Prato, M.; Monticelli, O. Biobased system composed of electrospun sc-PLA/POSS/cyclodextrin fibers to remove water pollutants. *ACS Sustain. Chem. Eng.* **2015**, *3*, 2917–2924. [CrossRef]
68. Wu, H.Q.; Kong, J.H.; Yao, X.Y.; Zhao, C.Y.; Dong, Y.L.; Lu, X.H. Polydopamine-assisted attachment of β-cyclodextrin on porous electrospun fibers for water purification under highly basic condition. *Chem. Eng. J.* **2015**, *270*, 101–109. [CrossRef]
69. Molnar, K.; Jedlovszky-Hajdu, A.; Zrinyi, M.; Jiang, S.; Agarwal, S. Poly (amino acid)-based gel fibers with pH responsivity by coaxial reactive electrospinning. *Macromol. Rapid Commun.* **2017**, *38*, 1700147. [CrossRef]
70. Sharma, P.R.; Chattopadhyay, A.; Sharma, S.K.; Hsiao, B.S. Efficient removal of UO22+ from water using carboxycellulose nanofibers prepared by the nitro-oxidation method. *Ind. Eng. Chem. Res.* **2017**, *56*, 13885–13893. [CrossRef]
71. Agarwal, S.; Jiang, S.; Chen, Y. Progress in the field of water-and/or temperature-triggered polymer actuators. *Macromol. Mater. Eng.* **2019**, *304*, 1800548. [CrossRef]
72. Wang, Q.; Yu, D.G.; Zhou, S.Y.; Li, C.; Zhao, M. Fabrication of amorphous electrospun medicated-nanocomposites using a Teflon-based concentric spinneret, e. *Polymer* **2018**, *18*, 3–11.
73. Jiang, S.; Chen, Y.; Duan, G.; Mei, C.; Greiner, A.; Agarwal, S. Electrospun nanofiber reinforced composites: A review. *Polym. Chem.* **2018**, *9*, 2685–2720. [CrossRef]
74. Ding, C.; Fang, H.; Duan, G.; Zou, Y.; Chen, S.; Hou, H. Investigating the draw ratio and velocity of an electrically charged liquid jet during electrospinning. *RSC Adv.* **2019**, *9*, 13608–13613. [CrossRef]

© 2020 by the authors. Licensee MDPI, Basel, Switzerland. This article is an open access article distributed under the terms and conditions of the Creative Commons Attribution (CC BY) license (http://creativecommons.org/licenses/by/4.0/).

polymers

MDPI

Article

The Effect of Natural Additives on the Composting Properties of Aliphatic Polyesters

Malgorzata Latos-Brozio and Anna Masek *

Institute of Polymer and Dye Technology, Lodz University of Technology, ul. Stefanowskiego 12/16, 90-924 Lodz, Poland; malgorzata.latos-brozio@dokt.p.lodz.pl
* Correspondence: anna.masek@p.lodz.pl

Received: 8 July 2020; Accepted: 14 August 2020; Published: 19 August 2020

Abstract: Compounds of plant origin are used with polymers as functional additives. However, these substances often have biological (antimicrobial) activity. The bactericidal and fungicidal properties of natural additives can affect the composting process of biodegradable polymers. The scientific novelty of the manuscript is the investigation of the effect of the addition of herbal antimicrobial functional substances on the composting process of green polymers. The aim of the study is to analyze composting processes of biodegradable polymers polylactide (PLA) and polyhydroxyalkanoate (PHA) containing β-carotene, juglone, morin, and curcumin. As part of the research, six-month composting of materials was performed. At time intervals of one month, the weight loss of samples, surface energy, colour change, mechanical properties, and carbonyl indices (based on FTIR spectroscopy) of composted materials were examined. The research results showed that the addition of selected plant substances slightly slowed down the process of polymer composting. Slower degradation of samples with plant additives was confirmed by the results of mechanical strength tests and the analysis of changes in carbonyl index (CI). The CI analysis showed that PLA and PHA containing a natural additive degrade a month later than reference samples. However, PLA and PHA polyesters with β-carotene, juglone, morin, and curcumin were still very biodegradable.

Keywords: biodegradable polyesters; plant substances; composting

1. Introduction

Biodegradable polymers such as polyhydroxyalkanoates (PHA) and polylactide (PLA) are becoming more popular due to their unique properties. Aliphatic polyesters PLA and PHA are easily composted and biodegradable, therefore they are an excellent alternative to commercial plastic packaging, especially for single use [1,2]. Biodegradation is the process of degradation caused by biological activity, particularly by enzymatic action, leading to a meaningful change in the chemical structure of the exposed polymeric material and resulting in the production of carbon dioxide, water, mineral salts (mineralization), and also new microbial cellular constituents (biomass). Biodegradable materials are plastics that undergo degradation due to the action of naturally occurring microorganisms, such as bacteria, fungi, and algae [3–6].

Nowadays, the development of green additives and fillers is increasing more and more to reduce the environmental impact of polymers. Natural compounds are increasingly used as functional substances in polymeric materials [7–19]. Phytocompounds, β-carotene, juglone, morin, and curcumin, were selected for this study.

Β-carotene is a plant pigment from the carotenoids group. Carotenoids have been shown to act as antioxidants by quenching photosensitizers, interacting with singlet oxygen and scavenging peroxygen radicals. Moreover, these compounds exhibit a significant antimicrobial effect [20–22]. Another selected phytocompound, juglone, is a naphthoquinone type dye reported to be found in plants of the family *Juglandaceae* [23]. Juglone has been tested for phytotherapeutic applications due to its antiviral, antibacterial, and antifungal properties [23–27]. The next chosen natural compound was morin, belonging to the flavonoids group. Morin, like other flavonoids, has strong antioxidant properties and antibacterial and antifungal activity [28,29]. Curcumin was the last phytocompound introduced into polymers. It has been shown that this natural spice and common food colorant has, among other attributes, significant antifungal effect [30,31].

Natural substances (β-carotene, juglone, morin, and curcumin) have been used in the authors' other manuscripts as stabilizers and indicators of aging time of polymers. However, in the cited publications, the influence of plant additives on the degradation processes of polymeric materials was not investigated [32–37]. Selected natural compounds, in addition to strong antioxidant and dyeing properties, also exhibit antimicrobial activity. The biological activity of compounds of plant origin may have a negative impact on the processes occurring during the biodegradation of polymers. Antimicrobial properties of phytocompounds can potentially slow down polymer composting and degradation processes.

Few literature references concern the analysis of the influence of functional plant substances on the degradation processes of eco-friendly polymers. In literature data, only Moraczewski et al. examined the effect of introducing natural anti-aging compounds of plant origin on polylactide composting [38]. Coffee, cocoa, and cinnamon extracts were added to polylactide in amounts of 0.5; 5, and 10 wt%. Samples were composted for 7, 14, 21, or 28 days. The authors showed that these plant extracts did not have negative effects on the industrial composting process. Coffee and cinnamon extracts even accelerated and intensified biodegradation of polylactide [38].

The aim of this study is to examine the effect of the addition of plant substances with antimicrobial activity on the composting process of biodegradable aliphatic polyesters. So far, the influence of β-carotene, juglone, morin, and curcumin on the composting of PLA and PHA has not been tested. The research presented in this manuscript develops scientific knowledge and helps to rectify the current dearth of information in the literature.

2. Materials and Methods

2.1. Reagents

The biodegradable polymeric materials used in this study were polylactide (PLA) and polymer P(3,4HB) 2001 from the polyhydroxyalkanoate group of polymers (PHA). Polylactide (PLA), Ingeo^{TM} Biopolymer 4043D PLA, was obtained from Nature Works^{TM} (Minnetonka, MN, USA) and had the properties: T_g = 55–60 °C, Tm = 145–160 °C, and melt flow index MFI = 6 g/10 min. Polyester (PHA) was produced by Simag Holdings LTD (Hong Kong, China) and had properties: P(3,4HB) containing 12 mol% 4-hydroxybutyrate, the average M_w was approximately 520 kDa, MVR = 15–20 g/10 min (assay conditions: temperature 170 °C, nominal load 2.16 kg), and a density of 1.25 g/cm^3.

Phytochemicals: β-carotene (≥93%; Mw 536.87 g/mol; melting point range 178–179 °C), juglone (5-hydroxy-1,4-naphthoquinone, 97%; Mw 174.15 g/mol; melting point range 161–163 °C), morin (hydrate, ≥90%; Mw 302.24 g/mol; melting point range 299–300 °C), and curcumin (≥94%; Mw 368.38 g/mol; melting point 183 °C) were introduced into the polyester materials. All natural additives were purchased from Sigma-Aldrich (Darmstadt, Germany). Structural formulae of polymers and polyphenols are shown in Figure 1.

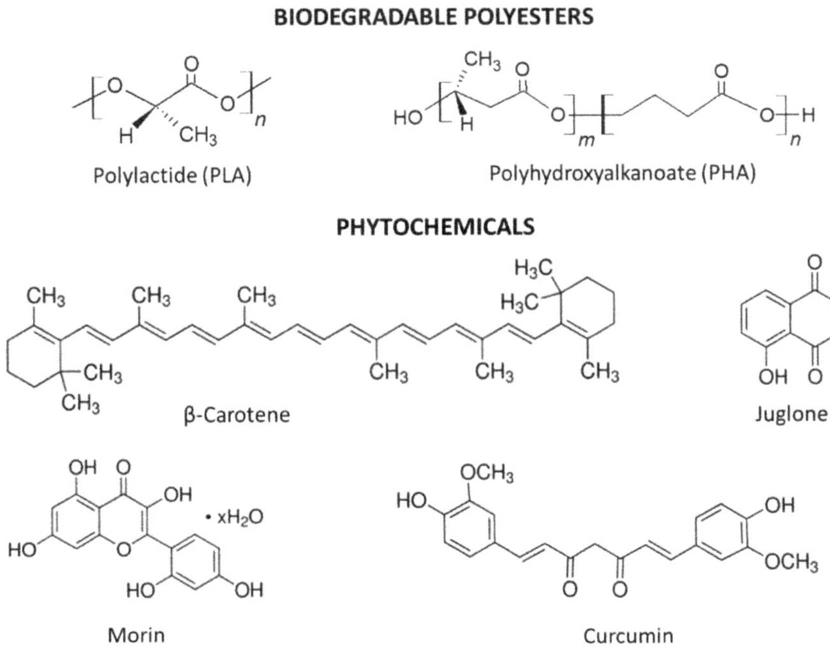

Figure 1. Structure of biodegradable polyesters and phytochemicals.

2.2. Method of Preparation of PLA and PHA with Phytochemicals

Dried (12 h, 50 °C) granulates of polymers PLA and PHA were mixed with 1 part by weight of β-carotene, juglone, morin, or curcumin and extruded using a laboratory extruder. Strip samples with a thickness of 1.6–1.8 mm were obtained. The temperature of the working chamber of the extruder was 180 °C for PLA (feed zone temperature 25 °C; cylinder zone temperature 190 °C; nozzle zone temperature 180 °C) and 160 °C for PHA (feed zone temperature 25 °C; cylinder zone temperature 165 °C; nozzle zone temperature 160 °C), screw rotation speed was 40 rpm, and extrusion pressure was 17 atm.

2.3. Method for Composting Polyester Samples

Polymer samples were placed in a ceramic vessel filled with garden soil with high peat content (soil pH 5.5–6.5) and left for 6 months. During the composting, a constant temperature of 30 °C and humidity (60%) were maintained and monitored. Samples were removed from the soil at intervals of 1 month for half a year. After removing soil residue with a brush, the samples were dried to constant weight and weighed, and then subjected to further analysis. The weight losses [%] were calculated based on the changes in the mass of the samples.

2.4. Measurement Methods

2.4.1. Surface Free Energy

A goniometer OEC 15EC (DataPhysics Instruments GmbH, Filderstadt, Germany) was utilized to determine surface free energy of the polymers before and after composting. Surface free energy was calculated by the method of Owens, Wendt, Rabel, and Kaelble (OWRK) using software module

SCA 20. Polar and disperse contributions to the surface energy and surface tension were combined by forming the sum of both parts, leading to (1) and (2):

$$\sigma_l = \sigma_l^d + \sigma_l^p \quad (1)$$

$$\sigma_S = \sigma_S^d + \sigma_S^p \quad (2)$$

where σ_l^d and σ_l^p represent the disperse and polar parts of the liquid, while σ_S^d and σ_S^p stand for the respective contributions of the solid.

Surface energy measurements were done based on the determination of contact angle. The measurements of contact angle were made for liquids with different polarities: distilled water, diiodomethane, and ethylene glycol. During the determination of surface energy on each of the three samples of one material, 10 contact angles were measured for each of the three liquids.

2.4.2. Change of Color

The change of color determinations were carried out using a CM-3600d spectrophotometer (Konica Minolta Sensing, Osaka, Japan). Color measurements were performed to measure the color change of the polymeric materials before and after composting. The result of the test is the color as described in the CIE-Lab space and the color in a system of three coordinates: L, a and b, where L is the lightness parameter (maximum value of 100, representing a perfectly reflecting diffuser, minimum value of zero representing the color black), a is the axis of red–green, and b is the axis of yellow–blue. The a and b axes have no specific numerical limits. The change of color, dE*ab, was computed according to Equation (3):

$$\mathrm{dE}*\mathrm{ab} = \sqrt{(\Delta a^2) + (\Delta b^2) + (\Delta L^2)} \quad (3)$$

Visual changes of color of samples before and after composting were recorded using a camera.

For change of color determination, 3 samples of each type of the polymeric material were prepared for each time interval of composting. Spectrophotometric determinations were done at 5 measuring points on each of the 3 control samples.

2.4.3. Mechanical Properties

The mechanical properties tests presented in the manuscript were static tensile test. Mechanical properties of polymers were determined using a Zwick Roell Z005 test machine (Zwick Roell, Ulm, Germany) before and after composting. To determine mechanical properties, six control samples were cut out from extruded strips with a thickness of 1.6–1.8 mm and length of 150 mm. The measurement conditions were: a preload of 0.1 N and a test speed of 50 mm/min. The parameters: T_{Fmax}, the maximum tensile stress [MPa], E_{Fmax}, the elongation at break for maximum tensile stress [%], σ, the tensile strength [MPa], and ε, the total elongation at break (just before the destruction of the sample) [%] were measured.

2.4.4. Fourier Transform Infrared Spectroscopy (FTIR)

The measurements were carried out using a Thermo Scientific Nicolet 6700 FT-IR spectrometer (Thermo Fisher Scientific, Waltham, MA, USA) equipped with a diamond accessory, Smart Orbit ATR, for analyzing samples in the wave number range from 4000–400 cm^{-1}. Samples were placed at the output of infrared beams. As the result of the study, oscillatory spectra were obtained, the analysis of which allows determination of the functional groups with which the radiation interacted.

For each of the analyzed samples, based on the FT-IR spectrum, a carbonyl index (CI) was calculated, according to Equation (4):

$$CI = \frac{I_{C=O}}{I_{C-H}} \quad (4)$$

where:

$I_{C=O}$—the intensity of the peak corresponding to the carbonyl groups C = O [-],
I_{C-H}—the intensity of the peak corresponding to the aliphatic carbon chains [-].
Carbonyl index (CI) is a measure of the number of carbonyl groups in the tested samples generated during composting.

3. Results and Discussion

Measuring the weight loss of a material after a certain composting time is the easiest and the most popular method for testing the approximate degree of biodegradation. In the process, some of the degraded material is converted into water, CO_2 and other products, which results in a reduction of its mass. The biodegradation efficiency depends, not only on the conditions of composting, but also on the dimensions of the polymer sample.

Figure 2A,B shows the weight loss of PLA and PHA samples over 6 months of composting. After the first month of PLA composting, slight weight losses were found. The weight loss of the reference sample was 0.02% and of the samples containing natural additives from 0% (PLA/morin) to 0.08% (PLA/β-carotene). During the next four months of composting, the weight loss for PLA ranged from 0.14 to 0.38%. For samples containing plant substances, a weight loss from 0.10% to 0.84% was found. The greatest weight loss occurred after the sixth month of composing. After the last stage of composting, the weight loss of PLA was 2.84% and PLA with natural compounds from 0.60% (PLA/juglone) to 3.22% (PLA/β-carotene). For PHA samples, the weight loss after the first month of composting was 0.48% (reference sample) and from 0.04 to 0.36% for samples with natural additives. As a function of time, there was a gradual increase in weight loss of all samples, the greatest for the reference PHA (months 2–5: weight change from 0.54 to 3.91%) and for samples containing juglone (months 2–5: weight loss range from 0.56 to 3.35%). As in the case of PLA samples, the composition based on PHA also showed the greatest weight loss after 6 months of composting (PHA 5.64%, PHA with additives 0.40–3.67%). Higher mass losses of samples made using PHA polymer may result from the structure of the polymer matrix. As demonstrated by SEM studies published in another manuscript [39], the PHA samples were porous, while the PLA samples were compact and smooth. The high porosity of PHA materials can cause increased adsorption and absorption of water present during composting, as well as increased availability of material for microorganisms.

The behavior of PLA and PHA samples with curcumin and juglone seems to be interesting. PLA samples with juglone showed the lowest weight loss, while PHA samples with this plant additive showed weight losses similar to the reference PHA. Similarly, for the samples with curcumin–PLA/curcumin had the highest weight loss (greater than the reference PLA), and PHA/curcumin–low, similar to the samples with morin and β-carotene. This behavior of the samples may be related to the miscibility and solubility of specific natural additive in the polymer matrix of PLA or PHA. Moreover, the polarity of the PLA and PHA polymer matrices is different. This is evidenced by the surface energy values of polymers (PLA 42.05 mN/m^2; PHA 37.63mN/m^2). Due to the different polarity of the materials, the interactions of curcumin and juglon with polymers may be different. This may be the reason for a different effect of additives on the weight loss of PLA and PHA samples.

Reference samples of both polymers were more compostable than samples containing additives of plant origin. Due to fungicidal and bactericidal properties, polymer compositions with the addition of phytochemicals may show lower susceptibility to composting and biodegradation. Despite the lower compostability, the polyester materials were still degradable under composting conditions.

In the compost environment (elevated temperature, high humidity, presence of microorganisms), there are favorable conditions for the occurrence of two mechanisms of degradation of polymeric materials, i.e., hydrolytic degradation (a chemical process that causes chemical bonds in the polymer molecule to break) and enzymatic degradation (actions of microorganisms). In practice, the distribution of polymers under composting conditions is complex, which makes it impossible to identify only one mechanism responsible for this process. Degradation leads to a reduction in the molecular weight of all high molecular components of the polymer material, which in the case of one-component material, is associated with its distribution. In multi-component polymer materials, degradation of one of the components of a composite can only cause the loss of its cohesiveness and dispersion of other components in the environment, not complete degradation.

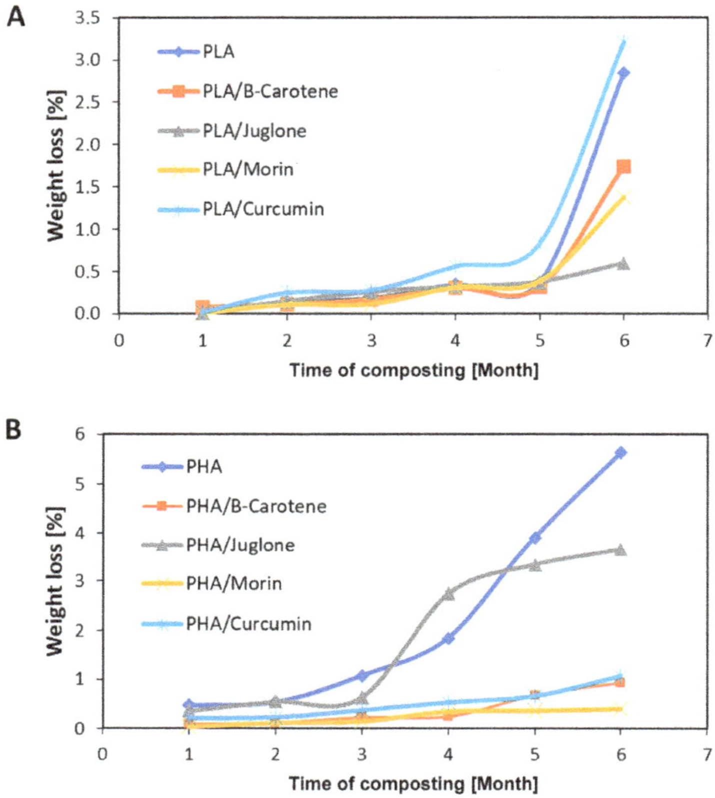

Figure 2. The weight loss of PLA (**A**) and PHA (**B**) samples over 6 months of composting.

The second step in the study was to examine the color change of the samples after composting (Figures 3 and 4). The change of color is the first sign of degradation of polymeric materials. Samples of both polyesters with the addition of substances of plant origin were characterized by a very pronounced color change, larger than the reference samples. When the color change factor dE*ab > 5, the colors are perceived as completely different. The largest color changes were found for samples containing B-carotene and juglone. Both B-carotene and juglone are substances of intense color and are successfully used as natural dyes, e.g., B-carotene is widely applied in the food industry. Juglone and substances from the quinone group of dyes are used, among other things, for dyeing fabrics. Composting samples can cause changes in the structure of the polymer matrix but can also initiate changes in the structure of substances of plant origin, e.g., cracking of C-C and C-H bonds, which will result in degradation of the plant additive and, as a consequence, a clear change in the color of the polymer material. Natural substances were added to various polymer matrices that differed from polarity, miscibility, affinity for specific additives. Therefore, the color change trends between PLA and PHA samples may be different. Visual changes of color of the samples during composting are shown in the photographs in Figures 3A and 4A. Visual changes in the samples of both polyesters correspond to the spectrophotometric determination of the dE*ab coefficient.

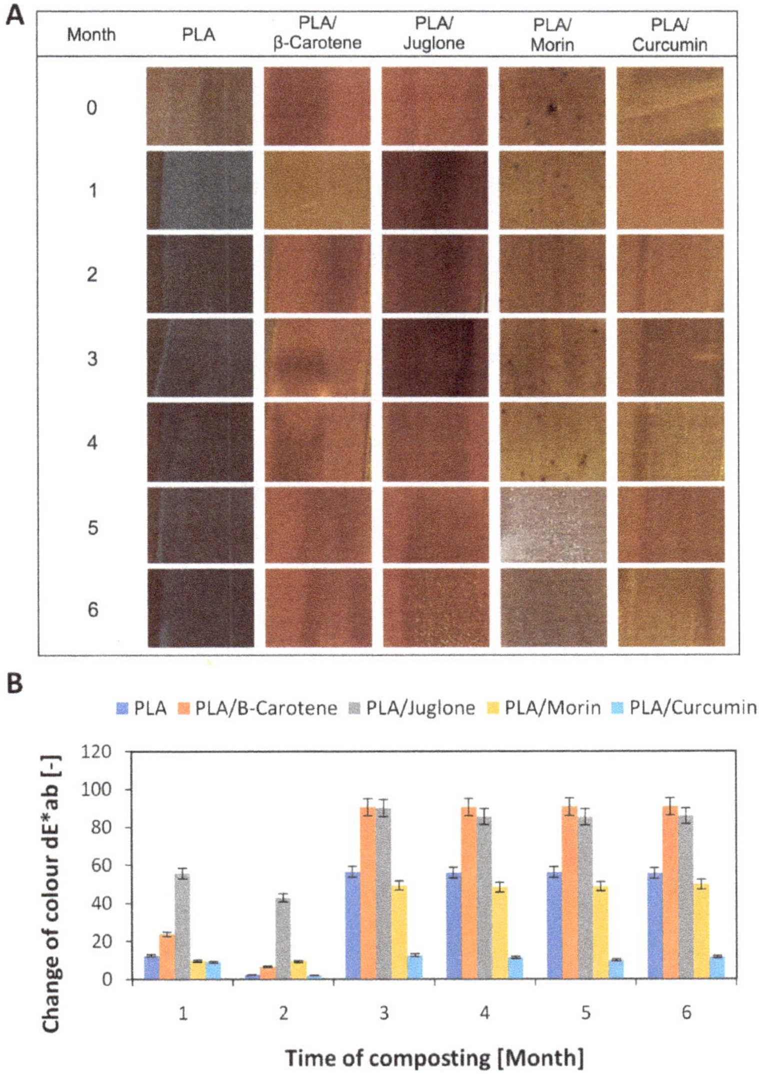

Figure 3. Visual changes of PLA samples (**A**) over 6 months of composting, confirmed by spectrophotometric determination of changes of color (**B**) in CIE-Lab space system.

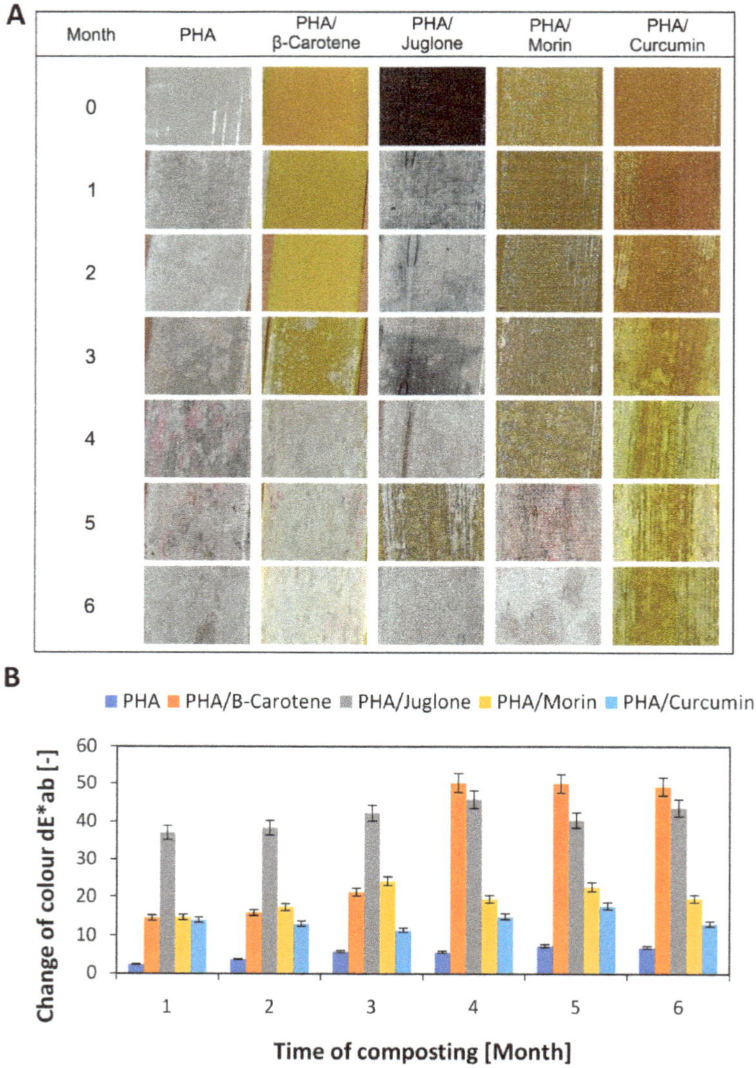

Figure 4. Visual changes of PHA samples (**A**) over 6 months of composting, confirmed by spectrophotometric determination of changes of color (**B**) in CIE-Lab space system.

An important parameter for polymeric materials is their surface energy, calculated on the basis of contact angle (Figure 5). The average contact angle for water for PLA and PLA with natural additives was 73°. A contact angle value below 90° indicates that the material has hydrophilic nature, and hence good compostability. Samples based on PHA polymer also had a hydrophilic surface character—the average contact angle for water was 72° (Tables 1 and 2). The hydrophilic nature of both polyesters determines their good susceptibility to composting processes. For materials based on polylactide and polyhydroxyalkanoate, it was not found that the addition of substances of natural origin significantly affected the change in the surface energy of reference samples. Only the PLA/morin and PHA/juglone samples showed slightly lower surface energy compared to reference samples. The lower surface energy of the PLA/morin sample, compared to the standard PLA, may be due to the heterogeneity of the material. Morin dissolves poorly in polylactide. The heterogeneous structure of the sample may have reduced its surface energy compared to the reference sample. For materials made using PHA polymer, lower surface energy was found for the PHA/juglone sample relative to the PHA reference sample. The decrease in sample surface energy may be due to the juglone interaction with the PHA polymer. Similar behavior of juglone in PHA has been described elsewhere [33].

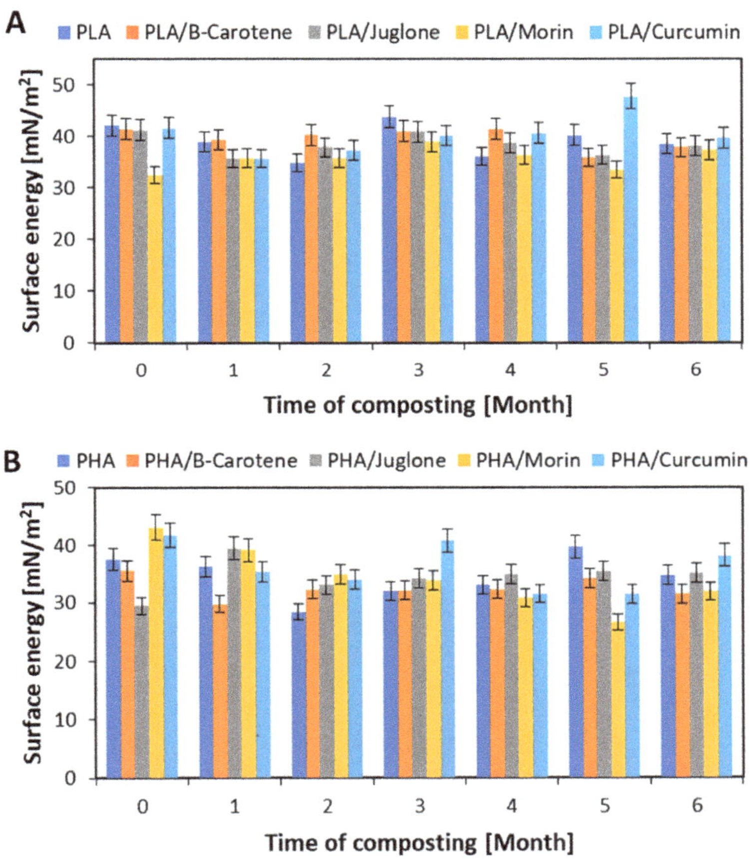

Figure 5. Changes in surface energy of PLA (**A**) and PHA (**B**) samples during composting.

The reason for changes in the surface energy of samples during composting may be a change in the morphology of the materials. Composting, especially PHA, caused visible degradation of the samples' surfaces (delicate cracks, roughness), which undoubtedly affected the changes in the surface energy values.

Plant substances, due to their fungicidal and bactericidal properties, can reduce the susceptibility of polymer compositions to composting and biodegradation. However, the analysis of the surface energy changes of the samples of both polyesters suggested that the surface energy values of the samples during composting were close to those for polyesters without plant substances, and indicated good degradability of the materials analyzed.

Table 1. Contact angle obtained using water, diiodomethane, and ethylene glycol for PLA samples during 6 months of composting.

Liquid	Contact Angle during 6 Months of Composting [°]						
	0	1	2	3	4	5	6
				PLA			
Water	61.3	77.1	70.8	76.8	75.6	75.2	66.8
Diiodomethane	36.7	36.8	53.7	27.6	47.3	38.7	42.2
Ethylene glycol	44.8	55.7	51.1	47.4	51.0	45.3	51.5
				PLA/B-Carotene			
Water	69.9	80.5	73.4	76.7	77.9	81.2	78.5
Diiodomethane	35.3	39.8	38.2	38.9	32.9	45.4	38.2
Ethylene glycol	44.3	50.4	45.6	41.0	51.2	56.7	59.2
				PLA/Juglone			
Water	74.5	73.7	68.7	68.9	76.8	85.2	77.9
Diiodomethane	37.2	48.0	46.9	39.9	39.8	45.1	40.2
Ethylene glycol	42.3	50.9	46.7	40.4	50.9	58.2	53.7
				PLA/Morin			
Water	83.2	83.0	74.0	75.5	72.3	84.5	85.8
Diiodomethane	52.8	51.8	44.3	42.9	49.1	51.9	44.4
Ethylene glycol	58.8	44.1	57.7	43.6	48.2	56.4	55.8
				PLA/Curcumin			
Water	77.6	84.4	70.0	75.1	78.8	74.2	76.7
Diiodomethane	32.0	51.2	46.3	34.9	34.6	38.1	35.1
Ethylene glycol	51.5	46.5	49.0	52.2	53.0	51.5	54.9

Table 2. Contact angle obtained using water, diiodomethane, and ethylene glycol for PHA samples during 6 months of composting.

Liquid	Contact Angle during 6 Months of Composting [°]						
	0	1	2	3	4	5	6
				PHA			
Water	70.0	67.2	89.2	82.8	74.7	74.5	80.2
Diiodomethane	50.1	53.9	57.7	55.4	52.2	37.2	47.7
Ethylene glycol	41.4	48.1	70.6	55.8	58.0	49.9	51.1
				PHA/B-Carotene			
Water	81.8	78.7	87.9	78.3	78.3	79.1	90.4
Diiodomethane	35.1	65.0	56.7	62.0	62.0	52.9	57.6
Ethylene glycol	76.6	53.5	53.4	45.2	45.2	49.3	50.6
				PHA/Juglone			
Water	83.0	60.1	84.8	73.4	83.5	76.0	75.4
Diiodomethane	57.1	56.6	49.8	49.3	50.0	52.7	53.3
Ethylene glycol	64.9	44.3	62.8	57.4	52.5	44.3	49.8
				PHA/Morin			
Water	70.7	83.1	71.8	77.9	80.1	87.6	88.4
Diiodomethane	32.2	40.3	43.8	51.4	60.5	60.8	52.3
Ethylene glycol	40.0	52.4	62.8	54.3	53.5	71.9	54.5
				PHA/Curcumin			
Water	65.9	70.3	76.6	60.8	80.1	81.6	80.4
Diiodomethane	34.3	57.1	58.2	44.5	56.0	54.4	38.1
Ethylene glycol	45.5	44.3	42.1	44.0	57.1	60.3	45.2

In the next stage of research, the effect of composting on the mechanical properties of selected polylactide and polyhydroxyalkanoate samples was analyzed. Table 3 summarizes the mechanical properties of PLA samples before and after 6 months of composting. In the first step of composting (after 1 month), polymer compositions PLA and PLA/curcumin showed an initial increase in the parameter T_{Fmax} [MPa] and σ [MPa] (maximum stress at break; tensile strength). Such results may indicate an increase in the crystalline phase content of samples after composting. After subsequent composting stages, PLA and PLA samples with curcumin showed a slow decrease in mechanical properties. For the PLA/juglone sample, the largest increase in Ts_{max} and σ was observed after 2 months of composting. In subsequent months, a decrease of the mechanical properties was noted. After 6 months of composting, it was found that the PLA reference had definitely lower values of all mechanical parameters than the samples with juglone and curcumin. For example, the σ value of the PLA sample was 36.2 MPa and for the PLA/juglone and PLA/curcumin samples were 55.2 MPa and 50.2 MPa, respectively. These results suggested that the addition of plant substances retards the composting of polylactide-based samples.

Table 3. Mechanical properties of samples based on PLA over 6 months of composting (T_{Fmax} the maximum stress transferred by the sample [MPa], E_{Fmax} the elongation at break for maximum tensile stress [%], σ the tensile strength [MPa], and ε the total elongation at break [%]).

Sample	Time of Composting [Month]	T_{Fmax} [MPa]	E_{Fmax} [%]	σ [MPa]	ε [%]
PLA	0	44.9	6.4	39.6	8.1
	1	98.3	3.8	83.3	4.4
	2	40.5	3.6	37.6	3.9
	3	45.0	4.8	42.7	4.8
	4	38.1	4.0	37.9	4.0
	5	54.1	3.8	52.5	3.8
	6	37.0	2.9	36.2	2.9
PLA/Juglone	0	49.9	4.9	48.1	5.3
	1	46.7	3.7	56.3	3.8
	2	59.3	3.8	58.3	4.0
	3	52.1	4.8	51.3	4.9
	4	51.5	4.3	51.4	4.3
	5	52.9	4.8	51.2	4.8
	6	57.1	4.7	55.2	4.8
PLA/Curcumin	0	54.1	5.1	47.1	5.9
	1	97.2	4.3	19.3	18.1
	2	45.4	3.6	42.7	4.7
	3	53.0	4.8	51.2	4.9
	4	49.3	5.4	47.4	5.5
	5	58.5	4.9	57.9	5.3
	6	51.9	4.6	50.2	4.6

The degradation of biodegradable polymers, especially polylactide, is very specific. According to literature data [40], depending on the degradation conditions, there are different changes in the polylactide crystallinity. These changes also depend on the degree of crystallinity of the polymer. Changes in the polylactide crystallinity are part of the changes associated with the degradation of the polymer material, not a separate process. According to literature, when modelling chain cleavage induced crystallization in biodegradable PLLAs, it can be assumed that the crystal growth occurs much faster than the hydrolysis reaction. In semi-crystalline PLLAs, the amorphous polymer chains entrapped by the spherulites degrade much faster than the free amorphous polymer chains outside the spherules. The research results shown in publication [40] suggest that, during polylactide degradation, there is an increase in polymer crystallinity that accompanies the simultaneous decrease of its molar mass. Crystallization of polymers is, therefore, part of the degradation process, which is why there was an initial increase in mechanical properties during the composting study that may be associated with an increase in the degree of crystallinity of the samples.

Table 4 shows the mechanical properties of PHA-based samples before and after composting. Analysis of changes in the mechanical properties of PHA samples showed that the samples are clearly degraded after 3 months of composting. The significant decomposition of samples prevented the determination of the mechanical properties of all materials based on PHA. The samples were very brittle and cracked. As in the case of PLA samples, an initial increase in the degree of crystallinity of the PHA samples was observed (higher σ [MPa] tensile strength values), indicating the onset of material degradation. After subsequent composting time intervals, a clear decrease in the mechanical property parameters was observed. Polyester materials in which PHA was used as the polymer matrix were characterized by greater compostability than PLA samples.

Table 4. Mechanical properties of PHA-based samples over 6 months of composting (T_{Fmax} the maximum stress transferred by the sample [MPa], E_{Fmax} the elongation at break for maximum tensile stress [%], σ the tensile strength [MPa], and ε the total elongation at break [%]).

Sample	Time of Composting [Month]	T_{Fmax} [MPa]	E_{Fmax} [%]	σ [MPa]	ε [%]
PHA	0	29.7	4.2	19.3	6.9
	1	21.1	2.3	21.1	2.3
	2	18.6	2.5	18.5	2.5
	3	10.9	2.3	10.5	2.3
	4	*	*	*	*
	5	*	*	*	*
	6	*	*	*	*
PHA/Juglone	0	32.5	3.8	6.4	5.8
	1	18.7	2.2	18.7	2.2
	2	11.9	1.6	11.9	1.6
	3	9.2	1.8	9.9	1.8
	4	*	*	*	*
	5	*	*	*	*
	6	*	*	*	*
PHA/Curcumin	0	30.0	4.4	5.99	6.1
	1	17.3	1.9	17.2	1.9
	2	4.09	1.4	3.87	1.5
	3	7.75	0.9	7.75	0.9
	4	*	*	*	*
	5	*	*	*	*
	6	*	*	*	*

* Significant degradation of the samples prevented the analysis of mechanical properties.

During biodegradation, changes occur in the chemical structure of polymers that can be monitored by infrared spectroscopy (FT-IR). Degradation processes can cause detachment of substituents (giving signals in the IR spectrum) or cracking of C-C and C-H bonds in the main chain with the simultaneous formation of carbonyl, peroxide, and hydroxyl groups. If the polymer contains functional groups characterized by absorption in a specific range, by recording absorption spectra and analyzing changes in individual bands, it is possible to determine the degree of polymer degradation. The ratio of the intensity of a given band undergoing change to the bandwidth of a group not subject to change in the course of biodegradation is calculated. This ratio is called the group's index (e.g., carbonyl index) and can be expressed as a percentage. Figure 6 summarizes the carbonyl indices of selected PLA and PHA samples calculated on the basis of FTIR spectra.

During composting, an increase in carbonyl indices of all samples was observed. For the PLA reference sample, an increase in the carbonyl index was found after 3 months of composting, while for the PLA/morin sample it was after 4 months of composting. The increase in carbonyl indices indicated a greater degree of degradation of the polymer compositions. The addition of morin to polylactide slowed down the composting process—a clear change in the carbonyl index was observed a month later than in the reference sample.

The PHA polyester reference sample had higher carbonyl indices than the PLA sample, which indicated a greater susceptibility to degradation of the polyhydroxyalkanoate polymer. Significant structural changes, indicating the degradation of the PHA sample, were visible after the first month of composting, which translated into a more than 100% increase in the carbonyl index compared to the PHA/morin sample. Similar to the samples made using polylactide, for PHA samples also it was found that the addition of a plant-derived substance, morin, stabilized the polyester composition and retarded the composting process (slow increase of carbonyl index, pronounced after the fourth month of composting).

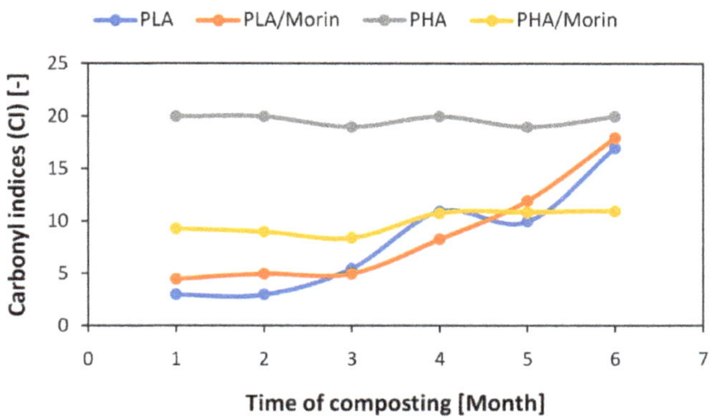

Figure 6. Carbonyl indices (CI) of selected PLA and PHA samples during composting.

The appearance of OH hydroxyl groups in the FTIR spectra may indicate the hydrolytic degradation. However, in the spectra of the analyzed samples after composting, bands in range 3650–3450 cm^{-1} (corresponding to OH group) were not present. This may mean that enzymatic degradation was the dominant degradation mechanism of biodegradable materials.

4. Conclusions

Samples based on polyester from the group of polyhydroxyalkanoates, PHA, were more susceptible to composting, as evidenced by higher weight losses of samples and clear changes in carbonyl indexes calculated on the basis of FTIR spectra, as well as analysis of mechanical properties. Higher susceptibility of PHA samples to composting and biodegradation may be the result of the porous structure of the material, which ensured increased adsorption and absorption of water (hydrolytic degradation), as well as greater availability of material for microorganisms. Slower degradation was observed for PLA samples with a compact and smooth structure.

Polymer compositions with the addition of phytochemicals with antimicrobial properties were slightly less susceptible to composting and biodegradation, which was assessed by determining the weight loss of samples, carbonyl indices, and surface energy. Despite the lower susceptibility to composting, polyester materials with natural substances were still very degradable.

As a result of composting, not only the polymer matrix, but also plant additives underwent structural changes. Degradation of substances of natural origin caused, for example, the oxidation and cracking of their C-C and C-H bonds, which resulted in clear change of their color, as well as visual change throughout the polymeric materials. Composting can be an effective form of recycling polyester materials containing plant substances (β-carotene, juglone, morin, and curcumin) with biological activity.

Author Contributions: Conceptualization, M.L.-B. and A.M.; methodology, A.M.; software, M.L.-B.; validation, M.L.-B., and A.M.; formal analysis, A.M.; investigation, M.L.-B.; resources, M.L.-B.; data curation, A.M.; writing—original draft preparation, M.L.-B.; writing—review and editing, A.M.; visualization, M.L.-B.; supervision, A.M.; project administration, A.M.; funding acquisition, A.M. All authors have read and agreed to the published version of the manuscript.

Funding: This study was supported by the National Centre for Research and Development (NCBR) project LIDER/32/0139/L-7/15/NCBR/2016.

Conflicts of Interest: The authors declare no conflict of interest.

References

1. Chollet, B.; Lopez-Cuesta, J.-M.; Laoutid, F.; Ferry, L. Lignin Nanoparticles as A Promising Way for Enhancing Lignin Flame Retardant Effect in Polylactide. *Materials* **2019**, *12*, 2132. [CrossRef]
2. Musioł, M.; Jurczyk, S.; Sobota, M.; Klim, M.; Sikorska, W.; Zięba, M.; Janeczek, H.; Rydz, J.; Kurcok, P.; Johnston, B.; et al. (Bio)Degradable Polymeric Materials for Sustainable Future—Part 3: Degradation Studies of the PHA/Wood Flour-Based Composites and Preliminary Tests of Antimicrobial Activity. *Materials* **2020**, *13*, 2200. [CrossRef] [PubMed]
3. Iovino, R.; Zullo, R.; Rao, M.A.; Cassar, L.; Gianfreda, L. Biodegradation of poly(lactic acid)/starch/coir biocomposites under controlled composting conditions. *Polym. Degrad. Stabil.* **2008**, *93*, 147–157. [CrossRef]
4. Kalea, G.; Aurasa, R.; Singha, S.P.; Narayanb, R. Biodegradability of polylactide bottles in real and simulated composting conditions. *Polym. Test.* **2007**, *26*, 1049–1061. [CrossRef]
5. Nurul Fazita, M.R.; Jayaraman, K.; Bhattacharyya, D.; Hossain, M.S.; Mohamad Haafiz, M.K.; Abdul Khalil, H.P.S. Disposal Options of Bamboo Fabric-Reinforced Poly(Lactic) Acid Composites for Sustainable Packaging: Biodegradability and Recyclability. *Polymers* **2015**, *7*, 1476–1496. [CrossRef]
6. Jurczyk, S.; Musioł, M.; Sobota, M.; Klim, M.; Hercog, A.; Kurcok, P.; Janeczek, H.; Rydz, J. (Bio)degradable Polymeric Materials for Sustainable Future—Part 2: Degradation Studies of P(3HB-co-4HB)/Cork Composites in Different Environments. *Polymers* **2019**, *11*, 547. [CrossRef]
7. Kirschweng, B.; Tatraaljai, D.; Foldes, E.; Pukanszky, B. Natural antioxidants as stabilizers for polymers. *Polym. Degrad. Stabil.* **2017**, *145*, 25–40. [CrossRef]
8. Hussain, T.; Tausif, M.; Ashraf, M. A review of progress in the dyeing of eco-friendly aliphatic polyester based polylactic acid fabrics. *J. Clean. Prod.* **2015**, *108*, 476–483. [CrossRef]
9. Singh, S.; Gaikwad, K.K.; Lee, J.S. Anthocyanin—A Natural Dye for Smart Food Packaging Systems. *Korean J. Packag. Sci. Technol.* **2018**, *24*, 167–180. [CrossRef]
10. Tichoniuk, M.; Radomska, N.; Cierpiszewski, R. The Application of Natural Dyes in Food Freshness Indicators Designed for Intelligent Packaging. *Stud. Oeconomica Posnaniensia* **2017**, *5*, 19–34. [CrossRef]
11. Doudin, K.; Al-Malaika, S.; Sheena, H.H.; Tverezovskiy, V.; Fowler, P. New genre of antioxidants from renewable natural resources: Synthesis and characterisation of rosemary plant-derived antioxidants and their performance in polyolefins. *Polym. Degrad. Stabil.* **2016**, *130*, 126–134. [CrossRef]
12. Grigsby, W.J.; Bridson, J.H.; Schrade, C. Modifying biodegradable plastics with additives based on condensed tannin esters. *J. Appl. Polym. Sci.* **2015**, *132*, 41626. [CrossRef]
13. Gordobil, O.; Egües, I.; Llano-Ponte, R.; Labidi, J. Physicochemical properties of PLA lignin blends. *Polym. Degrad. Stabil.* **2014**, *108*, 330–338. [CrossRef]
14. Nwakaudu, A.A.; Nwakaudu, M.S.; Owuamanam, C.I.; Iheaturu, N.C. The Use of Natural Antioxidant Active Polymer Packaging Films for Food Preservation. *Appl. Signals Rep.* **2015**, *2*, 38–50.
15. Sanches-Silva, A.; Costa, D.; Albuquerque, T.G.; Buonocore, G.G.; Ramos, F.; Castilho, M.C.; Machado, A.V.; Costa, H.S. Trends in the use of natural antioxidants in active food packaging: A review. *Food Addit. Contam. Part. A* **2014**, *31*, 374–395. [CrossRef] [PubMed]
16. Medina-Jaramillo, C.; Ochoa-Yepes, O.; Bernal, C.; Famá, L. Active and smart biodegradable packaging based on starch and natural extracts. *Carbohydr. Polym.* **2017**, *176*, 187–194. [CrossRef] [PubMed]
17. Lopez-Rubio, A.; Lagaron, J.M. Improvement of UV stability and mechanical properties of biopolyesters through the addition of β-carotene. *Polym. Degrad. Stabil.* **2010**, *95*, 2162–2168. [CrossRef]
18. Lionetto, F.; López-Muñoz, R.; Espinoza-González, C.; Mis-Fernández, R.; Rodríguez-Fernández, O.; Maezzoli, A. A Study on Exfoliation of Expanded Graphite Stacks in CandelillaWax. *Materials* **2019**, *12*, 2530. [CrossRef]
19. Andreotti, S.; Franzoni, E.; Degli Esposti, M.; Fabbri, P. Poly(hydroxyalkanoate)s-Based Hydrophobic Coatings for the Protection of Stone in Cultural Heritage. *Materials* **2018**, *11*, 165. [CrossRef]
20. Keceli, T.M.; Erginkaya, Z.; Turkkan, E.; Kaya, U. Antioxidant and Antibacterial Effects of Carotenoids Extracted from Rhodotorula glutinis Strains. *Asian J. Chem.* **2013**, *25*, 42–46. [CrossRef]
21. Manimala, M.R.A.; Murugesan, R. In vitro antioxidant and antimicrobial activity of carotenoid pigment extracted from *Sporobolomyces* sp. isolated from natural source. *J. Nat. Appl. Sci.* **2014**, *6*, 649–653. [CrossRef]

22. Hayashi, M.; Naknukool, S.; Hayakawa, S.; Ogawa, M.; Nimatulah, A.-B.A. Enhancement of antimicrobial activity of a lactoperoxidase system by carrot extract and b-carotene. *Food Chem.* **2012**, *130*, 541–546. [CrossRef]
23. Ali, S.; Hussain, T.; Nawaz, R. Optimization of alkaline extraction of natural dye from henna leaves and its dyeing on cotton by exhaust method. *J. Clean Prod.* **2009**, *17*, 61–66. [CrossRef]
24. Tan, D.T.C.; Osman, H.; Mohamad, S.; Kamaruddin, A.H. Synthesis and antibacterial activity of juglone derivatives. *J. Chem. Chem. Eng.* **2012**, *6*, 8489.
25. Montenegro, R.C.; Araujo, A.J.; Molina, M.T.; Filho, J.D.B.M.; Rocha, D.D.; Lopéz-Montero, E.; Goulart, M.O.F.; Bento, E.S.; Alves, A.P.N.N.; Pessoa, C.; et al. Cytotoxic activity of naphthoquinones with special emphasis on juglone and its 5-methyl derivative. *Chem. Biol. Interact.* **2010**, *184*, 439–448. [CrossRef] [PubMed]
26. Oliveira, I.; Sousa, A.; Ferreira, I.C.; Bento, A.; Estevinho, L.; Pereira, J.A. Total phenols, antioxidant potential and antimicrobial activity of walnut (*Juglans regia* L.) green husks. *Food Chem. Toxicol.* **2008**, *46*, 23262331. [CrossRef]
27. Bonjar, G.S.; Aghighi, S.; Nik, A.K. Antibacterial and antifungal survey in plants used in indigenous herbal medicine of south east regions of Iran. *J. Biol. Sci.* **2004**, *4*, 405412.
28. Kopacz, M.; Woznicka, E.; Gruszecka, J. Antibacterial Activity of Morin and Its Complexes with La(III), Gd(III) and Lu(III) Ions. *ACTA Pol. Pharm.* **2005**, *62*, 65–67.
29. Cushnie, T.P.T.; Lamb, A.J. Antimicrobial activity of flavonoids. *Int. J. Antimicrob. Agents* **2005**, *26*, 343–356. [CrossRef]
30. Ou, J.-L.; Mizushina, Y.; Wang, S.-Y.; Chuang, D.-Y.; Nadar, M.; Hsu, W.-L. Structure–activity relationship analysis of curcumin analogues on anti-influenza virus activity. *FEBS J.* **2013**, *280*, 5829–5840. [CrossRef]
31. Neelofar, K.; Shreaz, S.; Rimple, B.; Muralidhar, S.; Nikhat, M.; Khan, L.A. Curcumin as a promising anticandidal of clinical interest, Curcumin as a promising anticandidal of clinical interest. *Can. J. Microbiol.* **2011**, *57*, 204–210. [CrossRef] [PubMed]
32. Masek, A.; Latos-Brozio, M. The Effect of Substances of Plant Origin on the Thermal and Thermo-Oxidative Ageing of Aliphatic Polyesters (PLA, PHA). *Polymers* **2018**, *10*, 1252. [CrossRef] [PubMed]
33. Latos, M.; Masek, A.; Zaborski, M. The potential of juglone as natural dye and indicator for biodegradable polyesters. *Proc. Inst. Mech. Eng. L J. Mater. Des. Appl.* **2019**, *233*, 276–285. [CrossRef]
34. Masek, A.; Latos, M.; Piotrowska, M.; Zaborski, M. The potential of quercetin as an effective natural antioxidant and indicator for packaging materials. *Food Packag. Shelf Life.* **2018**, *16*, 51–58. [CrossRef]
35. Masek, A. Flavonoids as Natural Stabilizers and Color Indicators of Ageing for Polymeric Materials. *Polymers* **2015**, *7*, 1125–1144. [CrossRef]
36. Masek, A.; Chrzescijanska, E.; Diakowska, K.; Zaborski, M. Application of β-carotene, a natural flavonoid dye, to polymeric materials as a natural antioxidant and determination of its characteristics using cyclic voltammetry and FTIR spectroscopy. *Int. J. Electrochem. Sci.* **2015**, *10*, 3372–3386.
37. Latos-Brozio, M.; Masek, A. The application of natural food colorants as indicator substances in intelligent biodegradable packaging materials. *Food Chem. Toxicol.* **2020**, *135*, 11097. [CrossRef]
38. Moraczewski, K.; Malinowski, R.; Sikorska, W.; Karasiewicz, T.; Stepczynska, M.; Jagodzinski, B.; Rytlewski, P. Composting of Polylactide Containing Natural Anti-Aging Compounds of Plant Origin. *Polymers* **2019**, *11*, 1582. [CrossRef]
39. Latos-Brozio, M.; Masek, A. Biodegradable Polyester Materials Containing Gallates. *Polymers* **2020**, *12*, 677. [CrossRef]
40. Gleadall, A.; Pan, J.; Atkinson, H. A simplified theory of crystallisation induced by polymer chain scissions for biodegradable polyesters. *Polym. Degrad. Stabil.* **2012**, *97*, 1616–1620. [CrossRef]

© 2020 by the authors. Licensee MDPI, Basel, Switzerland. This article is an open access article distributed under the terms and conditions of the Creative Commons Attribution (CC BY) license (http://creativecommons.org/licenses/by/4.0/).

polymers

MDPI

Article

Wastewater Treatment by a Polymeric Bioflocculant and Iron Nanoparticles Synthesized from a Bioflocculant

Nkosinathi Goodman Dlamini [1,*], Albertus Kotze Basson [1] and Rajasekhar VSR Pullabhotla [2,*]

1 Department of Biochemistry and Microbiology, University of Zululand, Private Bag X1001, KwaDlangezwa 3886, South Africa; BassonA@unizulu.ac.za
2 Department of Chemistry, University of Zululand, Private Bag X1001, KwaDlangezwa 3886, South Africa
* Correspondence: nathidlamini03@gmail.com (N.G.D.); PullabhotlaV@unizulu.ac.za (R.V.S.R.P.); Tel.: +27-359-026-155 (R.V.S.R.P.)

Received: 22 June 2020; Accepted: 14 July 2020; Published: 21 July 2020

Abstract: Wastewater remains a global challenge. Various methods have been used in wastewater treatment, including flocculation. The aim of this study was to synthesize iron nanoparticles (FeNPs) using a polymeric bioflocculant and to evaluate its efficacy in the removal of pollutants in wastewater. A comparison between the efficiencies of the bioflocculant and iron nanoparticles was investigated. A scanning electron microscope (SEM) equipped with an energy-dispersive X-ray analyzer (EDX) and Fourier transform-infrared (FT-IR) spectroscopy were used to characterize the material. SEM-EDX analysis revealed the presence of elements such as O and C that were abundant in both samples, while FT-IR studies showed the presence of functional groups such as hydroxyl (–OH) and amine (–NH_2). Fe nanoparticles showed the best flocculation activity (FA) at 0.4 mg/mL dosage as opposed to that of the bioflocculant, which displayed the highest flocculation activity at 0.8 mg/mL, and both samples were found to be cation-dependent. When evaluated for heat stability and pH stability, FeNPs were found thermostable with 86% FA at 100 °C, while an alkaline pH of 11 favored FA with 93%. The bioflocculant flocculated poorly at high temperature and was found effective mostly at a pH of 7 with over 90% FA. FeNPs effectively removed BOD (biochemical oxygen demand) and COD (chemical oxygen demand) in all two wastewater samples from coal mine water and Mzingazi River water. Cytotoxicity results showed both FeNPs and the bioflocculant as nontoxic at concentrations up to 50 μL.

Keywords: biosafety; flocculation; removal efficiency; wastewater

1. Introduction

Approximately 90% of wastewater is discharged untreated into water bodies in developing countries (Corcoran) [1]. The aquatic ecosystem is threatened by this as edible and drinkable water become contaminated [2]. Colloids are heterogeneous matter characterized by kinetically non-labile and thermodynamically instable characteristics. Colloids, organic, and inorganic pollutants in water are a major concern of this era. Colloids have a tendency of not settling under gravity in a solution [3]. Both organic and inorganic hazardous pollutants, including derivatives of phenols and dyes released from different industries, have turned out to be a global problem [4,5]. Textile industries are one of the largest sources that are contributing to the pollution of water. This is due to the application of different chemicals throughout the textile processing [6,7]. Untreated effluent discharge from the textile processing results in highly toxic wastewater [8]. This effluent contains high levels of chemical oxygen demand (COD) and biochemical oxygen demand (BOD) and is highly turbid. The release of this untreated effluent to sea, lakes, or rivers affect the environment badly [9]. In developing countries, close to 10% of the population dies due to waterborne infections as well as cancer caused by untreated industrial effluents in water [10]. Hence, treatment and removal of the pollutants that are present in water bodies are necessary, though it is never an easy task.

Several techniques have been employed to treat the effluents and to remove toxic compounds from the water [10,11]. The methods include constructed wetlands, membrane filtration, hybrid ion exchange materials and electrocoagulation, etc. All these water treatment technologies play a substantial role in the treatment of effluents from industries. However, the major downside of these techniques is that they are either very expensive or produce immense amounts of sludge [12].

Of late, secondary metabolites (bioflocculants) produced by microorganisms during growth are viewed as the possible solution to water treatment. These flocculants are favored due to their environmental friendliness, biodegradability, and nontoxicity [11], and they cause no environmental harm and can remove heavy metals from wastewater [13]. Ugbenyen and Okoh [14] stated that chemicals that stimulate flocculation by aggregation of colloids and other suspended particles, forming a floc, are called flocculants. Both organic and inorganic flocculants have been used in the purification of water in various industries. This includes organic synthetic polymers, inorganic aluminium, and ferric salt [15]. Natural flocculants have also been used in various downstream processes such as treatment of wastewater, purification of potable water, and fermentation and food industries [16]. Commonly, the flocculants categories are: synthetic organic flocculants, which include polyacrylamide derivatives; inorganic flocculants, which include polyaluminium chloride; and naturally occurring flocculants, which include the secondary secretion (bioflocculants) from microorganisms [17].

Flocculants of chemical nature have been used widely in the process of flocculation due to the cost effectiveness and flocculating efficiency [18]. Nonetheless, some environmental and health concerns have been raised through their usage due to the monomers of these flocculants being reported as toxic to humans, and aluminium salts being associated with Alzheimer's disease [11,19]. Therefore, researchers in the world have shifted focus in the application of these biodegradable, environmentally friendly flocculants to replace chemically synthesized flocculants. Despite all these interesting properties of biodegradability and environmental-friendliness, natural flocculants have the disadvantages of low shelf life, are very expensive to produce, have low yield, and have minimal flocculation activity [18]. Therefore, to overcome these shortcomings, we investigate the application of bioflocculant-synthesized nanoparticles in comparison to chemical synthetic flocculant (ferric chloride) and bioflocculant.

Hence, in the present study, we report the synthesis of iron nanoparticles using a polymeric bioflocculant, and its application in wastewater treatment in comparison to a bioflocculant and biosafety evaluation.

2. Materials and Methods

2.1. Production Medium Chemicals

All reagents for production media used were obtained from Sigma-Aldrich (St. Louis, MO, USA). The standard production medium as described by Zhang, et al. [20] was followed. A litre of the filtered sea water was used together with the following reagents: glucose (20.0 g), KH_2PO_4 (2.0 g), K_2HPO_4 (5.0 g), $(NH_4)_2SO_4$ (0.2 g), NaCl (0.1 g), CH_4N_2O (0.5 g), $MgSO_4$ (0.2 g), and yeast extract (0.5 g).

2.2. Extraction and Purification of the Bioflocculant

The bacteria used were previously isolated from the sediment sample from Sodwana Bay in the Province of KwaZulu-Natal in South Africa (28°450′ S 31°540′ E) and identified as *Alcaligenes faecalis* HCB2 [11]. Bioflocculant extraction was achieved following a method as described by Dlamini et al. [21]. Firstly, 1 L of the production medium was prepared and autoclaved at 121 °C for 15 min. Subsequently, 1% in (50 mL) inoculum was added and the medium incubated in a shaker at 165 rpm for 72 h at 30 °C, and after incubation, the medium was centrifuged at 8000 rpm at 4 °C for 30 min. This was done in order to remove cells and insoluble substances. The supernatant was transferred into a clean container and 1 L of distilled water and 2 L of ethanol were added to the supernatant, agitated, and the solution was stored at 4 °C for 12 h. Later, the precipitate formed was vacuum-dried and 100 mL of distilled water was added. A mixture of chloroform and *n*-butyl (5:2 *v*/*v*) was also added and the mixture was left to stand for 12 h at room temperature [21].

2.3. Synthesis of the Iron Nanoparticles (FeNPs)

To synthesize iron nanoparticles, a green-approach method was adopted [22]. A metal precursor for the synthesis of iron nanoparticles (FeNPs) used was iron sulphate ($FeSO_4$). Briefly, 0.5 g of pure bioflocculant was dissolved in 0.2 M ($FeSO_4$), and to prevent agglomeration of nanoparticles, 10 mL of 5.0 M sodium hydroxide (NaOH) solution was added. The mixture was left overnight at room temperature and nanoparticles formation was confirmed by physical observation, i.e., color change and characterization. Subsequently, the mixture was centrifuged at 5000 rpm at 4 °C for 15 min to harvest the synthesized nanoparticles and the resulting precipitate was vacuum-dried at 25 °C for 24 h [22].

2.4. Characterization of the Bioflocculant and Iron Nanoparticles

2.4.1. Morphology and Element Analysis

A scanning electron microscope (SEM, JEOL, USA, Inc., Peabody, MA, USA) and energy-dispersive X-ray spectroscopy (EDX, JEOL, USA, Inc., Peabody, MA, USA) were used to evaluate morphology and elements in FeNPs and the bioflocculant.

2.4.2. Functional Groups Analysis

Fourier transform-infrared (FT-IR, Bruker, Gauteng, South Africa) spectroscopy was used to identify and confirm the functional groups present in FeNPs and the bioflocculant (Tensor 27, Bruker FT-IR spectrophotometer, Bruker, Gauteng, South Africa).

2.5. Determination of Flocculation Activity

The process in which mediation of flocculants is achieved in the presence of microorganisms or biodegradable macromolecular flocculants released by microorganisms is called bioflocculation.

Kaolin clay was used as the test material in this study; 4.0 g in 1 litre distilled water was prepared. Kaolin clay solution (50 mL) was added into a 250 mL conical flask, and thereafter, 2.0 mL (0.2 mg/mL) solution of the bioflocculant or iron nanoparticles was added and 3.0 mL $CaCl_2$ (1.0 g/L) solution was also added. The mixtures were shaken for 1 min and transferred to 100 mL graduated measuring cylinders. The mixture was left to stand for 5 min before the supernatant was taken for analysis [23]. The following equation was used to calculate the flocculation activity:

Flocculation activity

$$FA\ \% = \frac{[A - B]}{A} \times 100 \quad (1)$$

where A is the optical density of a control at 550 nm and B is the optical density of a sample at 550 nm. The flocculation mechanism for the bioflocculant is bridging [11].

2.6. Optimization of the Flocculation Efficiency of the Bioflocculant and FeNPs

2.6.1. Evaluation of Flocculation Activity of the Bioflocculant and FeNPs

To evaluate the most effective dosage, different concentrations were prepared (0.2, 0.4, 0.6, and 0.8 mg/mL) by dissolving the bioflocculant and FeNPs in distilled water to obtain respective concentrations. A liter of kaolin solution was prepared using distilled water (4 g/L), after which, 100 mL of kaolin solution, 2 mL of the bioflocculant or FeNPs, and 3 mL of 1% $CaCl_2$ were transferred into a 300 mL conical flask. The mixture was vigorously shaken for 1 min before being transferred into a measuring cylinder (100 mL) and allowed to settle for 5 min at room temperature. This procedure was also followed for the control, where 2 mL of nanoparticles were replaced by 2 mL distilled water. The clear top layer of the supernatant was pipetted into a cuvette to determine the flocculation activity. A UV-Visible spectrophotometer was used to measure the optical density (OD_{550nm}). All experiments were conducted in triplicates. Equation (1) above was used to calculate the flocculation activity. The most effective dosage was used for the subsequent experiment and kaolin clay used as test material.

2.6.2. Effect of Cations on Flocculating Activity

Different salts were used to ascertain cation effect on flocculation activity, solutions were used to replace 1% $CaCl_2$, and the salts used were comprised of monovalent (LiCl and NaCl), divalent ($MgCl_2$ and $CaCl_2$), and trivalent ($FeCl_3$) at the same concentration. The control was maintained without cations. To measure the flocculating activity, the above procedure was used to evaluate cation effect on flocculation activity.

2.6.3. Effect of pH and Temperature on Flocculating Activity

A solution of NaOH (1.0 M) or HCl (1.0 M) was used whenever necessary to adjust pH in a range (3 to 11). The flocculation activity was assessed using the previously described method. Both the bioflocculant and FeNPs were subjected to high temperatures (50–100 °C) in a water bath for 30 min to determine thermostability, after which the flocculation activity was calculated using the method described above.

2.7. *Wastewater Treatment*

To assess removal efficiency (RE), coal mine wastewater and Mzingazi River water samples were collected and autoclaved at 121 °C for 15 min to ensure that no microorganisms were present to interfere with experimentation. The samples were collected from Tendele Coal Mine and Mzingazi River in KwaZulu Natal, RSA. Following the method described by Maliehe, Basson, and Dlamini [11], COD and BOD removal was evaluated. A UV-Vis spectrophotometer Pharo 300 Spectroquant® was used at 680 nm for the RE measurement. The removal efficiency (RE) of the pollutants was calculated by the following equation:

$$RE\ (\%) = \frac{C_i - C_f}{C_i} \times 100 \quad (2)$$

where: C_i is the initial value before treatment with the bioflocculant and nanoparticles and C_f is the value after treatment.

2.8. *Cytotoxicity of the Bioflocculant and Iron Nanoparticles*

A method described by Daniels and Singh [24] was adopted to evaluate cytotoxicity of the bioflocculant and nanoparticles using human embryonic kidney (HEK 293) and breast cancer cells (MCF-7). Cells with cell suspensions of 1×10^5 cells/mL concentrations were platted on 96-well-plate. Using a tenfold serial dilution method, the cells were seeded with different concentrations of nanoparticles (25–100 µg/µL). After 48 h of incubation, media containing 1% of fetal bovine serum (FBS) were used for the administration of nanoparticles and the plates were returned to the incubator for 48 h. To ascertain cell viability, tetrazolium salt (Sigma) was added as an indicator after 48 h of incubation. Then, 15 µL of MTT (5 mg/mL) in phosphate-buffered saline (PBS) was added to each well and incubated at 37 °C for 4 h. After sucking off from the wells, the medium with MTT and the formed formazan crystals were dissolved in 100 µL of dimethyl sulfoxide (DMSO). The optical density of the solutions was measured at 570 nm using a microplate reader [24].

The % cell inhibition was determined using the following formula:

Cell viability (%) $= \frac{F_1}{F_0} \times 100$, where F_1 and F_0 are the final values obtained after and before treatment with the bioflocculant and nanoparticles, respectively.

2.9. Experimental, Software, and Statistical Analysis

All data was collected in triplicates and the error bars in the Figures show the standard deviations of the data. Data were subjected to one-way analysis of variances (ANOVA) using Graph Pad Prism™ 6.1. A significant level of $p < 0.05$ was used.

3. Results

3.1. FT-IT Spectra of the Bioflocculant and Iron Nanoparticles

Figure 1 below represents the functional groups present in the bioflocculant and iron nanoparticles. The peak at 3245 cm^{-1} (bioflocculant) and 3250 cm^{-1} (iron nanoparticles) indicates the presence of hydroxyl (–OH) and amine (–NH_2) functional groups in the sample.

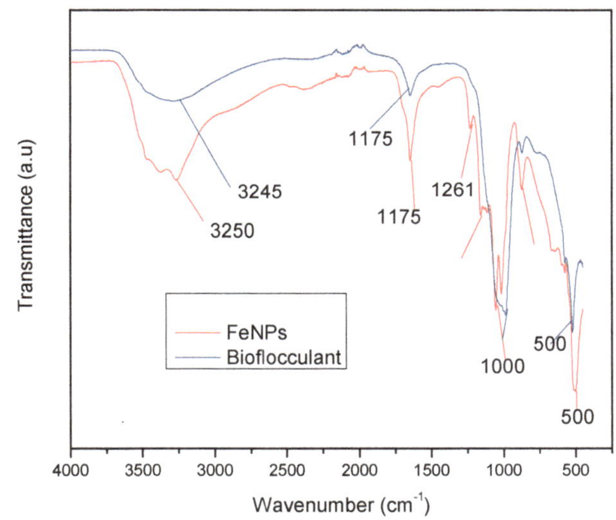

Figure 1. Fourier transform-infrared (FT-IR) spectra of the bioflocculant and iron nanoparticles.

3.2. The SEM Morphology of the Bioflocculant and Iron Nanoparticles

Figures 2 and 3 below represent the surface morphology of the bioflocculant and iron nanoparticles, respectively. The bioflocculant has the crystal-like morphology while the nanoparticles seem to have granular-like shapes.

Figure 2. SEM surface morphology of the bioflocculant.

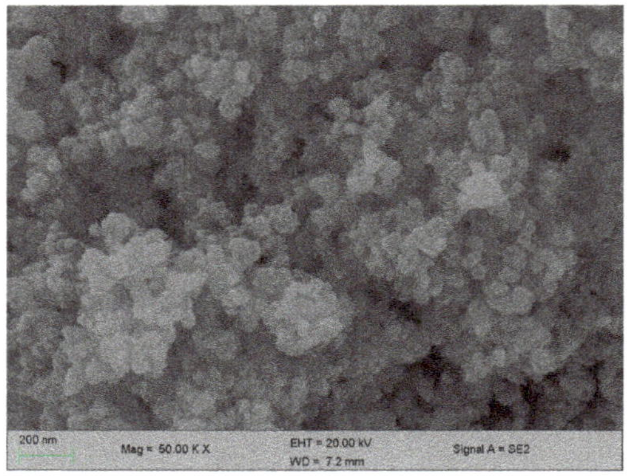

Figure 3. SEM surface morphology of the iron nanoparticles.

3.3. Elemental Composition of the Bioflocculant and Iron Nanoparticles

In Table 1 below, elements such as O, C, Mg, P, K, Ca, Fe, and Cu are present in the bioflocculant and iron nanoparticle samples. From both samples, oxygen and carbon account for over 50%, while iron and copper were only found to be present in the iron nanoparticles alone and absent in the bioflocculant.

Table 1. Energy-dispersive X-ray analysis (EDX) of the bioflocculant and iron nanoparticles.

Elements	Sample	
	Bioflocculant (wt.%)	**FeNPs (wt.%)**
C	13.21	12.39
O	55.25	47.94
Mg	13.35	1.12
P	16.00	13.43
K	0.14	0.24
Ca	2.04	7.33
Fe	-	17.31
Cu	-	0.30
Total	100.00	100.00

3.4. Dosage Concentration Effect on Flocculation

An adequate dosage is required for an efficient flocculation process. Fe nanoparticles showed the optimum flocculation activity (FA) at 0.4 mg/mL dosage as opposed to that of the bioflocculant, which displayed the highest flocculation activity at 0.8 mg/mL (Figure 4). The optimum dosage for each flocculant was then used for subsequent experiments.

3.5. Temperature Effect on Flocculation Activity

The FeNPs are more thermostable compared to the bioflocculant, as the flocculation activity is above 86% at 100 °C, while the significant drop in flocculation activity is observed with the increased temperature in the bioflocculant (Figure 5).

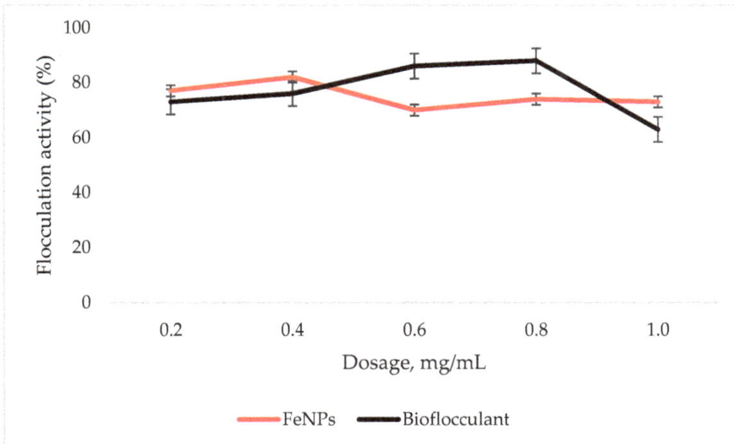

Figure 4. Dosage effect on flocculation activity. Error bars denote statistical significance at ($p < 0.05$).

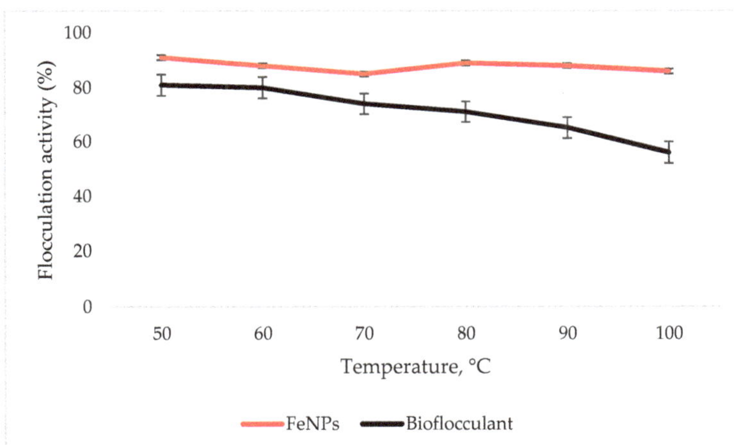

Figure 5. Temperature effect on flocculation activity. Error bars denote statistical significance at ($p < 0.05$).

3.6. Effect of pH on Flocculation Activity

Both the FeNPs and bioflocculant flocculate well in alkaline conditions, with FeNPs having the optimum flocculation activity at a pH of 11, while that of the bioflocculant is at a pH of 7 (Figure 6).

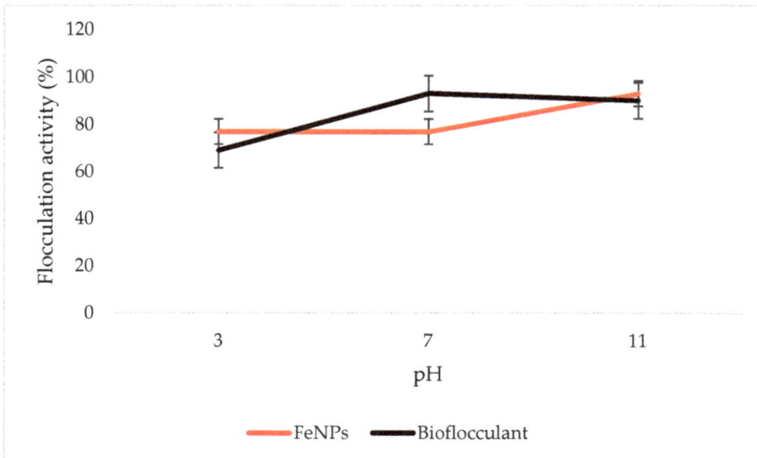

Figure 6. pH effect on flocculation activity. Error bars denote statistical significance at ($p < 0.05$).

3.7. Effect of Metal Ions on Flocculation Activity

The nanoparticles and bioflocculant flocculated poorly when the cation was not added, with 49% and 46% flocculation activity, respectively (Table 2).

Table 2. Cation effect on flocculation activity.

Cations	Flocculation Activity (%)	
	Bioflocculant	**FeNPs**
Control	49 ± 3.35	46 ± 2.03 [b]
Fe^{3+}	31 ± 3.15	85 ± 2.72 [a]
Mg^{2+}	63 ± 6.78	82 ± 1.53 [a]
Ca^{2+}	71 ± 5.42	82 ± 3.64 [a]
Li^{+}	75 ± 2.31	72 ± 1.15 [a]
Na^{+}	62 ± 7.28	72 ± 1.15 [a]

Different letters ([a] and [b]) denote statistical significance at ($p < 0.05$).

3.8. The Removal of COD and BOD

Table 3 represents removal efficiency by $FeCl_3$, FeNPs, and the bioflocculant; Fe nanoparticles were the most effective in reducing both COD and BOD compared to the other two flocculants.

Table 3. Chemical oxygen demand (COD) and biochemical oxygen demand (BOD) removal in wastewater by the bioflocculant and iron nanoparticles.

Flocculant	Types of Waste Water	Types of Pollutants in Water	Water Quality before Treatment (mg/L)	Water Quality after Treatment (mg/L)	Removal Efficiency (%)
FeNPs	Coal mine water	COD	842	204	76
		BOD	123.2	23	81
	Mzingazi river water	COD	3.300	1.700	48
		BOD	136	24	82
Bioflocculant	Coal mine water	COD	842	208	72
		BOD	123.2	77.88	59
	Mzingazi river water	COD	3.300	1.68	51
		BOD	136	72.08	53

3.9. Evaluation of Cytotoxicity of the FeNPs and Bioflocculant

In vitro cytotoxicity of both the FeNPs and bioflocculant were evaluated and the FeNPs were found nontoxic at low concentrations and the bioflocculant was nontoxic at all concentrations (Figures 7 and 8).

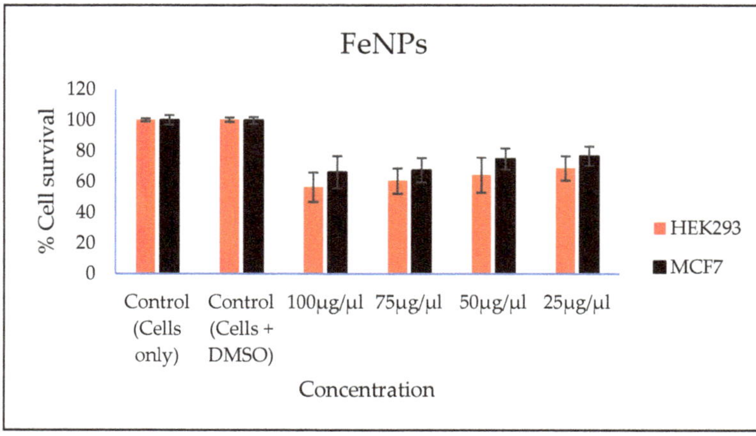

Figure 7. In vitro cytotoxicity effect of FeNPs nanoparticles on HEK293 and MCF7 cells.

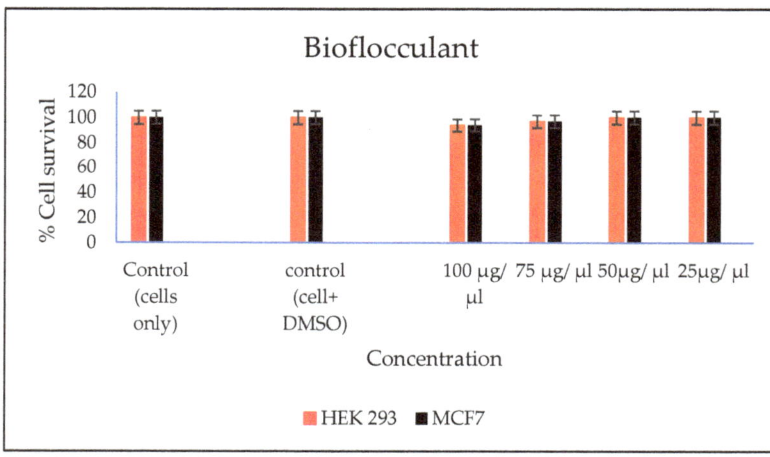

Figure 8. In vitro cytotoxicity effect of bioflocculant nanoparticles on HEK293 and MCF7 cells.

4. Discussion

The functional groups present in the molecular chains of the bioflocculant facilitate the binding capability of the bioflocculants [25]. The presence of the –OH group plays the significant role in reducing and stabilizing nanoparticles during synthesis [26]. Thermostability of the nanoparticles when subjected to heat further confirm the presence of the hydroxyl group. The flocculation process is influenced by the surface morphology of the flocculant and accounts for the effectiveness or poor efficiency of the flocculant [20]. In Figure 3, the crystal-like and granular morphology is observed. The change in the bioflocculant structure is the indication of the formation of nanoparticles in the synthesis. Furthermore, it can be noted that the nanoparticles have more surface area for pollutants absorption. Therefore, it can be deduced that the synthesis of nanoparticles does not only modify the surface structure, but it also increases the surface area on nanoparticles for particles flocculation and pollutants removal in water. The flexibility and stability of flocculants is brought about by the different elements present in the sample. In Table 1, elements such as O and C were found in the bioflocculant sample and account for a major percentage, as these elements form the backbone structure of the biomolecule. Furthermore, Mg, P, K, and Ca account for the production media that were used for the bioflocculant production. Similarly, the as-synthesized iron nanoparticles also had O and C, which account for 60.33%, and Fe was found to be the second highest present element at 17.31%, which indicates that the nanoparticles synthesis was successful. The copper grid, which was used during analysis, could account for 0.30% Cu present in the sample.

To effectively neutralize some of the negative charges on colloidal particles, an adequate dosage is required; if the dosage is insufficient, poor flocculation results [27]. Contrary to this, excess dosage may increase the viscosity, which results in poor flocculation activity [28]. As illustrated in Figure 4, the optimum flocculation activity was achieved at 0.4 mg/mL and 0.8 mg/mL for the nanoparticles and bioflocculant, respectively. An increase in flocculation activity was observed between 02–04 mg/mL for FeNPs, however, with the increase in dosage concentration to 0.6–1.0 mg/mL, the flocculation activity dropped a little and it remained consistent throughout. This could be due to the competition and repulsion of negatively charged kaolin particles, which in turn block binding sites. The low flocculation activity for the bioflocculant at 0.2–0.6 mg/mL may be due to the fact that low dosage did not permit bridging phenomena to occur effectively [11]. Both the bioflocculant and the FeNPs were subjected to different temperatures (50–100 °C) for 30 min in a water bath. As depicted in Figure 5, higher flocculation activity was observed at 50 °C with 91 and 81% for the nanoparticles and bioflocculant, respectively. The increase in temperature did not affect the flocculation process of the as-synthesized nanoparticles. The flocculation activity remained above 86%, suggesting that the nanoparticles are thermostable. The results are comparable with those of other studies [13,27,29], where heat could not affect the effectiveness of bioflocculants, indicating their thermostability. This could be attributed to the presence of the -OH group as indicated in Figure 1 above. The results are comparable with that of Sekelwa, et al. [30], where the presence of hydroxyl groups, evidenced by the IR spectra within the polymer, favored the possibility of hydrogen bonding with one or more water molecules.

Key factors that influence the flocculation process include pH. Flocculation activity may be affected by pH; it may alter flocculant status charge and surface characteristics of colloidal particles in suspension [27]. In Figure 6, the highest flocculation activity of 90% was achieved with FeNPs at a strong alkaline pH of 11. Nonetheless, the flocculation activity was still above 77% at a strong acidic pH of 3, suggesting that FeNPs can be applied in both acidic and alkaline conditions, but are most effective by using alkaline conditions. Contradictory to this, the flocculation activity of the bioflocculant was poor in acidic conditions with the optimum of 93% at a pH of 7. The poor performance in strong acidic conditions may be attributed to protein denaturation in the bioflocculant [11]. These findings suggest that the nanoparticles can be a suitable flocculant in coal mine waste, as the pH is mostly alkaline.

Residual negative net surface charge of the bioflocculant functional group is neutralized by cations, which in turn enhance the flocculation activity [31]. Various metal ion effects were evaluated on the as-synthesized nanoparticles and bioflocculant as shown in Table 2. The highest flocculation activity of 85% was observed when a trivalent cation (Fe^{3+}) was used as an enhancing metal ion. However, both the monovalent and divalent cations could still have enhanced the flocculation activity with the flocculation activity above 70%. Contrary to this, the nanoparticles flocculate poorly without the presence of the cation, suggesting that they are cation-dependent. In the bioflocculant, both the monovalent and divalent were found to be most effective, with Li^{+} being the highest flocculation activity at 75%. The least flocculation activity was observed when the trivalent cation (Fe^{3+}) was used. This conflicts the findings that suggest monovalent cations reduce the strength of the bonds and results in loose flocs, thus producing poor flocculation activity [32].

The higher amount of both COD and BOD is not good for the aquatic ecosystem. This condition results in the decrease of the amount of dissolved oxygen (DO), which in turn results in anaerobic conditions that are detrimental to higher aquatic life. Furthermore, a high amount of BOD in water signifies a high amount of nutrients, which may result in an algal bloom. In Table 3, different wastewaters were used to evaluate the effectiveness of FeNPs in comparison to a bioflocculant. Samples were analysed using a UV-Vis spectrophotometer Spectroquant® at 620 nm wavelength. The removal of COD and BOD was conducted using the 0.4 and 0.8 mg/mL for the FeNPs and bioflocculant, respectively, as these concentrations were found to be effective from optimization in Figure 4. The nanoparticles proved to be most effective when compared to both the bioflocculant and ferric chloride with BOD over 80%, while COD was 76% for coal mine wastewater and least effective on river water with just 48%. Contrary to this, the Actinomycete bioflocculant that was used in the wastewater treatment and removal of heavy metals by Agunbiade et al. performed below 70% for both COD and BOD removal efficacy [13]. The bioflocculant remained consistently poor in all the samples for BOD removal with just 50% efficacy. However, a remarkable improvement was observed in COD removal for coal mine wastewater by the bioflocculant with 72%, but it remained poor in the river water sample. Therefore, it can be deduced that FeNPs are a better flocculant compared to the bioflocculant and ferric chloride. Bioflocculants are generally nontoxic but they still need to be tested for biosafety reasons [33]. In Figure 7, nanoparticles were evaluated against human normal cells (HEK 293) and cancer cells (MCF7). As-synthesized nanoparticles are found to be nontoxic at low concentrations, as the cell survival was above 76% for both cells at 25–50 µL. With the increase in concentration, cell survival rates decrease, however, cell survival was still above 56%. It is therefore recommended that FeNPs should not be used at high concentrations, as it may result in cell toxicity. Contrarily, the bioflocculant proved to be nontoxic against both cells at the highest concentration of 100 µL, with the cell survival over 90%.

5. Conclusions

The sample as-synthesized nanoparticles and bioflocculant revealed the presence of the functional groups –OH and $–NH_2$, respectively. SEM-EDX indicated a huge percentage of O and C wt.% in both samples. FeNPs are most effective at low concentrations while the bioflocculant works best when the dosage is increased to 0.8 mg/mL. FeNPs are effective in all pH conditions and temperature ranges, while the bioflocculant was only effective at lower temperatures and neutral in weak alkaline conditions. Nanoparticles could remove effectively both COD and BOD in all water samples, while the bioflocculant and ferric chloride were seen to be less effective. FeNPs are nontoxic only at lower concentrations, while the bioflocculant is nontoxic even at higher concentrations. Therefore, FeNPs can be recommended as an alternative flocculant provided a lower concentration is maintained. For future prospects, more characterization should be conducted (XPS) to ascertain the oxidation state of the synthesized material. In addition, more characterization is necessary to establish the mechanism behind the formation of nanoparticles.

Author Contributions: Conceptualization, A.K.B. and R.V.S.R.P.; formal analysis, N.G.D. and R.V.S.R.P.; investigation, N.G.D.; supervision, A.K.B. and R.V.S.R.P.; writing—original draft, N.G.D.; writing—review and editing, R.V.S.R.P. All authors have read and agreed to the published version of the manuscript.

Funding: Rajasekhar Pullabhotla would like to acknowledge the National Research Foundation (NRF, South Arica) for the financial support in the form of the Incentive Fund Grant (Grant No: 103691) and Research Developmental Grant for Rated Researchers (112145).

Acknowledgments: Nkosinathi Dlamini would like to acknowledge the Council for Scientific and Industrial Research (CSIR, South Africa) for the financial assistance in the form of the Ph.D. bursary. The authors would like to acknowledge the Electron Microscopy Unit at the University of KwaZulu-Natal, Westville campus, for providing support by letting us use the SEM-EDX facilities for the characterization of nanomaterials. The authors would like to acknowledge Moganavelli Singh, University of KwaZulu-Natal, Westville campus and her team for assisting with cytotoxicity tests. Rajasekhar Pullabhotla would like to acknowledge the National Research Foundation (NRF, South Arica) for the financial support in the form of the Incentive Fund Grant (Grant No: 103691) and Research Developmental Grant for Rated Researchers (112145).

Conflicts of Interest: The authors declare no conflicts of interest.

References

1. Corcoran, E.; Nellemann, C.; Baker, E.; Bos, R.; Osborn, D.; Savelli, H. (Eds.) *Sick Water? The Central Role of Wastewater Management in Sustainable Development: A Rapid Response Assessment*; United Nations Environment Programme: Birkeland Trykkeri AS, Norway, 2010.
2. Yang, Z.; Liu, S.; Zhang, W.; Wen, Q.; Guo, Y. Enhancement of coal waste slurry flocculation by CTAB combined with bioflocculant produced by Azotobacter chroococcum. *Sep. Purif. Technol.* **2019**, *211*, 587–593. [CrossRef]
3. Santschi, P.H. Marine colloids, agents of the self-cleansing capacity of aquatic systems: Historical perspective and new discoveries. *Mar. Chem.* **2018**, *207*, 124–135. [CrossRef]
4. Ali, I.; Asim, M.; Khan, T.A. Low cost adsorbents for the removal of organic pollutants from wastewater. *J. Environ. Manag.* **2012**, *113*, 170–183. [CrossRef] [PubMed]
5. Qu, X.; Alvarez, P.J.; Li, Q. Applications of nanotechnology in water and wastewater treatment. *Water Res.* **2013**, *47*, 3931–3946. [CrossRef] [PubMed]
6. Verma, Y. Acute toxicity assessment of textile dyes and textile and dye industrial effluents using Daphnia magna bioassay. *Toxicol. Ind. Health* **2008**, *24*, 491–500. [CrossRef]
7. Samanta, A.K.; Agarwal, P. Application of natural dyes on textiles. *Indian J. Fibre Text. Res.* **2009**, *34*, 384–399.
8. Merzouk, B.; Madani, K.; Sekki, A. Using electrocoagulation–electroflotation technology to treat synthetic solution and textile wastewater, two case studies. *Desalination* **2010**, *250*, 573–577. [CrossRef]
9. Verma, A.K.; Dash, R.R.; Bhunia, P. A review on chemical coagulation/flocculation technologies for removal of colour from textile wastewaters. *J. Environ. Manag.* **2012**, *93*, 154–168. [CrossRef]
10. Tiwari, D.K.; Behari, J.; Sen, P. Application of nanoparticles in waste water treatment 1. *World Appl. Sci. J.* **2008**, *3*, 417–433.
11. Maliehe, T.S.; Basson, A.K.; Dlamini, N.G. Removal of Pollutants in Mine Wastewater by a Non-Cytotoxic Polymeric Bioflocculant from Alcaligenes faecalis HCB2. *Int. J. Environ. Res. Public Health* **2019**, *16*, 4001. [CrossRef]
12. Lu, H.; Wang, J.; Wang, T.; Wang, N.; Bao, Y.; Hao, H. Crystallization techniques in wastewater treatment: An overview of applications. *Chemosphere* **2017**, *173*, 474–484. [CrossRef]
13. Agunbiade, M.O.; Pohl, C.; Van Heerden, E.; Oyekola, O.; Ashafa, A. Evaluation of FreshWater Actinomycete Bioflocculant and Its Biotechnological Applications in Wastewaters Treatment and Removal of Heavy Metals. *Int. J. Environ. Res. Public Health* **2019**, *16*, 3337. [CrossRef] [PubMed]
14. Ugbenyen, A.; Okoh, A. Characteristics of a bioflocculant produced by a consortium of Cobetia and Bacillus species and its application in the treatment of wastewaters. *Water SA* **2014**, *40*, 140–144. [CrossRef]
15. Guo, H.; Hong, C.; Zhang, C.; Zheng, B.; Jiang, D.; Qin, W. Bioflocculants' production from a cellulase-free xylanase-producing Pseudomonas boreopolis G22 by degrading biomass and its application in cost-effective harvest of microalgae. *Bioresour. Technol.* **2018**, *255*, 171–179. [CrossRef] [PubMed]
16. Karthiga devi, K.; Natarajan, K.A. Production and characterization of bioflocculants for mineral processing applications. *Int. J. Miner. Process.* **2015**, *137*, 15–25. [CrossRef]
17. Sun, J.; Zhang, X.; Miao, X.; Zhou, J. Preparation and characteristics of bioflocculants from excess biological sludge. *Bioresour. Technol.* **2012**, *126*, 362–366. [CrossRef] [PubMed]
18. Abu Tawila, Z.; Ismail, S.; Dadrasnia, A.; Usman, M. Production and characterization of a bioflocculant produced by Bacillus salmalaya 139SI-7 and its applications in wastewater treatment. *Molecules* **2018**, *23*, 2689. [CrossRef]
19. Gao, J.; Bao, H.-Y.; Xin, M.-X.; Liu, Y.-X.; Li, Q.; Zhang, Y.-F. Characterization of a bioflocculant from a newly isolated Vagococcus sp. W31. *J. Zhejiang Univ. Sci. B* **2006**, *7*, 186–192. [CrossRef] [PubMed]
20. Zhang, Z.-Q.; Bo, L.; XIA, S.-Q.; Wang, X.-J.; Yang, A.-M. Production and application of a novel bioflocculant by multiple-microorganism consortia using brewery wastewater as carbon source. *J. Environ. Sci.* **2007**, *19*, 667–673. [CrossRef]
21. Dlamini, N.G.; Basson, A.K.; Pullabhotla, V.S.R. Optimization and Application of Bioflocculant Passivated Copper Nanoparticles in the Wastewater Treatment. *Int. J. Environ. Res. Public Health* **2019**, *16*, 2185. [CrossRef] [PubMed]

22. Dlamini, N.G.; Basson, A.K.; Shandu, J.S.E.; Mavuso, S.S.; Pullabhotla, V.S.R. Biosynthesis, Characterization, and Application of Iron Nanoparticles: In Dye Removal and as Antimicrobial Agent. *Water Air Soil Pollut.* **2020**, *231*, 1–10. [CrossRef]
23. Xia, X.; Lan, S.; Li, X.; Xie, Y.; Liang, Y.; Yan, P.; Chen, Z.; Xing, Y. Characterization and coagulation-flocculation performance of a composite flocculant in high-turbidity drinking water treatment. *Chemosphere* **2018**, *206*, 701–708. [CrossRef] [PubMed]
24. Daniels, A.N.; Singh, M. Sterically stabilized siRNA: Gold nanocomplexes enhance c-MYC silencing in a breast cancer cell model. *Nanomedicine* **2019**, *14*, 1387–1401. [CrossRef] [PubMed]
25. Xiong, Y.; Wang, Y.; Yu, Y.; Li, Q.; Wang, H.; Chen, R.; He, N. Production and characterization of a novel bioflocculant from Bacillus licheniformis. *Appl. Environ. Microbiol.* **2010**, *76*, 2778–2782. [CrossRef] [PubMed]
26. Mata, Y.; Torres, E.; Blazquez, M.; Ballester, A.; González, F.; Munoz, J. Gold (III) biosorption and bioreduction with the brown alga Fucus vesiculosus. *J. Hazard. Mater.* **2009**, *166*, 612–618. [CrossRef] [PubMed]
27. Li, X.-M.; Yang, Q.; Huang, K.; Zeng, G.-M.; Liao, D.-X.; Liu, J.-J.; Long, W.-F. Screening and characterization of a bioflocculant produced by *Aeromonas* sp. *Biomed. Environ. Sci. BES* **2007**, *20*, 274–278.
28. Wang, L.; Ma, F.; Qu, Y.; Sun, D.; Li, A.; Guo, J.; Yu, B. Characterization of a compound bioflocculant produced by mixed culture of Rhizobium radiobacter F2 and Bacillus sphaeicus F6. *World J. Microbiol. Biotechnol.* **2011**, *27*, 2559–2565. [CrossRef]
29. Gong, W.-X.; Wang, S.-G.; Sun, X.-F.; Liu, X.-W.; Yue, Q.-Y.; Gao, B.-Y. Bioflocculant production by culture of Serratia ficaria and its application in wastewater treatment. *Bioresour. Technol.* **2008**, *99*, 4668–4674. [CrossRef]
30. Sekelwa, C.; Anthony, U.M.; Vuyani, M.L.; Anthony, O.I. Characterization of a thermostable polysaccharide bioflocculant produced by Virgibacillus species isolated from Algoa bay. *Afr. J. Microbiol. Res.* **2013**, *7*, 2925–2938.
31. Manivasagan, P.; Kang, K.-H.; Kim, D.G.; Kim, S.-K. Production of polysaccharide-based bioflocculant for the synthesis of silver nanoparticles by *Streptomyces* sp. *Int. J. Biol. Macromol.* **2015**, *77*, 159–167. [CrossRef]
32. Wu, J.-Y.; Ye, H.-F. Characterization and flocculating properties of an extracellular biopolymer produced from a Bacillus subtilis DYU1 isolate. *Process. Biochem.* **2007**, *42*, 1114–1123. [CrossRef]
33. Pathak, M.; Devi, A.; Bhattacharyya, K.; Sarma, H.; Subudhi, S.; Lal, B. Production of a non-cytotoxic bioflocculant by a bacterium utilizing a petroleum hydrocarbon source and its application in heavy metal removal. *RSC Adv.* **2015**, *5*, 66037–66046. [CrossRef]

© 2020 by the authors. Licensee MDPI, Basel, Switzerland. This article is an open access article distributed under the terms and conditions of the Creative Commons Attribution (CC BY) license (http://creativecommons.org/licenses/by/4.0/).

polymers

MDPI

Article

Development of ZnO Nanoflake Type Structures Using Silk Fibres as Template for Water Pollutants Remediation

K. Jagajjanani Rao [1,*], Tarangini Korumilli [1], Akshaykumar KP [2], Stanisław Wacławek [3], Miroslav Černík [3,*] and Vinod V. T. Padil [3,*]

1 Department of Biotechnology, Vel Tech Rangarajan Dr.Sagunthala R&D Institute of Science and Technology, Chennai, Tamil Nadu 600062, India; drktarangini@veltech.edu.in

2 Tata Institute of Fundamental Research Hyderabad, Hyderabad Sy. No 36/P, Serilingampally Mandal, Hyderabad, Telangana 500107, India; kpakshay95@gmail.com

3 Department of Nanomaterials in Natural Sciences, Institute for Nanomaterials, Advanced Technologies and Innovation (CXI), Technical University of Liberec (TUL), Studentská 1402/2, 1 461 17 Liberec, Czech Republic; stanislaw.waclawek@tul.cz

* Correspondence: drjagajjananirao@veltech.edu.in (K.J.R.); miroslav.cernik@tul.cz (M.Č.); vinod.padil@tul.cz (V.V.T.P.); Tel.: +420-723372911 (V.V.T.P.)

Received: 9 April 2020; Accepted: 15 May 2020; Published: 18 May 2020

Abstract: We have fabricated ZnO nanoflake structures using degummed silk fibers as templates, via soaking and calcining the silk fibers bearing ZnO nanoparticles at 150 °C for 6 h. The obtained ZnO nanostructures were characterized using scanning electron microscopy (SEM), X-ray diffraction analysis (XRD), and UV-vis and fluorescence spectroscopic analysis. The size (~500–700 nm) in length and thicknesses (~60 nm) of ZnO nanoflakes were produced. The catalysis performances of ZnO nanoflakes on silk fibers (ZnSk) via photo-degradation of naphthalene (93% in 256 min), as well as Rose Bengal dye removal (~1.7 mM g^{-1}) through adsorption from aqueous solution, were practically observed. Further, ZnSk displayed superb antibacterial activity against the tested model gram-negative *Escherichia coli* bacterium. The produced ZnSk has huge scope to be used for real-world water contaminants remediation applications.

Keywords: silk fibers; ZnO; nanoflakes; photo-degradation; dye removal; antibacterial efficiency

1. Introduction

ZnO nanoparticles have significant technological applications, as they are well known for their catalytic, electrical, optoelectronic, antimicrobial properties, etc. [1]. They possess a large direct band gap of ~3.37 eV, and are used as functional devices in solar cells, and photocatalysts in remedial applications and sensor materials [2]. ZnO has vast areas of application, and a variety of nanostructures including nanoparticles, nanowires, nanorods, nanotubes, nanobelts, and other complex morphologies have been fabricated and reported [3]. Researchers suggest that nanostructures can be attached to host polymer materials such as porous resins, cellulose and silica, to minimize a hazard to humans and the environment derived from the uncontrolled release [4]. The nanoparticles fixed to the host materials make them bulkier and they can be easily removed and recovered after application, which makes them of potential use in removal of pollutants from the environment, e.g., to break down organochlorine pesticides, halogenated herbicides and azo dyes [5].

Taking a practical science point of view into consideration, recent research has been accelerated towards the development of novel composite materials using thermostable biopolymers (chitin, spongin, silk) and selected inorganic phases [6,7]. Using hydrothermal synthesis conditions primarily, the synergized concept uses extreme biomineralization and bioinspired materials chemistry for the generation of advanced and functional composite materials [6]. The name given to this approach and filed is "extreme biomimetics", which produces materials with hierarchical and nanostructural organizations, alongside three-dimensional (3D) architecture at micro- and macrolevels [8,9]. Reported studies show biological templates with 3D morphology have been used as a matrix to generate TiO_2, Zirconia, ZnO, hematite, germanium oxide composites, etc. [6,9,10]. For instance, spongin, a protein scaffold from marine demosponge, *Hippospongia communis* was used to produce TiO_2 3D composite structures using hydrothermal conditions (120 °C for 3 h) [6]. Likewise, lyophilized skeletons of marine sponges *Aplysina cauliformis* were used to produce novel chitin–GeO_2 photoluminescent nanocomposites, using a hydrothermal reaction with a temperature of 185 °C [9]. In another study, β-chitinous scaffolds isolated from marine cephalopod *Sepia officinalis* to synthesize a template for the in vitro formation of ZnO nanostructure composites at 70 °C, which are extreme for the used biological material [10]. The key step in the above mentioned studies is thermal stability of the biopolymers during hydrothermal synthesis and processing [6,9,10]. Among inorganic nanomaterials, ZnO structures and composites are unique with antibacterial activity, wide bandgap, photocatalytic properties, and are known for having numerous practical applications [8,10].

Template directed synthesis of ZnO materials is gaining significant interest in research communities, and various porous and non-porous materials of ZnO have been synthesized [2,5]. Polyethylene glycol, copolymer gels, polyvinylpyrrolidone, nanofibres, etc., templates have been used to produce porous materials, while various soft biological materials, like butterfly wings, bacteria, bamboo, cotton, wool, silk fibers, etc., have been used to produce ZnO networks and hybrid materials [2,11,12]. The selection of natural and renewable sources to synthesize nanomaterials is imperative to sustainable development, and can limit hazardous wastes. Materials from biological sources with distinctive structures are plentiful in nature and are inexpensive and environmentally friendly. Exploring these structures for inorganic-organic nanocomposites, in particular with ZnO composites, holds significant challenges that need to be addressed. Process simplicity, more than one preparation step, pre- and post-treatments, high-temperature dependency, and a selection of suitable templates, etc., are key components, which affect the process pathway from the laboratory to commercial application.

Natural fibers from silkworms are well known for their biomedical applications, and recent studies display their potential as templates for nanoparticle synthesis [13,14] Degummed silk fibers are rich in glycine, alanine, and serine functional groups, which interact with metal ion containing precursors and can assist in nanoparticle deposition [15]. The application of silk materials for water treatment is a new area of research still in its infancy. Challenges like functionalization and impregnation with nanomaterials for applications like toxic pollutants removal, oil/solvent-water separation, smart textiles etc., are the key thrust areas in this field [15,16] Silk materials are biocompatible, biodegradable, and mechanically strong materials, readily available and suitable for water treatment applications, as they pose no disposal problem in the environment after their subsequent exhaustive use [15]. There are limited studies where research groups used silk materials as a template, and as an assistive substrate for the synthesis and growth of ZnO particles [11,12,17]. These involve multistep procedures and the use of harmful chemicals like N, N-dimethyl formamide (DMF). ZnO nanocomposites for toxic dyes removal, contaminant remediation, and water treatment are of scientific interest, and the use of silk fibers for the production of ZnO composites with simple preparation steps is a challenging task.

Our work focused on introducing a facile template assisted synthesis strategy to attain flake type ZnO nanoparticles. In this study, a green template source (silk fibers) was utilized with zinc acetate precursor, with a reasonably low temperature annealing process, to attain ZnO nanoflakes on silk fibers (ZnSk). The synthesized ZnSk were used for pollutant remediation studies, such as naphthalene degradation and Ross Bengal (RB) dye removal from the aqueous system successfully. The stated chemicals were selected as model pollutants, as they have profound effects on the environment. For instance, the photo-degradation of naphthalene is slow, with photochemical half-life usually beyond 100 h [18], and RB, a Xanthan dye extensively used in printing industries, insecticides, etc., produces the residual dye with the potential to pollute water [19]. Furthermore, the observed leach out of ZnO from the ZnSk is low; consequently, the synthesized ZnSk structures with decent antimicrobial activity against the tested *E. coli* bacteria may be useful for large-scale wastewater treatment applications in real-time settings.

2. Materials and Methods

2.1. Degumming Step

Dried cocoons of *Bombyxmori* silkworm were gifted by Silkworm Culture and Reeling Unit, Chebrolu, Andhra Pradesh, India. Further degumming steps were adopted from a reported study [20]. In brief, the obtained cocoons without insects were peeled into pieces and treated 3 times with 0.5 wt % aqueous Na_2CO_3 solution at 100 °C for 30 min. The process was repeated until the silk fibers tuned to white from pale yellow and the degumming solution stayed clear. Next, the samples obtained were washed with deionized water and naturally dried to get degummed silks.

2.2. Synthesis of ZnO Nanoflakes on Silk Fibres (ZnSk)

ZnO nanostructures were obtained on silk fibers by modifying the approach presented in a reported study using zinc acetate as a starting material [21]. In a typical experiment, 30 mM $Zn(CH_3COO)_2{\cdot}2H_2O$ solution was prepared using a mixture of H_2O: methanol: ethanol in the volumetric ratio of 5:4:6. A small amount of acetic acid was added to adjust the pH value to ~4.8 to prevent the formation of hydroxides. Then, the solution was homogeneously mixed using a sonicator and 1 mL of it added to degummed silk fibroin of 56.8 mg. The wetted silk fiber was transferred to a glass crucible and kept in an oven at 150 °C for 6 h. The solution without the silk fiber was kept as a control with the above ascribed conditions. The obtained ZnSk structures were further analyzed.

2.3. Catalytic Study

The photocatalytic activity of the obtained ZnSk was investigated by means of naphthalene degradation experiment. Primarily, 80 mg of the ZnSk sample (contains 10 mg of Zn upon on analysis with Multiparameter Photometer) was added to 30 mL naphthalene solution of 43 mg/L in a 100 mL wide mouth glass flask. The investigation was performed at room temperature (34 °C), with continuous stirring (800 rpm) in a rotary shaker incubator. During incubation, the obtained suspensions were irradiated from top under UV lamp (11W), Philips PL-S, Hyderabad, India, a compact UVC source with intensity of 22 lux. Samples for analysis were taken periodically and the decomposition of naphthalene versus irradiation time was monitored by UV–visible spectrophotometer.

For reference measurements, silk fibers devoid of ZnO were also exposed to naphthalene, in the same way as the above stated experimental conditions, and monitored spectrally.

2.4. Rose Bengal (RB) Dye Removal

Adsorption experiments were carried out by maintaining a ratio of RB/(ZnSk or silk fibers) as 2.45 mg/g. An orbital incubator shaker (moving at 800 rpmREMI, Mumbai, India) was used for the study, and samples were monitored till 400 min. A pH of ~6.5 and a temperature of ~34 °C were used for investigations devoid of light. The concentration of residual RB was determined using a UV-visible spectrophotometer (UV-3600 Shimadzu, Kyoto, Japan), where the measurements were made at 545 nm, which corresponds to its maximum absorbance. Silk fibers devoid of ZnO are used as control. The change in RB dye amounts were calculated from the concentrations in solutions before and after dye adsorption. The mechanism of adsorption by silk and ZnO deposited fibers were estimated by using the pseudo-first and second-order kinetic models [22]. Additionally, the adsorption behavior of RB dye was estimated using Langmuir adsorption isotherm model. For isotherm studies, ~0.287 g of ZnSk or silk fibers were used in a 50 mL aqueous solution of RB (concentrations 0.005 to 0.03 mM, pH ~ 6.5). All the experiments were done using sterile 100 mL glass flasks, properly covered, and periodic samples were removed for UV-vis spectral analysis.

2.5. Antimicrobial Activity of ZnSk Composite

The produced ZnSk composite was tested for bactericidal activity against Gram negative *E. coli*, using the zone of inhibition (ZOI) method. Bacterial cell suspension of 0.1 mL (from 2% *w/v*) was plated uniformly, with the help of spreader on nutrient agar plates. Using a sterile cork borer, 10 mm wells were made on agar plates, and ZnSk composite (of 20 mg) was placed in the well. Here, degummed silk fibers were used as a control and agar plates with samples were then incubated at 37 °C for 24 h. The diameters of the resulting ZOI in mm due to microbial growth were measured using imageJ software on digital images, and the antibacterial activity was determined.

2.6. Analysis and Characterization

UV-visible spectra of the samples were observed using Cyberlab UV100, UV-vis spectrophotometer. The scanning electron microscopy (SEM) observations were performed on JEOL-2100F machine (JEOL India Pvt. Ltd, Kolkata, India). X-ray diffraction (XRD) measurements were done by X-ray diffractometer (Philips PW 1830 HT, Amsterdam, Netherlands), with an accelerating voltage of 35 kV, with a current of 30 mA. The photoluminescence (PL) spectra were measured using spectrofluorometer (Hitachi F-7000, Chiyoda, Tokyo, Japan. The Zn content in ZnSk was identified using HANNA HI83399 Multiparameter Photometer.

3. Results

3.1. Microscopic Analysis

Degummed silk fibers with and without the treatment with zinc precursor are shown in Figure 1. While the degummed fibers show smooth morphology (Figure 1a); treated fibers exhibited a layer of particle deposition, due to formation of ZnO nanoparticles throughout the silk fibers surface (Figure 1b). The close magnification of the silk fiber images reveals that the formed ZnO particles are mainly flake type structures with 500–700 nm in length and ~60 nm thicknesses (Figure 1c–e). The top and side views of the fibers show the thickness, orientation, and aggregation of the ZnO flake structures on the surface of silk fibers. At the images (Figure 1d,e), there were few pseudo spherical type structures (160–190 nm), along with ZnO flakes.

In this mode of synthesis, water is the most suitable oxidizing agent, whilst methanol and ethanol facilitate the fast transformation of the precursor mist into vapor form during the heating step, after an addition of the mixtures to the degummed silk fibers. The formation of nanoscale ZnO at 150 °C for 6 h may be described by the following equation:

$$Zn(CH_3COO)_2 + H_2O + \text{degummed silk fibers} \rightarrow \text{ZnSk composite} + 2CH_3COOH(g)$$

ZnO particle synthesis devoid of silk fibers exhibited no flake type structures and furthermore bigger ZnO particles with a high degree of polydispersity were formed (Figure 2a). This indicates that there is a role of silk fibers in directing the ZnO particles size and shape. Degummed silk fibers are rich in fibroin content, mainly made up of non-polar amino acids, like glycine and alanine, which sum up approximately 76% of their structure [23]. We believe that they secure the growth process of ZnO structures by harboring the initial nuclei formed from the ZnO precursor, and then acting as a growth sites to form ZnO flakes. This hypothesis can be confirmed microscopically in Figure 2b,c, where there exist some aggregations of pseudo smaller spherical particles forming flake type structures on the surface of the silk fibers. We believe that these are the points or places where there is a possibility for growth of ZnO nanoparticles into bigger structures. We assume that the observed ZnO nanoparticles are in between their growth step to form bigger flake type structures.

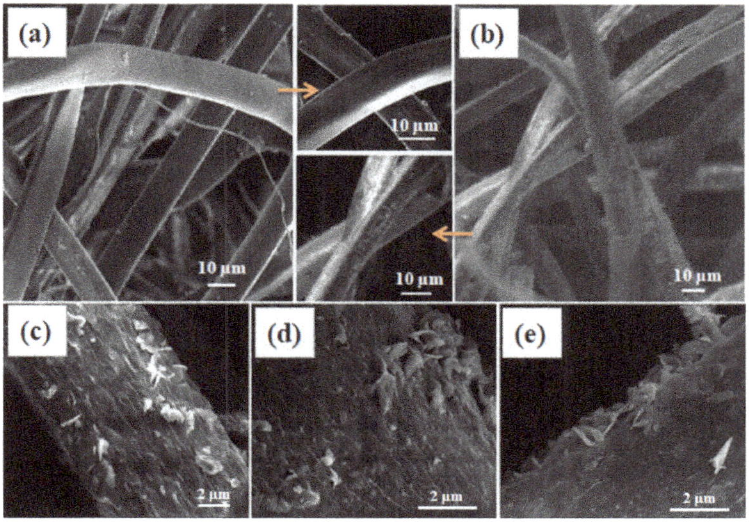

Figure 1. Scanning electron microscopy (SEM) images of (**a**) degummed silk fibers, (**b**) ZnO coated silk fibers. High magnification images from (**b**) are given in (**c**–**e**) where ZnO flake type structures are clearly visible.

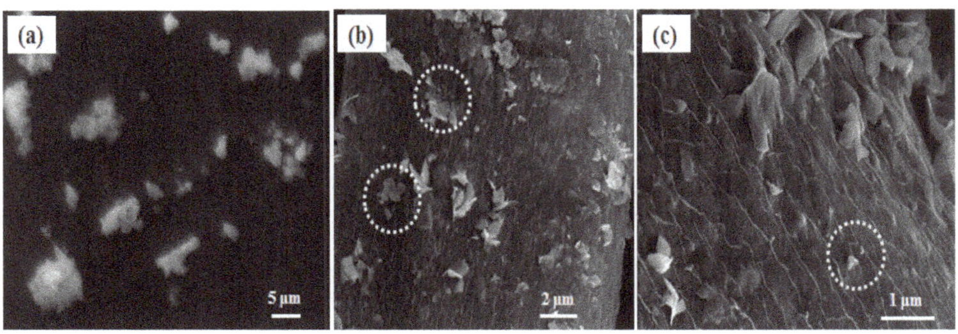

Figure 2. SEM image of (**a**) ZnO particles devoid of silk fibers; (**b**,**c**) shows ZnO nanoflakes on silk fiber (ZnSk) structures and the aggregations of small particles; dotted circular regions in white display aggregations of pseudo smaller spherical particles.

3.2. XRD Analysis

The X-ray diffraction spectra of the silk fibers and ZnSk composite are presented in Figure 3a. The silk fibers in Figure 3a exhibit strong peaks with 2θ value around 20.7°, this corresponds to the presence of crystalline domains in silk fibers [24,25]. The crystalline groups of silk include glycine-X repeats covering 94% of silk sequence. Here, X include alanine (65%), serine (23%), and tyrosine (9%) [24]. Coming to the diffraction pattern of ZnO (Figure 3a) deposited on silk fibers, 2θ values observed are to be at 30.26°, 32.91°, 36.08°, 47.08°, 56.79°, and the dashed lines in Figure 3a indicate different facets. The above sample data can be attributed to the hexagonal crystal system of ZnO, and matches near with JCPDS file no: 36-1451, 79-0206 [12,26]. Zn content in ZnSk composite identified to be 125 mg/g. This value is low and might be the reason for low intensity peaks from ZnSk samples.

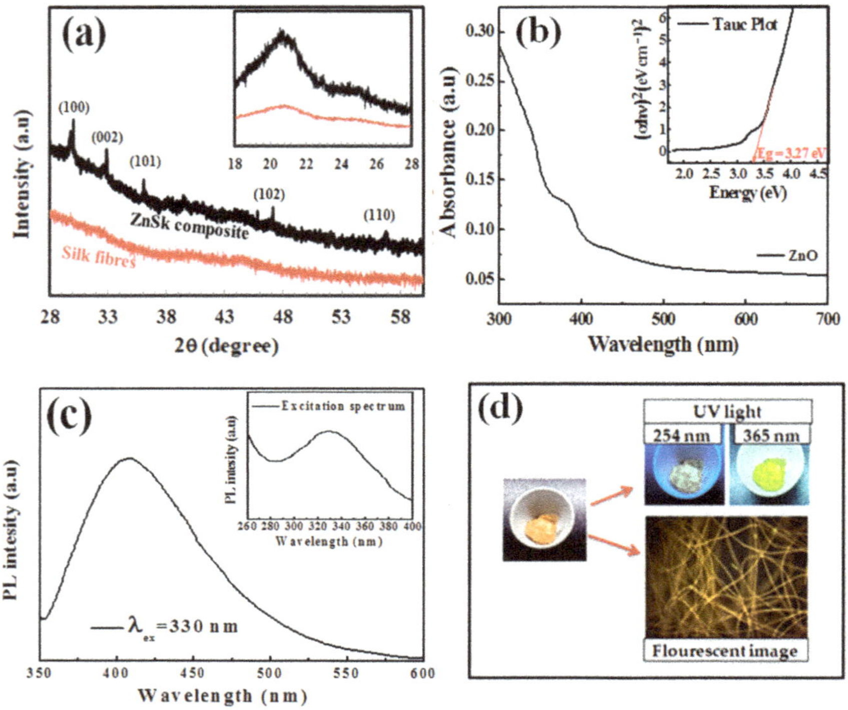

Figure 3. (**a**) X-ray diffraction (XRD) patterns of silk fibers and ZnSk composites. The hexagonal crystal system of ZnO with facets are indicated above the diffraction pattern of ZnSk composites, (**b**) UV-visible spectrum of ZnSk and Tauc plot depiction (inset), (**c**) fluorescence spectra of ZnSk and its excitation spectrum (inset), (**d**) visible emissions from ZnSk composites in a UV box and fluorescence microscopic observation with blue excitation.

3.3. UV-Vis Spectroscopy Analysis

Figure 3b shows the UV-visible spectrum of dispersed ZnSk in water. Only silk fiber suspension in water was taken as a reference here. The sample shows an absorption maximum in the UV region and the band gap energy was obtained using the observed spectrum. The optical band gap of the synthesized ZnO structures is calculated by using Tauc plot (Figure 3b insert), and was observed to be 3.27 eV. This value is near to a reported study with flake type structures of ZnO [27], and also matches with other studies as well [28,29].

Furthermore, the PL behaviour of ZnSk was observed using a spectrophotometer, and the measurements are shown in Figure 3c. The ZnSk exhibited strong and wide PL signal range from 380 to 500 nm, with the excitation light energy higher than the bandgap energy. The emission and excitation maxima of the ZnO coated silk fiber spectra are observed to be 330 nm and 420 nm, respectively (Figure 3c). The PL signal credited to excitonic PL and is a result of surface oxygen vacancies and defects of the ZnO nanoparticles [30]. The high intensity peak at 420 nm in particular attributed to band edge free excitons of ZnO nanostructures [30]. While the PL behavior from silk fibers is absent, the visible appearance and color changes of the ZnSk are observed using short (254 nm) and long (365 nm) UV irradiation lamp sources using a UV box (Figure 3d). The bright lemon green color emission at 365 nm wavelength exposure is clearly visible in the picture. The fluorescence emission of ZnSk was also visualized through a fluorescence microscope under blue excitation, and the image is shown in Figure 3d.

3.4. Degradation Study

Prior to degradation studies, the adsorption effect was monitored by exposing silk fibers and ZnSk composites to naphthalene for 30 min devoid of light. With no or negligible decrease in UV-vis spectrum of naphthalene at 274 nm, it was observed that naphthalene has no adsorption effect towards silk fibers or ZnSk composites.

Next, we studied degradation study, where Figure 4a shows the kinetics of the photocatalytic degradation of naphthalene solution by the ZnSk composites. The time dependent spectral behavior of naphthalene at absorption maximum (λ_{max}) of 274 nm is observed, with a decreasing trend in λ_{max} values and up to 93% degradation is observed in 256 min. Silk fibers devoid of ZnO did not displayed any change in λ_{max} value at 274 nm upon UV irradiation, indicating no effect on naphthalene degradation. The prepared ZnSk composites, after repeated use (three times), displayed a decreasing trend of naphthalene absorbance, and the average λ_{max} values at 274 nm with increasing time is shown in Figure 4a inset.

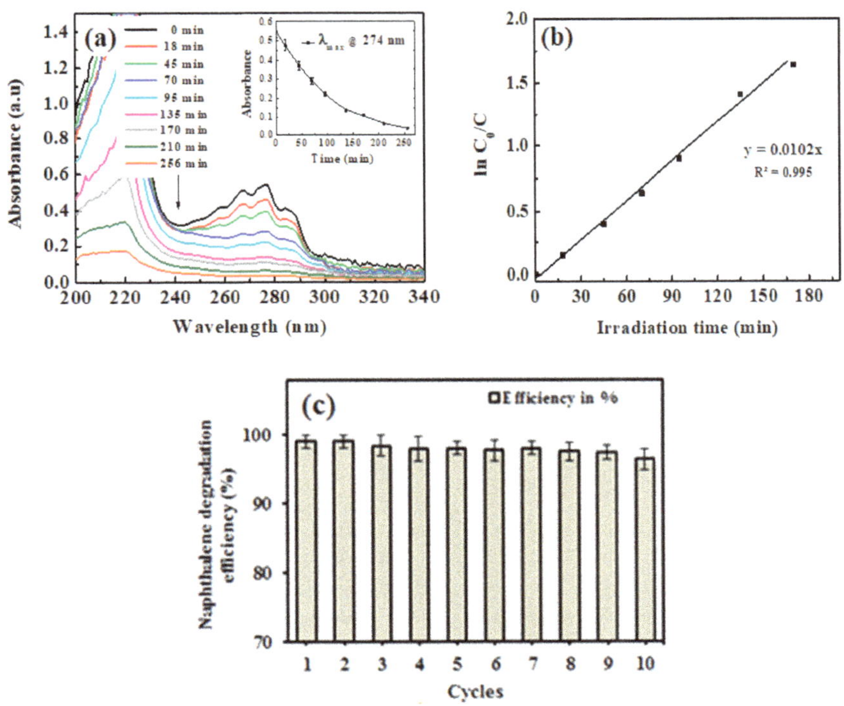

Figure 4. (**a**) UV-visible spectrum of naphthalene during degradation by ZnSk composite, with a plot of kinetic of absorbance change at 274 nm (inset), (**b**) first-order kinetic model fit to the degradation data. (**c**) Reusability of ZnSk composites for naphthalene degradation (80 mg ZnSk exposed to 43 mg/L naphthalene), and error bars represent standard deviation for three measurements.

The observed data, in accordance with the ln (C_0/C) vs irradiation time, fit better with the first-order kinetic model of the rate law (Figure 4b). The rate constant K form the slope of the observed trend line is 0.0102 min^{-1} and the half-life time of degradation is calculated to be 67.94 min. With a higher size ZnO nanoflake type structures, and with low deposition in terms of Zn (125 mg/gm of ZnSk) dose, the degradation behavior of naphthalene is comparable or superior to other reported studies with different nanoparticles (like TiO_2) and other approaches [31,32]. Furthermore, reports state that the photocatalytic activity of ZnO towards naphthalene is reasonably profound, and the degradation process is OH• radical mediated with quick end products [33,34]. The reaction can be effectively monitored using a simple UV-vis spectroscope, and the steps involved in the reaction are detailed below [33,34].

$$ZnO + hv \rightarrow e^- + h^+ \tag{1}$$

$$h^+ + H_2O \rightarrow H^+ + OH^{\cdot} \tag{2}$$

$$h^+ + OH^- \rightarrow OH^{\cdot} \tag{3}$$

$$e^- + O_2 \rightarrow O_2^- \tag{4}$$

$$2e^- + O_2 + 2H^+ \rightarrow H_2O_2 \tag{5}$$

$$e^- + H_2O_2 \rightarrow OH^{\cdot} + OH^- \tag{6}$$

In addition to the above, complementing our study and observations, the enhanced photocatalytic properties of anisotropic ZnO materials are attributed to the high surface area and the nanofeatures [33,34]. These structures result in decrease in chance of recombination for photo-excited electron–hole pairs. The excited electrons are captured by the oxygen vacancies on the surface, and thus restrain the recombination of electrons and holes [33,34]. The holes in the valence band of ZnO attack the surface hydroxyls and yield surface-bound OH• radicals, which participate in photocatalytic reactions effectively. The rate of generation of these OH• radicals is higher in anisotropic ZnO materials and thus promotes more photocatalytic activity. These preliminary results from our study (Figure 4) indicate that ZnSk composites could find promising potential in environmental and wastewater treatment applications.

The synthesized ZnSk structures have huge scope to be used in real time studies, as they are template based and firmly associated with silk fibers after repeated aqueous washings (observed Zn content is 118 mg/gm ZnSk with net loss in terms of Zn being ~4.8% after 12 washings). The application of ZnSk will prevent the leaching out of ZnO, and these structures can be recovered and reused easily, without any trouble. The consistent degradation ability (>97%) of ZnSk composites after 12 cycles (Figure 4c) further ensures its degradation efficiency, and these materials may be used for large scale remediation capability.

3.5. RB Removal/Adsorption Study

RB, an anionic dye, was used as a model pollutant to evaluate the adsorption properties of the produced ZnSk composite. As soon as the fibers were exposed to the RB solution, the rose color slowly disappeared, and the typical visible spectrum peak of RB at ~550 nm progressively decreased (Figure 5a). The rate of absorbance decrease was more significant for ZnSk composite compared to silk fibers, and the fibers became colored due to the RB adsorption.

Figure 5b shows the calculated adsorbed amount of RB for an experiment conducted up to 450 min at room temperature (32 °C). Even though the course of adsorption is similar for both of the experiments, the ZnSk composites exhibited markedly greater adsorption ability than the silk fibers devoid of ZnO. The adsorbed amount of RB was ~0.723 mg/g for silk fibers and ~1.770 mg/g for ZnSk. This value indicates the adsorption capacity of ZnSk is about ~2.45 times higher than silk fibers. The saturation time for RB adsorption was found to be ~120 min for both sorbents.

To investigate adsorption controlling mechanism, a pseudo-first order kinetic model was utilized to evaluate the experimental data [22]. The rate of adsorption based on the adsorption capacity in its linear form through pseudo-first order kinetic model is expressed as follows:

$$\log(q_e - q_t) = \log q_e - \frac{k_1}{2.303}t \tag{7}$$

Furthermore, the pseudo-second order model is presented as follows:

$$\frac{t}{q_t} = \frac{1}{k_2 q_e^2} + \frac{1}{q_e} t \tag{8}$$

$$h = k_2 q_e^2$$

where q_e (mg/g) is adsorption capacity, q_t is adsorption at time t and k_1 (min^{-1}), k_2 (g mg^{-1} min^{-1}) are pseudo-first and pseudo-second order rate constants, and h (mg g^{-1} min^{-1}) is the initial adsorption rate, corresponding to the second-order kinetic model.

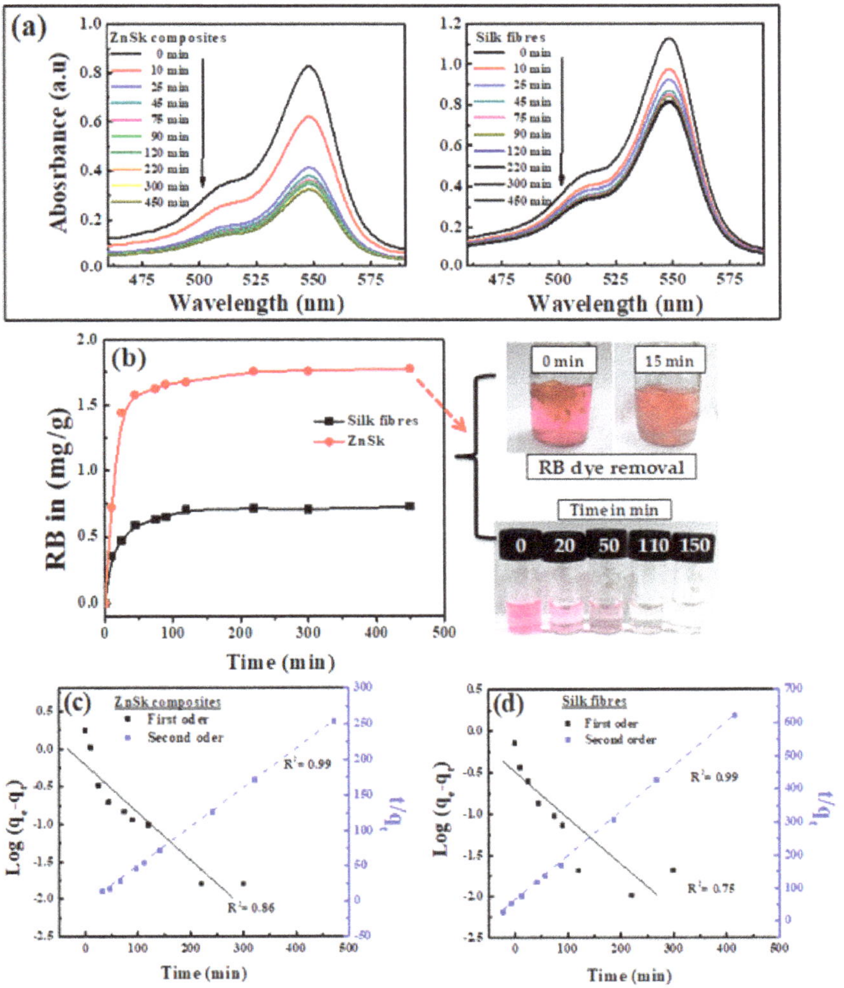

Figure 5. (**a**) UV-visible absorption spectra of Ross Bengal (RB) upon exposure to ZnSk composite and silk fibers. Initial concentration of RB used is 0.017 mM for ZnSk and 0.024 mM for silk fibers (for a 50 mL working volume solution). Weights of adsorbents used are 0.335 g (for ZnSk) and 0.477 g (for silk fiber). (**b**) The absorption threshold is shown in graph with RB color changes in the presence of ZnSk with time as a picture insert. Graphs from (**c**,**d**) correspond to first order and second order models for the adsorption of RB by ZnSk composites and silk fibers.

The adsorption kinetics was estimated using Equations (1) and (2) to get information on the adsorption behavior with respect to time. Batch adsorption data using the pseudo-first and second-order kinetic models are presented in Figure 5c,d for ZnSk and silk fibers individually. Correlation coefficient (R^2) values for the pseudo-first order kinetics are 0.86 and 0.75 for ZnSk and silk fibers respectively, while R^2 values for the second-order model are 0.99 and 0.99 for the same. A better correlation between the modelled and experimental data was observed using the pseudo-second order kinetics model. This favors chemisorption, and this behavior is the rate controlling factor for the adsorption of RB onto silk fibers and ZnSk composite used. The observed performance might be true, as the degummed silk fibers have the ability to fix anionic dyes through ionic bonding. Here, positive $-NH_2$ group of the fiber will bind to acid group of dye [35].

Furthermore, for examination of the adsorption behavior with different RB dye concentration, a quantitative Langmuir adsorption isotherm model was utilized to evaluate the experimental data. Langmuir equation in its linear form is usually expressed as:

$$\frac{C_e}{Q_e} = \frac{1}{Q_m K} + \frac{C_e}{Q_m} \quad (9)$$

where, Q_e is equilibrium dye concentration on adsorbent (mg g^{-1}), Q_m is monolayer capacity of the adsorbent (mg g^{-1}), K is adsorption constant (L mg^{-1}), and C_e is equilibrium dye concentration in solution (mg L^{-1}).

According to the above equation, a plot of C_e/Q_e versus C_e should be a straight line with a slope $1/Q_m$ and intercept $1/Q_mK$.

The equilibrium adsorption isotherm is of primary importance in the design of adsorption process. The Langmuir adsorption model is applied to the experimental data obtained from various RB concentrations and its adsorption (Figure 6a). A plot of C_e/Q_e vs. C_e is given in Figure 6b, and the calculated adsorption constant (K) values are 0.325 and 0.0497 (L mg^{-1}), and Q_m values are found to be ~1.706 and 0.679 mg g^{-1} for ZnSk and silk fibers, respectively. The observed Q_m values are promising and better than reported values using bottom ash type adsorbents [36] and Fe(III)–montmorillonite clay material [37]. The maximum RB adsorption achieved is ~0.04 mM g^{-1} in the former case, and it is 0.868 mg/g in the later reported study. The reusability of ZnSk composites are tested by exposing them to 0.006 mM RB with a ZnSk dose of 0.287 g in 50 mL solution (i.e., RD/ZnSk is 1 mg/g) for 12 cycles, and results are shown in Figure 6c. Values in the figure indicate a decent performance with ~97% RB removal ability after every wash cycle, whereas RB recovery efficiency was bit reduced after six cycles. This performance may be due to experimental error and Zn loss due to washing steps (as stated in Section 3.4). Overall, ZnSk composites are still better after 12 cycles of dye removal, and have the potential to be used in real time settings.

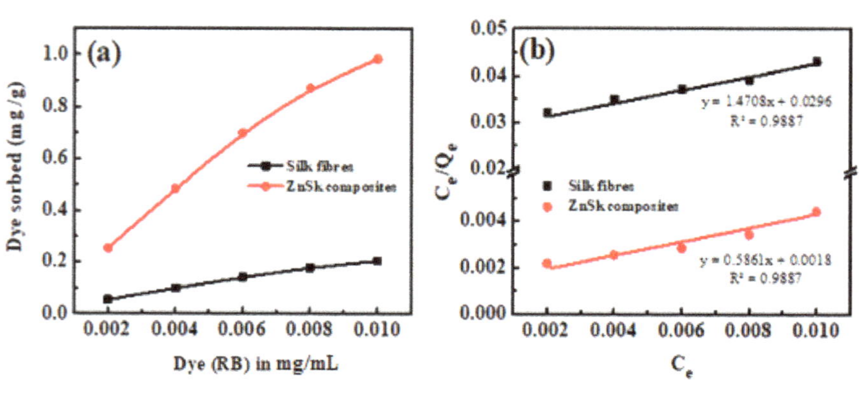

Figure 6. *Cont.*

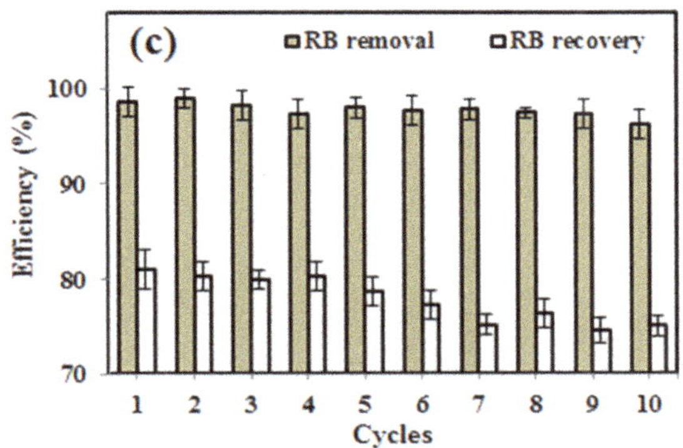

Figure 6. Using silk fibers and ZnSk composites, (**a**) RB adsorption vs. concentration and (**b**) linear plots showing the Langmuir adsorption isotherm parameters. (**c**) RB removal and recovery studies using ZnSk composites. Error bars are calculated standard deviation for three measurements.

3.6. Antimicrobial Activity of ZnO Composites

The most commonly used nanoparticles for water and wastewater treatment are Ag, TiO_2, and ZnO [38]. These particles with various sizes and shapes have huge scope as a new class of antimicrobial agents in the water disinfection and treatment fields. The applications of Ag and TiO_2 in real time settings are limited, due to cost and health effects. For instance, TiO_2 nanostructures could cause genetic mutations in human beings when applied in higher quantities [39]. In this scenario, being safe, stable, and non-toxic, nano ZnO is the principal choice, and has attracted interest as an antimicrobial agent, along with its exotic applications, which range from photocatalysis and sensors to biomedical applications [38–40].

The antimicrobial activity of as synthesized ZnSk composite was tested on *E. coli*, and the results are presented in Figure 7. ZOI results showed significant antimicrobial property by ZnSk composite, with the observed inhibition zone of 1.2 ± 0.1 mm, and silk fibers used as control are insensitive to tested *E. coli*. ZOI at ambient room temperature and at normal day light condition suggests that ZnSk composite have promising antimicrobial action. There are various mechanisms of antimicrobial activity of nanoscale ZnO are reported by research groups and key factors of toxicity include size, shape, reactive oxygen species (ROS) production, hydrogen peroxide generation, attachment to bacterial cell surfaces, etc. [1,38–40]. Coming to ZnSk composite in this study, ZnO nanoflakes are deposited steadily onto the silk fibers, and the most possible mechanism of toxicity may be due to generation of ROS. The release of ROS by ZnO nanoparticles will result in cause of oxidative stress, DNA damage, decomposition of the cell wall, membrane leakage of reducing sugars, proteins, etc. [38,40], which are identified reportedly.

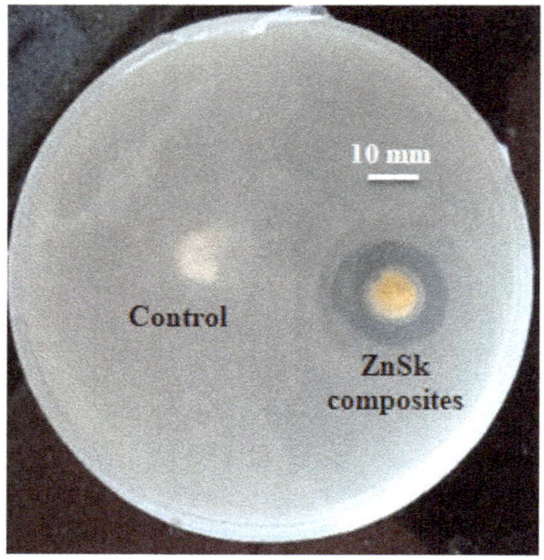

Figure 7. Zone of inhibition test showing antimicrobial effect of ZnSK composite. Degummed silk fibers are used as a control.

4. Conclusions

In this study, biopolymer, silk fiber was used as a template to synthesize ZnO nanoflakes with zinc acetate as a starting material. This biomimetic approach uses a simple hydrothermal process (with reaction temperature of 150 °C for 6 h) for the growth and formation of ZnO- silk fiber (ZnSk) composites. The ZnSk composite structure was visualized through the electron microscopy, and the results reveal the size of and morphology of the nanoflakes (~500–700 nm in length; ~60 nm thicknesses). The deposition of ZnO was confirmed by XRD, SEM, UV, and fluorescence studies. The efficiency of ZnSk composites in naphthalene degradation is demonstrated. The preliminary measurements of the photocatalytic property of ZnSk composites demonstrate more than 97% degradation ability, along with the reusability of the photocatalyst. Furthermore, taking the advantage of ZnSk composite's anionic dye adsorption capacity, we used RB for adsorption studies. The maximum RB adsorption capacity was found to be ~1.706 mg g^{-1} of ZnSk composite material, and the experimental data fitted better to the Langmuir model. In addition to the above, with a potent antimicrobial activity against the tested *E. coli* at ambient conditions, results suggest that the multifunctional use of the produced ZnSk composites could have huge scope to be used in environmental, industrial, and wastewater treatment applications.

Author Contributions: Conceptualization, K.J.R.; Investigation and Writing—original draft, K.J.R., T.K., and A.K.P.; Supervision, V.V.T.P.; Draft curation and Validation, V.V.T.P., S.W., and M.Č. All authors have read and agree to the published version of the manuscript.

Funding: This research received no external funding.

Acknowledgments: The authors would like to thank the University Innovation Cluster-Biotechnology, University of Rajasthan, Jaipur, India for their analytical support during project execution. The authors gratefully acknowledge financial support from the Ministry of Education, Youth and Sports in the Czech Republic under the "Project Hybrid Materials for Hierarchical Structures (HyHi, Reg. No. CZ.02.1.01/0.0/0.0/16_019/0000843), and Research Infrastructure NanoEnviCz, under Project No. LM2018124 supported by the Ministry of Education, Youth and Sports of the Czech Republic and European Union – European Structural and Investment Funds in the frames of Operational Program Research, Development and Education.

Conflicts of Interest: The authors of this study declare no conflict of interest.

References

1. Park, K.-H.; Han, G.D.; Neoh, K.C.; Kim, T.-S.; Shim, J.H.; Park, H.-D. Antibacterial activity of the thin ZnO film formed by atomic layer deposition under UV-A light. *Chem. Eng. J.* **2017**, *328*, 988–996. [CrossRef]
2. Lian, X.; Li, Y.; An, D.; Zou, Y.; Wang, Q.; Zhang, N. Synthesis of porous ZnO nanostructures using bamboo fibers as templates. *Mater. Sci.-Pol.* **2014**, *32*, 514–520. [CrossRef]
3. Umar, A.; Hahn, Y.-B. (Eds.) *Metal Oxide Nanostructures and Their Applications*; Nanotechnology book series; American Scientific Publ.: Los Angeles, CA, USA, 2010.
4. Khin, M.M.; Nair, A.S.; Babu, V.J.; Murugan, R.; Ramakrishna, S. A review on nanomaterials for environmental remediation. *Energy Environ. Sci.* **2012**, *5*, 8075–8109. [CrossRef]
5. Samadi, M.; Zirak, M.; Naseri, A.; Kheirabadi, M.; Ebrahimi, M.; Moshfegh, A.Z. Design and tailoring of one-dimensional ZnO nanomaterials for photocatalytic degradation of organic dyes: A review. *Res. Chem. Intermed.* **2019**, *45*, 2197–2254. [CrossRef]
6. Szatkowski, T.; Siwińska-Stefańska, K.; Wysokowski, M.; Stelling, A.L.; Joseph, Y.; Ehrlich, H.; Jesionowski, T. Immobilization of Titanium(IV) Oxide onto 3D Spongin Scaffolds of Marine Sponge Origin According to Extreme Biomimetics Principles for Removal of C.I. Basic Blue 9. *Biomimetics* **2017**, *2*, 4. [CrossRef]
7. Unterlass, M.M. Geomimetics and Extreme Biomimetics Inspired by Hydrothermal Systems—What Can We Learn from Nature for Materials Synthesis? *Biomimetics* **2017**, *2*, 8. [CrossRef]
8. Ehrlich, H. (Ed.) *Extreme Biomimetics*; Springer International Publishing: Cham, Switzerland, 2017.
9. Wysokowski, M.; Motylenko, M.; Beyer, J.; Makarova, A.; Stöcker, H.; Walter, J.; Galli, R.; Kaiser, S.; Vyalikh, D.; Bazhenov, V.V.; et al. Extreme biomimetic approach for developing novel chitin-GeO2 nanocomposites with photoluminescent properties. *Nano Res.* **2015**, *8*, 2288–2301. [CrossRef]
10. Wysokowski, M.; Motylenko, M.; Stöcker, H.; Bazhenov, V.V.; Langer, E.; Dobrowolska, A.; Czaczyk, K.; Galli, R.; Stelling, A.L.; Behm, T.; et al. An extreme biomimetic approach: Hydrothermal synthesis of β-chitin/ZnO nanostructured composites. *J. Mater. Chem. B* **2013**, *1*, 6469–6476. [CrossRef]
11. Hongfeng, L.; Jia, L.; Jun, W. Synthesis of Biomorphic ZnO Using Cotton as the Biotemplate. *Mater. Rev.* **2008**, S3.
12. Han, J.; Su, H.; Xu, J.; Song, W.; Gu, Y.; Chen, Y.; Moon, W.-J.; Zhang, D. Silk-mediated synthesis and modification of photoluminescent ZnO nanoparticles. *J. Nanopart. Res.* **2012**, *14*, 726. [CrossRef]
13. Shubha, P.; Gowda, M.L.; Namratha, K.; Shyamsunder, S.; Manjunatha, H.B.; Byrappa, K. Ex-situ fabrication of ZnO nanoparticles coated silk fiber for surgical applications. *Mater. Chem. Phys.* **2019**, *231*, 21–26. [CrossRef]
14. Gulrajani, M.L.; Gupta, D.; Periyasamy, S.; Muthu, S.G. Preparation and application of silver nanoparticles on silk for imparting antimicrobial properties. *J. Appl. Polym. Sci.* **2008**, *108*, 614–623. [CrossRef]
15. Gore, P.M.; Naebe, M.; Wang, X.; Kandasubramanian, B. Progress in silk materials for integrated water treatments: Fabrication, modification and applications. *Chem. Eng. J.* **2019**, *374*, 437–470. [CrossRef]
16. Zheng, K.; Zhong, J.; Qi, Z.; Ling, S.; Kaplan, D.L. Isolation of Silk Mesostructures for Electronic and Environmental Applications. *Adv. Funct. Mater.* **2018**, *28*, 1806380. [CrossRef]
17. Tomczak, M.M.; Gupta, M.K.; Drummy, L.F.; Rozenzhak, S.M.; Naik, R.R. Morphological control and assembly of zinc oxide using a biotemplate. *Acta Biomater.* **2009**, *5*, 876–882. [CrossRef] [PubMed]
18. Bertilsson, S.; Widenfalk, A. Photochemical degradation of PAHs in freshwaters and their impact on bacterial growth–influence of water chemistry. *Hydrobiologia* **2002**, *469*, 23–32. [CrossRef]
19. Naushad, M.; Al Othman, Z.A.; Awual, M.R.; Alfadul, S.M.; Ahamad, T. Adsorption of rose Bengal dye from aqueous solution by amberlite Ira-938 resin: Kinetics, isotherms, and thermodynamic studies. *Desalin. Water Treat.* **2016**, *57*, 13527–13533. [CrossRef]
20. Cai, L.; Shao, H.; Hu, X.; Zhang, Y. Reinforced and Ultraviolet Resistant Silks from Silkworms Fed with Titanium Dioxide Nanoparticles. *ACS Sustain. Chem. Eng.* **2015**, *3*, 2551–2557. [CrossRef]
21. Djelloul, A.; Bouzid, K.; Guerrab, F. Role of Substrate Temperature on the Structural and Morphological Properties of ZnO Thin Films Deposited by Ultrasonic Spray Pyrolysis. *Turk. J. Phys.* **2008**, *32*, 49–58.
22. Silvestri, D.; Mikšíček, J.; Wacławek, S.; Torres-Mendieta, R.; Padil, V.V.T.; Černík, M. Production of electrospun nanofibers based on graphene oxide/gum Arabic. *Int. J. Biol. Macromol.* **2019**, *124*, 396–402. [CrossRef]

23. Cao, T.-T.; Zhang, Y.-Q. Processing and characterization of silk sericin from Bombyx mori and its application in biomaterials and biomedicines. *Mater. Sci. Eng. C Mater. Biol. Appl.* **2016**, *61*, 940–952. [CrossRef] [PubMed]
24. Dong, Q.; Su, H.; Zhang, D. In situ depositing silver nanoclusters on silk fibroin fibers supports by a novel biotemplate redox technique at room temperature. *J. Phys. Chem. B* **2005**, *109*, 17429–17434. [CrossRef] [PubMed]
25. Zhang, H.; Li, L.-L.; Dai, F.-Y.; Zhang, H.-H.; Ni, B.; Zhou, W.; Yang, X.; Wu, Y.-Z. Preparation and characterization of silk fibroin as a biomaterial with potential for drug delivery. *J. Transl. Med.* **2012**, *10*, 117. [CrossRef] [PubMed]
26. Koutu, V.; Shastri, L.; Malik, M.M. Effect of NaOH concentration on optical properties of zinc oxide nanoparticles. *Mater. Sci.-Pol.* **2016**, *34*, 819–827. [CrossRef]
27. Khaghanpour, Z.; Naghibi, S. Perforated ZnO nanoflakes as a new feature of ZnO achieved by the hydrothermal-assisted sol–gel technique. *J. Nanostructure Chem.* **2017**, *7*, 55–59. [CrossRef]
28. Assi, N. Synthesis of ZnO-nanoparticles by microwave assisted sol-gel method and its role in photocatalytic degradation of food dye Tartrazine (Acid Yellow 23). *Int. J. Nano Dimens.* **2017**, *8*, 241–249.
29. Hammad, T.M.; Salem, J.K.; Harrison, R.G. Synthesis, Characterizaton, and Optical Properties of Y-Doped ZnO Nanaoparticles. *Nano* **2009**, *4*, 225–232. [CrossRef]
30. Liqiang, J.; Yichun, Q.; Baiqi, W.; Shudan, L.; Baojiang, J.; Libin, Y.; Wei, F.; Honggang, F.; Jiazhong, S. Review of photoluminescence performance of nano-sized semiconductor materials and its relationships with photocatalytic activity. *Sol. Energy Mater. Sol. Cells* **2006**, *90*, 1773–1787. [CrossRef]
31. Lair, A.; Ferronato, C.; Chovelon, J.-M.; Herrmann, J.-M. Naphthalene degradation in water by heterogeneous photocatalysis: An investigation of the influence of inorganic anions. *J. Photochem. Photobiol. Chem.* **2008**, *193*, 193–203. [CrossRef]
32. Jing, L.; Chen, B.; Zhang, B.; Zheng, J.; Liu, B. Naphthalene degradation in seawater by UV irradiation: The effects of fluence rate, salinity, temperature and initial concentration. *Mar. Pollut. Bull.* **2014**, *81*, 149–156. [CrossRef]
33. Xu, F.; Yuan, Z.-Y.; Du, G.-H.; Ren, T.-Z.; Bouvy, C.; Halasa, M.; Su, B.-L. Simple approach to highly oriented ZnO nanowire arrays: Large-scale growth, photoluminescence and photocatalytic properties. *Nanotechnology* **2006**, *17*, 588–594. [CrossRef]
34. Singh, P.; Mondal, K.; Sharma, A. Reusable electrospun mesoporous ZnO nanofiber mats for photocatalytic degradation of polycyclic aromatic hydrocarbon dyes in wastewater. *J. Colloid Interface Sci.* **2013**, *394*, 208–215. [CrossRef] [PubMed]
35. Chakraborty, J.N. 15-Dyeing with acid dye. In *Fundamentals and Practices in Colouration of Textiles*; Chakraborty, J.N., Ed.; Woodhead Publishing India: New Delhi, Indian, 2014; pp. 177–186.
36. Kumar Gupta, V.; Mittal, A.; Jhare, D.; Mittal, J. Batch and bulk removal of hazardous colouring agent Rose Bengal by adsorption techniques using bottom ash as adsorbent. *RSC Adv.* **2012**, *2*, 8381–8389. [CrossRef]
37. Vinuth, M.; Naik, H.S.B. Rapid Removal of Hazardous Rose Bengal Dye Using Fe(III)– Montmorillonite as an Effective Adsorbent in Aqueous Solution. *J. Environ. Anal. Toxicol.* **2016**, *06*. [CrossRef]
38. Dimapilis, E.A.S.; Hsu, C.-S.; Mendoza, R.M.O.; Lu, M.-C. Zinc oxide nanoparticles for water disinfection. *Sustain. Environ. Res.* **2018**, *28*, 47–56. [CrossRef]
39. Chakra, C.H.S.; Rajendar, V.; Rao, K.V.; Kumar, M. Enhanced antimicrobial and anticancer properties of ZnO and TiO2 nanocomposites. *3 Biotech* **2017**, *7*, 89. [CrossRef]
40. Tiwari, V.; Mishra, N.; Gadani, K.; Solanki, P.S.; Shah, N.A.; Tiwari, M. Mechanism of Anti-bacterial Activity of Zinc Oxide Nanoparticle Against Carbapenem-Resistant Acinetobacter baumannii. *Front. Microbiol.* **2018**, *9*. [CrossRef]

© 2020 by the authors. Licensee MDPI, Basel, Switzerland. This article is an open access article distributed under the terms and conditions of the Creative Commons Attribution (CC BY) license (http://creativecommons.org/licenses/by/4.0/).

polymers

MDPI

Article

A Polymeric Composite Material (rGO/PANI) for Acid Blue 129 Adsorption

Tomasz Kukulski [1], Stanisław Wacławek [2,*], Daniele Silvestri [2,*], Kamil Krawczyk [2], Vinod V. T. Padil [2], Ryszard Fryczkowski [1], Jarosław Janicki [1] and Miroslav Černík [2]

1 Institute of Textile Engineering and Polymer Materials, University of Bielsko-Biala, Willowa 2, 43-309 Bielsko-Biala, Poland; tkukulski@ath.bielsko.pl (T.K.); rfryczkowski@ath.bielsko.pl (R.F.); jjanicki@ath.bielsko.pl (J.J.)

2 Institute for Nanomaterials, Advanced Technologies and Innovation, Technical University of Liberec, Studentská 1402/2, 46117 Liberec 1, Czech Republic; kamil.krawczyk@tul.cz (K.K.); vinod.padil@tul.cz (V.V.T.P.); miroslav.cernik@tul.cz (M.Č.)

* Correspondence: stanislaw.waclawek@tul.cz (S.W.); daniele.silvestri@tul.cz (D.S.)

Received: 21 April 2020; Accepted: 28 April 2020; Published: 3 May 2020

Abstract: Over the years, polyaniline (PANI) has received enormous attention due to its unique properties. Herein, it was chosen to develop a new polymeric composite material: reduced graphene oxide/polyaniline (rGO/PANI). The composite was prepared by a simple and cost-effective fabrication method of formation by mixing and sonication in various conditions. The obtained materials were characterized and identified using various techniques such as scanning electron microscopy (SEM), Raman and ATR–FTIR spectroscopy, and X-ray diffraction (XRD). The objective of the paper was to confirm its applicability for the removal of contaminants from water. Water could be contaminated by various types of pollutants, e.g., inorganics, heavy metals, and many other industrial compounds, including dyes. We confirmed that the Acid Blue 129 dyes can be substantially removed through adsorption on prepared rGO/PANI. The adsorption kinetic data were modeled using the pseudo-first-order and pseudo-second-order models and the adsorption isotherm model was identified.

Keywords: graphene; nanocomposite; adsorption; Acid Blue 129

---

1. Introduction

Recently, graphene has been considered as a wonder material, especially after Prof. Geim and Novoselov gained Nobel prize for obtaining its stable form in 2010 [1]. They used mechanical exfoliation named today the 'Scotch tape method', which is considered as one of the physical 'top-down' approaches to get nanolayers. The other interesting way of obtaining graphene is by chemical methods. The first documented approach was made by Brodie in 1858 [2], whereas the next one by Staudenmaier in 1899 [3] and Hummers in 1958 [4], which nowadays is the most famous method for synthesis of graphene or its derivatives graphene oxides. Moreover, many modifications of this approach exist, for example Tour or Shi methods [5,6]. All the chemical methods are based on graphite oxidation by a strong oxidizer such as $KMnO_4$ or $KClO_3$, when graphene oxide (GO) is produced. It can be further deoxidized to reduced graphene oxide (rGO) by many reduction methods, such as chemical (e.g., hydrazine) [7,8], electrochemical [9], or thermal [10]. Graphene has extraordinary physical properties, owing to which, it has many potential applications in electronics, optics (as sensors), or as a component used in membranes and batteries [11–16]. Indeed, water treatment is a growing field for graphene-based materials [17–21].

Another interesting material that has recently brought the scientist´s attention (even though it was discovered over 150 years ago) is polyaniline (PANI) [22]. It is a conducting polymer with many advantages, e.g., simple synthesis, low cost, stability in environment, and easy doping/dedoping chemistry [23–25]. According to this it is used as protective coatings or supercapacitors [26,27]. These two remarkable materials can be bonded together to form the rGO/PANI composite. This composite is used often in supercapacitors applications [28]; however, recently it has been proposed as an excellent adsorbent [29], e.g., for heavy metals and dyes adsorption [30–32]. rGO due to the large specific surface area and PANI which is known for its facile synthesis, insolubility in water and stability, are potentially very good sorption materials [33,34]. The wastewater from the dyeing industry has many toxic properties and it has to be treated before the release to the environment [35]. Water and wastewater treatment are considered a vital branch of the environmental chemistry [36,37].

In this work we believed that the herein synthesized rGO/PANI composite can be used for efficient removal of the Acid Blue 129 (AB129) from water. The rGO/PANI composite was made by sonication and mixing at different temperatures. Taking into account the importance of the development of novel water treatment technologies and after successful trials for the removal of toxic metals from water by rGO/PANI composite [30], we are demonstrating herein first use of rGO/PANI for the adsorption of anionic dye from water. This study gives additional evidence of the possibility of using rGO/PANI for water and wastewater treatment.

2. Materials and Methods

2.1. Chemicals

Graphite (particle size <20 μm), polyaniline (PANI; emeraldine base, average molecular weight ~50,000), Acid Blue 129 (dye content 25%), hydrochloric acid (35 wt %), sulfuric acid (96 wt %), potassium permanganate (99.5 wt %), and hydrogen peroxide (30 wt %) were purchased from Sigma Aldrich (Saint Louis, MO, USA). Deionized water (18.2 MΩ·cm) was obtained by an ELGA purelab flex system (ELGA, Veolia Water, Marlow, UK) and was used in all experiments.

2.2. Preparation of Reduced Graphene Oxide

GO was synthesized by the modified Hummers' method described before [38]. Briefly, 0.5 L of H_2SO_4 was poured into a 5 L beaker and then 20 g of graphite was added. The mixture was stirred on a magnetic stirrer for 1 h, and then cooled down in an ice bath to 5 °C. After, 60 g of $KMnO_4$ was added in small batches, so the temperature of the reaction mixture would not exceed 35 °C. After adding $KMnO_4$, the reaction mixture was allowed to rest for 2 h, while the temperature was constantly controlled to keep it under 50 °C. In a subsequent step, 1 L of distilled water was added to the reaction mixture also in small batches, so the temperature of the reaction mixture was kept below 65 °C. Then, 0.8 L of distilled water with a temperature of 60 °C and 0.8 L of 3% H_2O_2 were added. GO was purified with 10% HCl and distilled water and finally centrifuged. Pure graphene oxide after drying was thermally reduced at 500 °C and nitrogen atmosphere.

2.3. Preparation of rGO/PANI Nanocomposite

There are only few works, which report green and effective sono-assisted synthesis of rGO-PANI composite [39–41]. In this work, we have extended these investigations to show the effect of various conditions on rGO bonding with PANI. PANI (3.88 g) was dispersed in 100 mL solution of 1 M HCl. Afterwards 0.5 g of rGO was added to this mixture with subsequent sonication for 15 mins. Thus created dispersions were mixed on a magnetic stirrer in various temperatures and periods (samples are named rGO/PANI-x,y; where x is temperature and y is time period, Table 1). The (expected) formation of amide bonds between rGO and PANI (as well as other chemical reactions), are time and temperature dependent. Therefore, the main aim of this part of the study was to investigate their influence on the rGO and PANI conjugation through the amide bond formation. The temperature range chosen and synthesis time were in accordance with the study of Sibilska et al. [42].

After mixing, samples in the form of powder were filtered under vacuum and purified by distilled water and 100 mL of 0.1 M NH_3 (two times) to reach neutral pH, in order to reach the PANI in the form of emeraldine base.

Table 1. Conditions for rGO/PANI composite preparation.

Time (h)	Temperature (°C)		
	0	50	100
1	rGO/PANI-0,1	rGO/PANI-50,1	rGO/PANI-100,1
8	rGO/PANI-0,8	rGO/PANI-50,8	rGO/PANI-100,8
24	rGO/PANI-0,24	rGO/PANI-50,24	rGO/PANI-100,24

2.4. Characterization Procedures

SEM (Phenom ProX, Thermo Scientific, Waltham, MA, USA) operating at an acceleration voltage 10 kV was used to analyze morphology and the structure of the composites. Before SEM analysis the samples were set onto aluminum holders, then, the sample holders were diffusion-coated with thin gold layer. Raman scattering was performed on micro Raman spectrometer (Raman DXR microscope, Thermo Scientific, Waltham, MA, USA) at 514 nm laser excitation of an argon laser with a spectral resolution of 1 cm^{-1} (full-width at half-maximum). XRD analysis measurements were taken using a URD 63 diffractometer (FPM-Seifert, Hamburg & Freiberg, Germany). The Cu Kα radiation was used at 40 kV and 30 mA. The monochromatization of the beam was achieved by a pulse height analyzer and nickel filter. Scintillation counter served as a detector. The searches were carried out in the range of angles from 4° to 60° at 0.1° pitch. Each diffraction curve was corrected for polarization, Lorentz factor and incoherent scattering. Attenuated total reflection—Fourier transform infrared spectroscopy (ATR–FTIR) spectra were obtained at 4000–7000 cm^{-1} (4 cm^{-1} resolution) utilizing a germanium ATR crystal (NICOLET IZ10, Thermo Scientific, Waltham, MA, USA) equipped with a horizontal ATR accessory (single reflection angle 45°). The surface areas of the composites were obtained using the BET (Brunauer–Emmett–Teller) technique (Autosorb iQ, Quantachrome Instruments, FL, USA). Absorption spectra were analyzed in a UV–vis spectrophotometer (Hach Lange DR 3900, Vancouver, WA, USA).

2.5. Quantum Chemical Analysis

The initial coordinates of the rGO/PANI composite were obtained with the Avogadro program (Open Molecules, Pittsburgh, PA, USA) [43]. All of the calculations were made with the Gaussian 16 software (Gaussian Inc. Wallingford CT, USA) [44]. The B3LYP/6-31G level of study was employed. Aniso-surface threshold value of 0.01 atomic units was set. The outputs were visualized with the Avogadro program.

2.6. Acid Blue 129 Adsorption Experiments

Adsorption tests were carried out in 100 mL beakers (50 mL dispersion of 25 mg rGO/PANI composite). The AB129 solutions (25 mg/L) were prepared by the addition of dye to the DI water. Samples were centrifuged at 14500 RPM and measured by the spectrophotometric technique with UV–vis spectrophotometer (Hach Lange DR 3900, Vancouver, WA, USA). Kinetic experiments were performed with all samples at room temperature and for 60 mins.

Kinetics models were used to determine the adsorption controlling mechanism, based on the experimental data. The pseudo-first-order kinetic model described the rate of the adsorption based on the adsorbed amounts. Its linear form is usually expressed as (Equation (1))

$$log(q_e - q_t) = logq_e - \frac{k_1}{2.303}t, \quad (1)$$

where q_e and q_t are the adsorption amounts at equilibrium and time t, respectively, and k_1 is the first-order rate constant. A plot of log (q_e − q_t) versus t gives k_1 as the slope and log q_e as the intercept value.

The pseudo-second-order kinetic model employed to adsorption kinetics was presented in [45] as

$$\frac{t}{q_t} = \frac{1}{k_2 q_e^2} + \frac{1}{q_e} t, \tag{2}$$

where k_2 is the appropriate rate constant. The plot of t/q_t versus t shows a linear relationship if the second-order kinetic is appropriate. Values of k_2 and q_e were calculated from the intercept and slope of the plots.

Langmuir adsorption isotherms were calculated based on batch tests according to equation

$$q_e = \frac{q_{max} k_l C_e}{1 + k_l C_e} \tag{3}$$

where q_e and q_{max} is the equilibrium and maximum adsorption amount, respectively, C_e the equilibrium concentration in solution and k_l is the Langmuir isotherm equilibrium constant related to free energy of adsorption. The adsorption was studied at a composite concentration of 0.5 g/L and different initial dye concentrations (3.1–50 mg/L).

The slope of the plot of C_e/q_e against C_e gives slope $1/q_{max}$, the intercept $1/q_{max}k_l$. The coherence between obtained results and the model-predicted values is indicated by the R^2 (determination coefficient).

3. Results and Discussion

3.1. Characterization

3.1.1. XRD

In Figure 1, XRD patterns of the rGO, PANI, and composites samples can be observed.

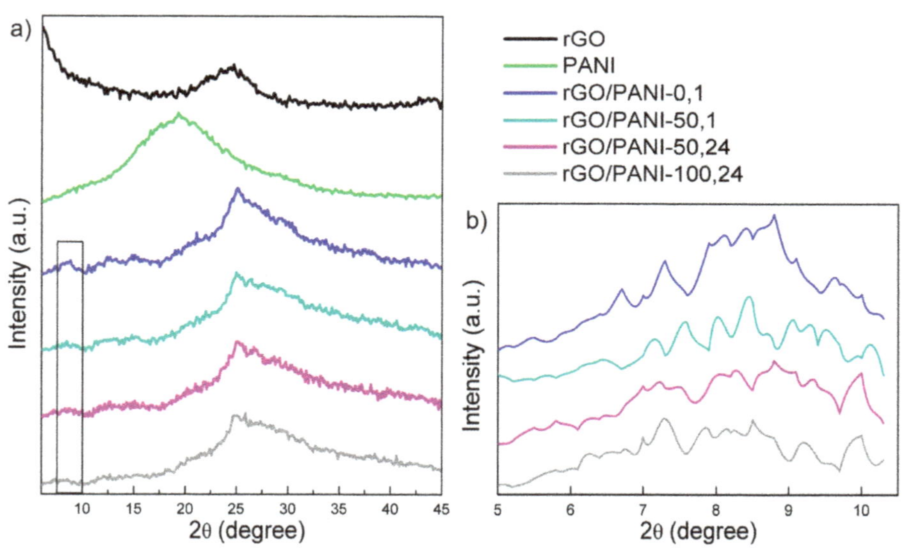

Figure 1. (**a**) Results of XRD diffractograms of (from the top) rGO, PANI; rGO/PANI-0,1; rGO/PANI-50,1; rGO/PANI-50,24 and rGO/PANI-100,24 samples. (**b**) Magnified view of the 2θ = 7°–8° peak.

As it could be seen, the XRD patterns of the pure rGO (Figure 1a) exhibit one typical peak at 2θ = 24° [46]. As for the bare PANI sample, one broad peak at 2θ = 20° could be observed. The XRD pattern of the rGO/PANI composite indicates a protonation of PANI. This could be evidenced by the shifting and division of the original PANI peak at 2θ = 20° to two peaks at 25° and 30°(Figure 1a). Moreover, in the same spectrum, a new peak appears at 7°–8° and the share of the amorphous component grows, it is visible as a broad halo around 30° [47,48]. According to these results, we could speculate about permanent bonding between the rGO and PANI. This could be further confirmed by the ATR–FTIR results presented in the subsequent subsection.

Furthermore, in Figure 1b comparison of the XRD spectra in the low-angle region was shown. The peak at the angle of 2θ = 7°–8° could be correlated to the more uniform dispersion of PANI on rGO [49] as well as to the protonation level of PANI (higher protonation state relating to a higher peak intensity [47]). From this, it can be concluded that PANI not only organizes itself under the influence of rGO but also the protonated = NH^{+}– groups can attract and immobilize negatively charged species [50], such as anionic dyes. Therefore, rGO-PANI-0,1 sample, having the highest intensity of 2θ = 7°–8° peak is believed to also have high potential of adsorbing Acid Blue 129 as well as other sulfonic/anionic dyes. However, in order to confirm this statement, it had to be validated whether PANI is undeniably covalently bound to rGO. This was assessed by ATR–FTIR and discussed in the next subsection.

3.1.2. ATR–FTIR

ATR–FTIR analysis was performed to confirm functional groups (and changes) present in PANI, rGO and the composite (Figure 2). This method is suitable for observing materials showing high absorption in the infrared range as the tested materials.

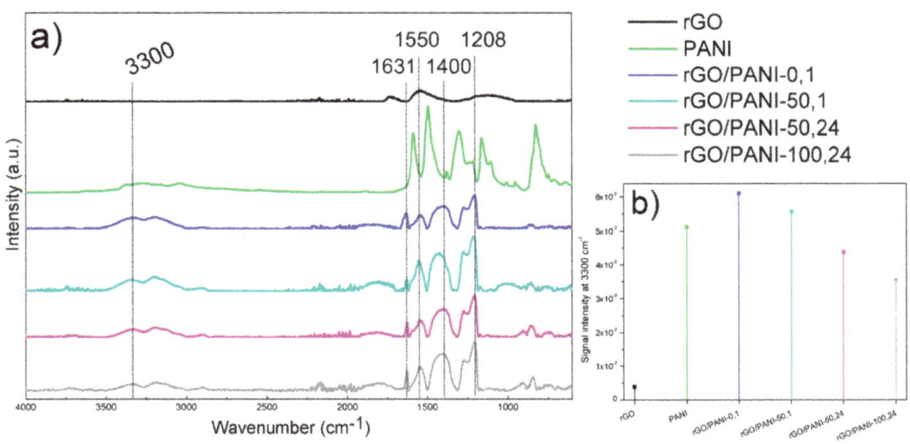

Figure 2. (**a**) Results of ATR–FTIR analysis of (from the top) rGO; PANI; rGO/PANI-0,1; rGO/PANI-50,1; rGO/PANI-50,24 and rGO/PANI-100,24 samples. (**b**) Comparison of peak intensities at 3300 cm^{-1}.

The rGO spectrum shows residual oxygen group remaining after reduction at the wavenumber of 1730 and 1130 cm^{-1}, C=O (carbonyl group) and C–OH, respectively. The peak at 1530 cm^{-1} corresponds to stretching C=C. The PANI spectrum shows characteristic peaks from benzenoid rings at 1587 and 1494 cm^{-1}, C–N band of an aromatic amine and N=Q=N vibration (Q- quinoid ring) at 1301 and 1166 cm^{-1}, respectively.

After the reaction of rGO with PANI, the peaks from benzenoid rings shift to the lower energy at 1629 and 1546 cm^{-1}, whereas at 1400 cm^{-1} a new peak characteristic for amide C–N bands appeared with a simultaneous signal increase typical for N–H stretching vibrations in a secondary amide (Figure 2b) at 3300 cm^{-1} [51]. Moreover, the peak present in the rGO/PANI spectrum at 1207 cm^{-1} is also characteristic for C–N stretching. The observed bands confirm the formation of chemical bonds between PANI and rGO. These types of interactions have a decisive impact on the electronic structure of the resulting composite, but also affect the supramolecular structure which has a significant impact on sorption properties, as well as the adsorption model of ionic compounds.

Due to the highest intensity of the peak at 3300 cm^{-1} and at 7°–8° given by FTIR and XRD analyses respectively, and the time spent for sample preparation (only 1 h of synthesis in a low temperature), rGO/PANI-0,1 sample has been selected for further investigations.

3.1.3. Morphology and Electron Distribution

The morphology of rGO and rGO/PANI composite is shown in Figure 3a,b, respectively.

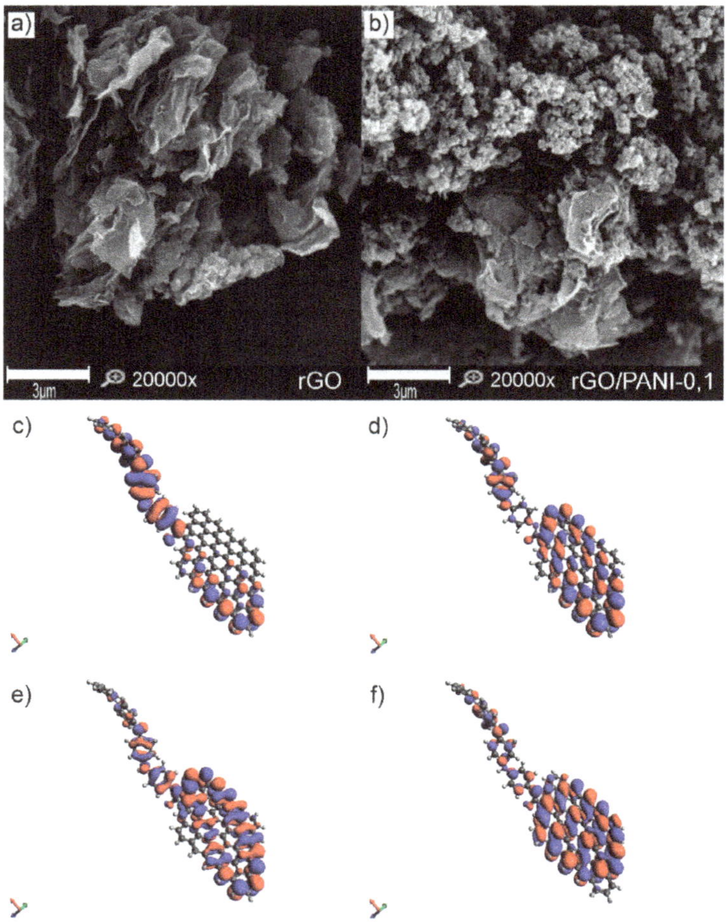

Figure 3. SEM images of (**a**) rGO, (**b**) rGO/PANI-0,1 composite (scale bar represents 3 µm), (**c**) LUMO, (**d**) HOMO, (**e**) LUMO-1, and (**f**) HOMO-1 (c, d, e and f figures are not in scale).

It is visible that the pure rGO sample is composed of the typical graphene sheets. Plates of rGO have a clean and smooth surface. After mixing rGO with PANI, a composite was form, which comprised of rGO flakes covered by PANI aggregated spheres (Figure S1) with an extensive specific surface (Figure 3b). Composites have a similar morphology regardless of the conditions of formation

(Figure S2). A similar structure of the composite rGO/PANI was observed by Yang et al. [52], who has synthesized it for supercapacitors application. However, in the reported method herein, PANI with a higher molecular weight was used, which can affect the adsorption and aggregation process.

Moreover, for better visualization of the rGO/PANI, a model of this composite was created with its electron distribution computed (Figure 3c–f). The lowest unoccupied molecular orbitals (LUMO) and highest occupied molecular orbitals (HOMO) of the rGO/PANI (Figure 3c,d) demonstrate large pi-pi conjugated system of connected by amide bond PANI molecule with the reduced graphene oxide. Moreover, as shown in Figure 3e,f, the extended pi-pi conjugation in the entire system was also confirmed by the LUMO-1 and the HOMO-1, indicating that amide groups play a significant role in bridging these two components [40]. Such extended system can be beneficial for adsorption of, e.g., contaminants with aromatic groups as reported by Peng et al. [53].

3.1.4. Raman Analysis

Figure 4a–c present the Raman spectra of rGO, PANI and rGO/PANI, respectively.

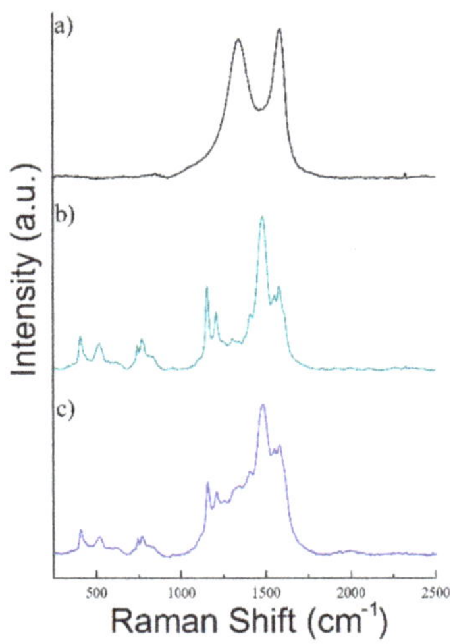

Figure 4. Raman spectra of (**a**) rGO, (**b**) PANI, and (**c**) rGO/PANI-0,1 composite.

As shown in Figure 4a rGO sample exhibits a typical Raman spectrum composed of two characteristic peaks D (at 1344 cm^{-1}) and G (at 1582 cm^{-1}) [54,55]. Moreover, the ratio of I_D/I_G bands was determined to be 0.94 in this sample, which is a typical value for rGO [56–59]. Nonetheless, the spectrum of PANI has shown C=C stretching in the quinonoid ring at 1548 cm^{-1}, C=N stretching vibration at 1483 cm^{-1}, C–N stretching vibrations of diverse benzenoid at 1210 cm^{-1} and C–H bending of the quinonoid ring peak at 1158 cm^{-1}. rGO/PANI composite has shown peaks at 1210, 1338, 1488, 1548, and 1583 cm^{-1} which could be found in the materials characterized individually. The 1483 cm^{-1} band shift may indicate a change in the area of the C=N bonds in PANI, which confirms the formation of the composite.

3.2. Adsorption of a Model Dye

AB129 was used as a model dye pollutant to assess rGO/PANI adsorption properties. After the composite was exposed to AB129, the blue color of the original solution slowly diminished. Typical Vis peak of AB129 at 630 nm wavelength progressively decreased over time. Furthermore, the rGO/PANI composite became blue due to the dye adsorption onto the composite surface. To characterize the interaction of the dye with adsorbents, important parameters must be determined such as adsorption capacity, isotherms, kinetics.

Adsorption experiment was carried out in a given ratio of the composite in water under room temperature for one hour. Adsorption kinetic for all prepared samples of the composite and the AB129 dye is shown in Figure 5.

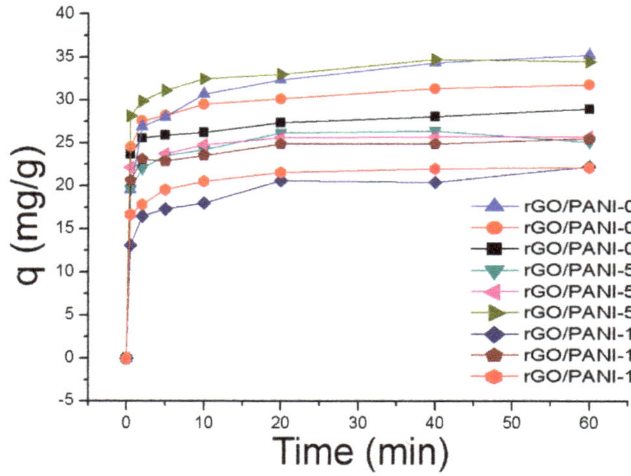

Figure 5. Adsorption of AB129 on rGO/PANI composites (conditions: 25 mg/L of AB129, 25 mg/50 mL of rGO/PANI, room temperature), the error bar was <5%.

The equilibrium sorption capacity, q_e was obtained in 40 mins. The results show that the equilibrium adsorbed concentration q_e of AB129 range from 22.15 to 35.26 mg/g depending on type of composite. The lower values were reached by samples, which were synthesized at a higher temperature. The best result of q_e was obtained by the rGO/PANI-0,1 composite (Figure 5) and therefore all further studies were performed only for this sample. To obtain the information on the adsorption rate and mechanism, adsorption kinetics were fitted with both pseudo-first and pseudo-second kinetic adsorption model (Figure 6). The R^2 of the pseudo-second-order model (R^2 = 0.999) is higher than one for the pseudo-first-order model (R^2 = 0.939) which is in consensus with previous observations [60]. This result was also checked on the other composite samples, where in most cases the pseudo-second kinetic adsorption model fitted the data better (Table S1). This model also includes chemisorption which may be a rate-limiting step [61].

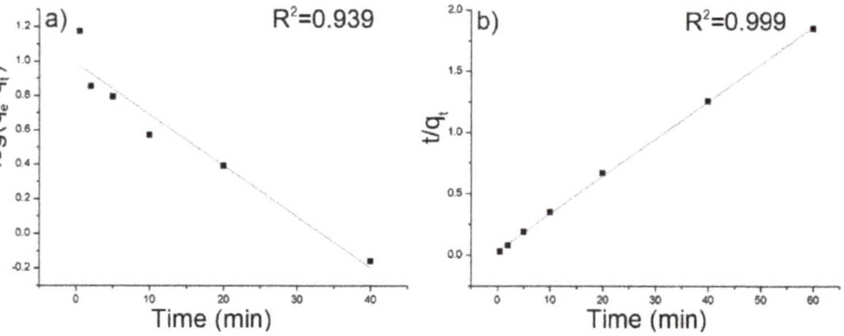

Figure 6. (**a**) Pseudo-first-order and (**b**) pseudo-second-order kinetic model for the adsorption of AB129 on rGO/PANI-0,1. The solid lines present fits of the pseudo-first and pseudo-second order kinetic model (conditions: 25 mg/L of AB129, 25 mg/50 mL of rGO/PANI, room temperature).

Equilibrium adsorption data determined for five different initial concentrations of dye were linearized to get parameters of Langmuir isotherm [62] (Figure 7).

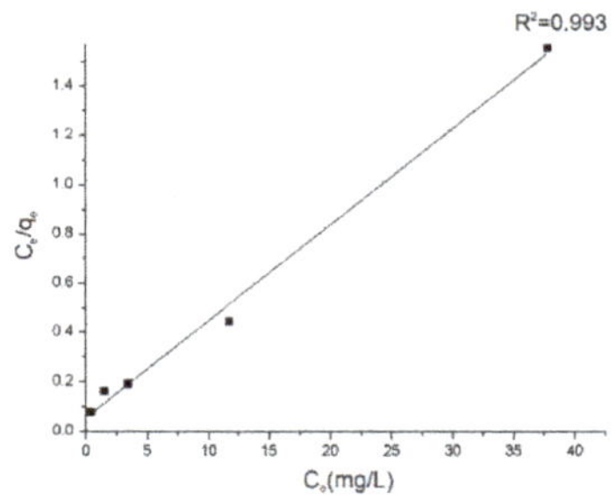

Figure 7. Langmuir isotherm model plot for the adsorption on rGO/PANI-0,1 (contact time: 60 mins).

The Langmuir isotherm is a satisfactory model for describing the correlation between the amounts of AB129 adsorbed by the composite and its equilibrium concentration in the solution. The maximum adsorption capacity (q_{max}) was calculated to be 25.57 mg/g, based on the fitting result where R^2 was 0.975. The main mechanisms involved in the adsorption of dyes onto the rGO/PANI composites as reported in many other studies are electrostatic attraction, physisorption, and complexation [29].

The obtained results are presented in Table 2 and were compared with the materials reported in the literature. In comparison to the other materials, rGO/PANI-0,1 composite is fast and effective adsorbent for AB129 removal from aqueous solution.

Table 2. Comparison of different adsorbents for Acid Blue 129 removal.

Adsorbent	q_{max} (mg/g)	Adsorbent Concentration (g/L)	Equilibrium Time (min)	Specific Surface Area (BET) (m^2/g)	Reference
Activated carbon cloth	61.64	1.4	500	1870	[63]
Almond shell	11.95	16	14	-	[64]
CuO-NP-AC	65.36	0.9	20–25	-	[65]
HCl-Modified Bentonite	13.8	3.33	55	87	[66]
Magnesium-Modified Bentonite	10.8	3.33	40	1310 *	[67]
Iron oxide/carbon nanocomposites	83.42	1	120	695	[68]
rGO/PANI-1,0	25.57	0.5	40	36	This work

* Langmuir surface area.

4. Conclusions

In this research, rGO/PANI composites were determined to be effective adsorbers for Acid Blue 129 dye. The composites were successfully prepared by a novel way relying on mechanical mixing at various times and temperatures. A total of nine composites were tested, from which the rGO/PANI-0,1 composite (synthesized in the lowest temperature) exhibited superior supramolecular and chemical structure that improved its adsorption properties.

SEM analysis shown structure of the composite, in which PANI particles were distributed on the rGO surface. Raman and ATR–FTIR spectroscopy confirmed chemical bonding between rGO and PANI. It was found that the amide bonds primarily were responsible for conjugation of rGO and PANI.

The results demonstrated that the rGO/PANI composite is an efficient adsorbent for the AB129. The maximum capacity of this material was determined (using Langmuir isotherm) to be 25.57 mg/g, whereas the adsorption–desorption equilibrium was reached after 40 mins. The adsorption process followed the pseudo-second-order kinetic model. This study proves that the rGO/PANI composite can be successfully used for the adsorption of Acid Blue 129 and other contaminants with similar chemical structure.

Supplementary Materials: The following are available online at http://www.mdpi.com/2073-4360/12/5/1051/s1, Figure S1: SEM image of PANI. Figure S2: SEM images of (a) rGO/PANI-50,1 (b) rGO/PANI-50,24, (c) rGO/PANI-100,24 composites. Table S1: R^2 values for kinetics experiments of rGO/PANI composite for pseudo-first-order (1st) and pseudo-second-order (2nd) kinetic model.

Author Contributions: Conceptualization, T.K. and S.W.; Methodology, T.K. and K.K.; Investigation, T.K. and K.K.; Writing—original draft preparation, T.K. and V.V.T.P.; Writing—review and editing, D.S., S.W., M.Č., J.J., and R.F.; Supervision, S.W. All authors have read and agreed to the published version of the manuscript.

Funding: This research was supported National Centre for Research and Development in Poland (POIR.04.01.02-00-0062/16) financed by the European Regional Development Fund (Operational Program Intelligent Development 2014–2020, 4.1.2). The authors would also like to acknowledge the Ministry of Education, Youth and Sports in the Czech Republic under the "Inter Excellence – Action programme" within the framework of project "Exploring the role of ferrates and modified nano zero-valent iron in the activation process of persulfates" (registration number LTAUSA18078) and the Research Infrastructures NanoEnviCz (Project No. LM2015073). This work was also supported by the Ministry of Education, Youth and Sports of the Czech Republic and the European Union - European Structural and Investment Funds in the frames of Operational Programme Research, Development and Education - project Hybrid Materials for Hierarchical Structures (HyHi, Reg. no. CZ.02.1.01/0.0/0.0/16_019/0000843).

Conflicts of Interest: The authors declare no conflict of interest.

References

1. Novoselov, K.S.; Geim, A.K.; Morozov, S.V.; Jiang, D.; Zhang, Y.; Dubonos, S.V.; Grigorieva, I.V.; Firsov, A.A. Electric field in atomically thin carbon films. *Science* **2004**, *306*, 666–669. [CrossRef]
2. Brodie, B.C. On the Atomic Weight of Graphit. *R. Soc. Lond.* **1858**, *149*, 423–429.
3. Staudenmaier, L. Verfahren zur Darstellung der Graphitsäure. *Ber. der Dtsch. Chem. Ges.* **1899**, *32*, 1394–1399. [CrossRef]
4. Hummers, W.S.; Offeman, R.E. Preparation of Graphitic Oxide. *J. Am. Chem. Soc.* **1958**, *80*, 1334–1339. [CrossRef]
5. Marcano, D.C.; Kosynkin, D.V.; Berlin, J.M.; Sinitskii, A.; Sun, Z.; Slesarev, A.; Alemany, L.B.; Lu, W.; Tour, J.M. Improved synthesis of graphene oxide. *ACS Nano* **2010**, *4*, 4806–4814. [CrossRef] [PubMed]
6. Chen, J.; Yao, B.; Li, C.; Shi, G. An improved Hummers method for eco-friendly synthesis of graphene oxide. *Carbon* **2013**, *64*, 225–229. [CrossRef]
7. Stankovich, S.; Dikin, D.A.; Piner, R.D.; Kohlhaas, K.A.; Kleinhammes, A.; Jia, Y.; Wu, Y.; Nguyen, S.T.; Ruoff, R.S. Synthesis of graphene-based nanosheets via chemical reduction of exfoliated graphite oxide. *Carbon* **2007**, *45*, 1558–1565. [CrossRef]
8. Pei, S.; Cheng, H.M. The reduction of graphene oxide. *Carbon* **2012**, *50*, 3210–3228. [CrossRef]
9. Zhou, M.; Wang, Y.; Zhai, Y.; Zhai, J.; Ren, W.; Wang, F.; Dong, S. Controlled synthesis of large-area and patterned electrochemically reduced graphene oxide films. *Chem. A Eur. J.* **2009**, *15*, 6116–6120. [CrossRef]
10. Becerril, H.A.; Mao, J.; Liu, Z.; Stoltenberg, R.M.; Bao, Z.; Chen, Y. Evaluation of solution-processed reduced graphene oxide films as transparent conductors. *ACS Nano* **2008**, *2*, 463–470. [CrossRef]
11. Georgakilas, V.; Tiwari, J.N.; Kemp, K.C.; Perman, J.A.; Bourlinos, A.B.; Kim, K.S.; Zboril, R. Noncovalent Functionalization of Graphene and Graphene Oxide for Energy Materials, Biosensing, Catalytic, and Biomedical Applications. *Chem. Rev.* **2016**, *116*, 5464–5519. [CrossRef] [PubMed]
12. Tan, C.; Cao, X.; Wu, X.J.; He, Q.; Yang, J.; Zhang, X.; Chen, J.; Zhao, W.; Han, S.; Nam, G.H.; et al. Recent Advances in Ultrathin Two-Dimensional Nanomaterials. *Chem. Rev.* **2017**, *117*, 6225–6331. [CrossRef] [PubMed]

13. Ruan, K.; Guo, Y.; Tang, Y.; Zhang, Y.; Zhang, J.; He, M.; Kong, J.; Gu, J. Improved thermal conductivities in polystyrene nanocomposites by incorporating thermal reduced graphene oxide via electrospinning-hot press technique. *Compos. Commun.* **2018**, *10*, 68–72. [CrossRef]
14. Sang, L.; Hao, W.; Zhao, Y.; Yao, L.; Cui, P. Highly aligned graphene oxide/waterborne polyurethane fabricated by in-situ polymerization at low temperature. *E-Polymers* **2018**, *18*, 75–84. [CrossRef]
15. Yang, H.; Liu, S.; Cao, L.; Jiang, S.; Hou, H. Superlithiation of non-conductive polyimide toward high-performance lithium-ion batteries. *J. Mater. Chem. A* **2018**, *6*, 21216–21224. [CrossRef]
16. Liao, X.; Ye, W.; Chen, L.; Jiang, S.; Hou, H.; Jiang, S.; Wang, G.; Zhang, L. Flexible hdC-G reinforced polyimide composites with high dielectric permittivity. *Compos. Part A Appl. Sci. Manuf.* **2017**, *101*, 50–58. [CrossRef]
17. Subramani, A.; Jacangelo, J.G. Emerging desalination technologies for water treatment: A critical review. *Water Res.* **2015**, *75*, 164–187. [CrossRef]
18. Kah, M.; Sigmund, G.; Xiao, F.; Hofmann, T. Sorption of ionizable and ionic organic compounds to biochar, activated carbon and other carbonaceous materials. *Water Res.* **2017**, *124*, 673–692. [CrossRef]
19. Yuan, X.; Wang, H. Adsorptive removal of methylene blue by rhamnolipid-functionalized graphene oxide from wastewater. *Water Res.* **2014**, *67*, 330–344.
20. Wei, Y.; Zhang, Y.; Gao, X.; Ma, Z.; Wang, X.; Gao, C. Multilayered graphene oxide membrane for water treatment: A review. *Carbon* **2018**, *139*, 964–981. [CrossRef]
21. Venkateshaiah, A.; Silvestri, D.; Ramakrishnan, R.K.; Wacławek, S.; Padil, V.V.T.; Černík, M.; Varma, R.S. Gum Kondagogu/Reduced Graphene Oxide Framed Platinum Nanoparticles and Their Catalytic Role. *Molecules* **2019**, *24*, 3643. [CrossRef] [PubMed]
22. Geniès, E.M.; Boyle, A.; Lapkowski, M.; Tsintavis, C. Polyaniline: A historical survey. *Synth. Met.* **1990**, *36*, 139–182. [CrossRef]
23. Zhang, J.; Han, J.; Wang, M.; Guo, R. Fe_3O_4/PANI/MnO_2 core-shell hybrids as advanced adsorbents for heavy metal ions. *J. Mater. Chem. A* **2017**, *5*, 4058–4066. [CrossRef]
24. Liu, Y.; Song, L.; Du, L.; Gao, P.; Liang, N.; Wu, S.; Minami, T.; Zang, L.; Yu, C.; Xu, X. Preparation of polyaniline/emulsion microsphere composite for efficient adsorption of organic dyes. *Polymers* **2020**, *12*, 167. [CrossRef] [PubMed]
25. Muhammad, A.; Shah, A.U.H.A.; Bilal, S. Effective Adsorption of Hexavalent Chromium and Divalent Nickel Ions from Water through Polyaniline, Iron Oxide, and Their Composites. *Appl. Sci.* **2020**, *10*, 2882. [CrossRef]
26. Mirmohseni, A.; Oladegaragoze, A. Anti-corrosive properties of polyaniline coating on iron. *Synth. Met.* **2000**, *114*, 105–108. [CrossRef]
27. Eftekhari, A.; Li, L.; Yang, Y. Polyaniline supercapacitors. *J. Power Sources* **2017**, *347*, 86–107. [CrossRef]
28. Male, U.; Modigunta, J.K.R.; Huh, D.S. Design and synthesis of polyaniline-grafted reduced graphene oxide via azobenzene pendants for high-performance supercapacitors. *Polymer* **2017**, *110*, 242–249. [CrossRef]
29. Zare, E.N.; Motahari, A.; Sillanpää, M. Nanoadsorbents based on conducting polymer nanocomposites with main focus on polyaniline and its derivatives for removal of heavy metal ions/dyes: A review. *Environ. Res.* **2018**, *162*, 173–195. [CrossRef]
30. Li, R.; Liu, L.; Yang, F. Preparation of polyaniline/reduced graphene oxide nanocomposite and its application in adsorption of aqueous Hg(II). *Chem. Eng. J.* **2013**, *229*, 460–468. [CrossRef]
31. Yang, Y.; Wang, W.; Li, M.; Wang, H.; Zhao, M.; Wang, C. Preparation of PANI grafted at the edge of graphene oxide sheets and its adsorption of Pb(II) and methylene blue. *Polym. Compos.* **2018**, *39*, 1663–1673. [CrossRef]
32. Ameen, S.; Seo, H.K.; Shaheer Akhtar, M.; Shin, H.S. Novel graphene/polyaniline nanocomposites and its photocatalytic activity toward the degradation of rose Bengal dye. *Chem. Eng. J.* **2012**, *210*, 220–228. [CrossRef]
33. Ansari, R.; Mosayebzadeh, Z. Application of polyaniline as an efficient and novel adsorbent for azo dyes removal from textile wastewaters. *Chem. Pap.* **2011**, *65*, 1–8. [CrossRef]
34. Stoller, M.D.; Park, S.; Zhu, Y.; An, J.; Ruoff, R.S. Graphene-Based Ultracapacitors. *Nano Lett.* **2008**, *8*, 3498–3502. [CrossRef] [PubMed]
35. Yaseen, D.A.; Scholz, M. Textile dye wastewater characteristics and constituents of synthetic effluents: A critical review. *Int. J. Environ. Sci. Technol.* **2019**, *16*, 1193–1226. [CrossRef]
36. Wacławek, S.; Lutze, H.V.; Grübel, K.; Padil, V.V.T.; Černík, M.; Dionysiou, D.D. Chemistry of persulfates in water and wastewater treatment: A review. *Chem. Eng. J.* **2017**, *330*, 44–62. [CrossRef]

37. Wacławek, S.; Černík, M.; Dionysiou, D.D. *The Development and Challenges of Oxidative Abatement for Contaminants of Emerging Concern*; Springer: Singapore, 2020; pp. 131–152.
38. Sieradzka, M.; Fryczkowski, R.; Biniaś, D.; Biniaś, W.; Janicki, J. A facile approach to obtaining PVDF/graphene fibers and the effect of nanoadditive on the structure and properties of nanocomposites. *Polym. Test.* **2019**, *81*, 106229. [CrossRef]
39. Wu, Q.; Xu, Y.; Yao, Z.; Liu, A.; Shi, G. Supercapacitors based on flexible graphene/polyaniline nanofiber composite films. *ACS Nano* **2010**, *4*, 1963–1970. [CrossRef]
40. An, J.; Liu, J.; Zhou, Y.; Zhao, H.; Ma, Y.; Li, M.; Yu, M.; Li, S. Polyaniline-grafted graphene hybrid with amide groups and its use in supercapacitors. *J. Phys. Chem. C* **2012**, *116*, 19699–19708. [CrossRef]
41. Ansari, M.O.; Yadav, S.K.; Cho, J.W.; Mohammad, F. Thermal stability in terms of DC electrical conductivity retention and the efficacy of mixing technique in the preparation of nanocomposites of graphene/polyaniline over the carbon nanotubes/polyaniline. *Compos. Part B Eng.* **2013**, *47*, 155–161. [CrossRef]
42. Sibilska, I.; Feng, Y.; Li, L.; Yin, J. Trimetaphosphate Activates Prebiotic Peptide Synthesis across a Wide Range of Temperature and pH. *Orig. Life Evol. Biosph.* **2018**, *48*, 277–287. [CrossRef] [PubMed]
43. Hanwell, M.D.; Curtis, D.E.; Lonie, D.C.; Vandermeersch, T.; Zurek, E.; Hutchison, G.R. Avogadro: An advanced semantic chemical editor, visualization, and analysis platform. *J. Cheminform.* **2012**, *4*, 17. [CrossRef] [PubMed]
44. Cardenas, L.; Macleod, J.; Lipton-Duffin, J.; Seifu, D.G.; Popescu, F.; Siaj, M.; Mantovani, D.; Rosei, F. Reduced graphene oxide growth on 316L stainless steel for medical applications. *Nanoscale* **2014**, *6*, 8664–8670. [CrossRef] [PubMed]
45. Thekkae Padil, V.V.; Filip, J.; Suresh, K.I.; Wacławek, S.; Černík, M. Electrospun membrane composed of poly[acrylonitrile-co-(methyl acrylate)-co-(itaconic acid)] terpolymer and ZVI nanoparticles and its application for the removal of arsenic from water. *RSC Adv.* **2016**, *6*, 110288–110300. [CrossRef]
46. Aradhana, R.; Mohanty, S.; Nayak, S.K. Comparison of mechanical, electrical and thermal properties in graphene oxide and reduced graphene oxide filled epoxy nanocomposite adhesives. *Polymer* **2018**, *141*, 109–123. [CrossRef]
47. Łuzny, W.; Hasik, M. Structural properties of polyaniline protonated with heteropolyacids. *Solid State Commun.* **1996**, *99*, 685–689.
48. Pouget, J.P.; Józefowicz, M.E.; Epstein, A.J.; Tang, X.; MacDiarmid, A.G. X-ray Structure of Polyaniline. *Macromolecules* **1991**, *24*, 779–789. [CrossRef]
49. Gao, W.; Sun, X.; Niu, H.; Song, X.; Li, K.; Gao, H.; Zhang, W.; Yu, J.; Jia, M. Phosphomolybdic acid functionalized covalent organic frameworks: Structure characterization and catalytic properties in olefin epoxidation. *Microporous Mesoporous Mater.* **2015**, *213*, 59–67. [CrossRef]
50. Jin, L.; Chai, L.; Ren, L.; Jiang, Y.; Yang, W.; Wang, S.; Liao, Q.; Wang, H.; Zhang, L. Enhanced adsorption-coupled reduction of hexavalent chromium by 2D poly(m-phenylenediamine)-functionalized reduction graphene oxide. *Environ. Sci. Pollut. Res.* **2019**, *26*, 31099–31110. [CrossRef]
51. Villar-Rodil, S.; Paredes, J.I.; Martínez-Alonso, A.; Tascón, J.M.D. Atomic Force Microscopy and Infrared Spectroscopy Studies of the Thermal Degradation of Nomex Aramid Fibers. *Chem. Mater.* **2001**, *13*, 4297–4304. [CrossRef]
52. Yang, C.; Zhang, L.; Hu, N.; Yang, Z.; Su, Y.; Xu, S.; Li, M.; Yao, L.; Hong, M.; Zhang, Y. Rational design of sandwiched polyaniline nanotube/layered graphene/polyaniline nanotube papers for high-volumetric supercapacitors. *Chem. Eng. J.* **2017**, *309*, 89–97. [CrossRef]
53. Peng, B.; Chen, L.; Que, C.; Yang, K.; Deng, F.; Deng, X.; Shi, G.; Xu, G.; Wu, M. Adsorption of Antibiotics on Graphene and Biochar in Aqueous Solutions Induced by π-π Interactions. *Sci. Rep.* **2016**, *6*, 31920. [CrossRef] [PubMed]
54. Wang, C.; Zhao, M.; Li, J.; Yu, J.; Sun, S.; Ge, S.; Guo, X.; Xie, F.; Jiang, B.; Wujcik, E.K.; et al. Silver nanoparticles/graphene oxide decorated carbon fiber synergistic reinforcement in epoxy-based composites. *Polymer* **2017**, *131*, 263–271. [CrossRef]
55. Duan, G.; Fang, H.; Huang, C.; Jiang, S.; Hou, H. Microstructures and mechanical properties of aligned electrospun carbon nanofibers from binary composites of polyacrylonitrile and polyamic acid. *J. Mater. Sci.* **2018**, *53*, 15096–15106. [CrossRef]
56. Ferrari, A.; Robertson, J. Interpretation of Raman spectra of disordered and amorphous carbon. *Phys. Rev. B Condens. Matter Mater. Phys.* **2000**, *61*, 14095–14107. [CrossRef]

57. Loryuenyong, V.; Totepvimarn, K.; Eimburanapravat, P.; Boonchompoo, W.; Buasri, A. Preparation and Characterization of Reduced Graphene Oxide Sheets via Water-Based Exfoliation and Reduction Methods. *Adv. Mater. Sci. Eng.* **2013**, *2013*, 923403. [CrossRef]
58. Zhou, S.; Zhou, G.; Jiang, S.; Fan, P.; Hou, H. Flexible and refractory tantalum carbide-carbon electrospun nanofibers with high modulus and electric conductivity. *Mater. Lett.* **2017**, *200*, 97–100. [CrossRef]
59. Vatankhah, A.R.; Hosseini, M.A.; Malekie, S. The characterization of gamma-irradiated carbon-nanostructured materials carried out using a multi-analytical approach including Raman spectroscopy. *Appl. Surf. Sci.* **2019**, *488*, 671–680. [CrossRef]
60. Silvestri, D.; Mikšíček, J.; Wacławek, S.; Torres-Mendieta, R.; Padil, V.V.T.; Černík, M. Production of electrospun nanofibers based on graphene oxide/gum Arabic. *Int. J. Biol. Macromol.* **2019**, *124*, 396–402. [CrossRef]
61. Boyd, G.E.; Adamson, A.W.; Myers, L.S. The Exchange Adsorption of Ions from Aqueous Solutions by Organic Zeolites. II. Kinetics[1]. *J. Am. Chem. Soc.* **1947**, *69*, 2836–2848. [CrossRef]
62. Langmuir, I. The constitution and fundamental properties of solids and liquids. part i. solids. *J. Am. Chem. Soc.* **1916**, *38*, 2221–2295. [CrossRef]
63. Hoda, N.; Bayram, E.; Ayranci, E. Kinetic and equilibrium studies on the removal of acid dyes from aqueous solutions by adsorption onto activated carbon cloth. *J. Hazard. Mater.* **2006**, *137*, 344–351. [CrossRef] [PubMed]
64. Fat'hi, M.R.; Asfaram, A.; Hadipour, A.; Roosta, M. Kinetics and thermodynamic studies for removal of Acid Blue 129 from aqueous solution by almond shell. *J. Environ. Health Sci. Eng.* **2014**, *12*, 62. [CrossRef]
65. Nekouei, F.; Nekouei, S.; Tyagi, I.; Gupta, V.K. Kinetic, thermodynamic and isotherm studies for acid blue 129 removal from liquids using copper oxide nanoparticle-modified activated carbon as a novel adsorbent. *J. Mol. Liq.* **2015**, *201*, 124–133. [CrossRef]
66. Ullah, Z.; Hussain, S.; Gul, S.; Khan, S.; Bangash, F.K. Use of HCl-modified bentonite clay for the adsorption of Acid Blue 129 from aqueous solutions. *Desalin. Water Treat.* **2016**, *57*, 8894–8903. [CrossRef]
67. Hussain, S.; Ullah, Z.; Gul, S.; Khattak, R.; Kazmi, N.; Rehman, F.; Khan, S.; Ahmad, K.; Imad, M.; Khan, A. Adsorption characteristics of magnesium-modified bentonite clay with respect to acid blue 129 in aqueous media. *Polish J. Environ. Stud.* **2016**, *25*, 1947–1953. [CrossRef]
68. Ianoş, R.; Păcurariu, C.; Muntean, S.G.; Muntean, E.; Nistor, M.A.; Nižňanský, D. Combustion synthesis of iron oxide/carbon nanocomposites, efficient adsorbents for anionic and cationic dyes removal from wastewaters. *J. Alloys Compd.* **2018**, *741*, 1235–1246. [CrossRef]

© 2020 by the authors. Licensee MDPI, Basel, Switzerland. This article is an open access article distributed under the terms and conditions of the Creative Commons Attribution (CC BY) license (http://creativecommons.org/licenses/by/4.0/).

polymers

MDPI

Article

PET Fibers Modified with Cloisite Nanoclay

Janusz Fabia [1,*], Andrzej Gawłowski [1], Monika Rom [1,*], Czesław Ślusarczyk [1], Anna Brzozowska-Stanuch [2] and Marta Sieradzka [1]

[1] Institute of Textile Engineering and Polymer Materials, University of Bielsko-Biala, Willowa 2, 43-309 Bielsko-Biala, Poland; agawlowski@ath.bielsko.pl (A.G.); cslusarczyk@ath.bielsko.pl (C.Ś.); msieradzka@ath.bielsko.pl (M.S.)

[2] BOSMAL Automotive Research and Development Institute Ltd., Sarni Stok 93,43-300 Bielsko-Biala, Poland; Anna.Brzozowska-Stanuch@bosmal.com.pl

* Correspondence: jfabia@ath.bielsko.pl (J.F.); mrom@ath.bielsko.pl (M.R.)

Received: 29 February 2020; Accepted: 26 March 2020; Published: 1 April 2020

Abstract: The alternative method of reducing the flammability of polyethylene terephthalate (PET) fibers, analogous to dyeing of PET fibers with dispersed dyes in a high-temperature bath, was proposed. A commercial organophilic montmorillonite Cloisite®15A (C15A) was applied as a flame retardant. The aim of the presented work was to evaluate the effectiveness of the introduced modifier and the improvement of the flame-retardant properties of PET fibers by limiting oxygen index (LOI) and thermogravimetric analysis (TGA) measurements. Evolved gas analysis (EGA) by spectrometric method (FTIR) during coupled thermogravimetric analysis (TGA) was applied in order to confirm no increase in the toxicity of volatile degradation products released from burning modified fibers. The nanocomposite nature of modified fibers was confirmed based on the structural parameters of the fibers determined using wide-angle X-ray scattering (WAXS) and small angle X-ray scattering (SAXS) X-ray diffraction methods.

Keywords: PET fibers; Cloisite; flame retardancy; LOI; TGA; FTIR; SEM; WAXS; SAXS

1. Introduction

Polyethylene terephthalate (PET) fibers are produced in very large quantities worldwide, due to their relatively low cost. Moreover, the fibers are characterized by very good mechanical properties, high resistance to physical and chemical agents. Therefore, they are very widely used as fibers for textiles, technical products (e.g., cords for tires, safety belts) and the production of articles intended for interior furnishings of public facilities or car upholstery. Due to such a wide range of applications, products from PET fibers are facing very strict requirements in terms of fire safety. Standard PET fibers are not very resistant to fire and large amounts of smoke are emitted during their combustion. In addition, the so-called dripping effect is observed; detached drops of molten polymer burn and promote the spread of fire. Therefore, imparting flame-retardant properties and resistance to dripping to PET fibers is of particular interest to researchers [1]. The flame-retardant modification of PET fibers can be achieved in various ways. Most often, a properly ground (micro- or nano-) anti-pyrene is added to the polymer melt [2–4], but also a surface application of the flame retardant is employed during the impregnation process [5–9]. The chemical method is used less often. An example here may be the introduction of the aryl ether monomer (2,2′- (4,4′ - (1,4-phenylenebis (oxy)) bis (4,1-phenylene)) bis (oxy) diethanol—PBPBD) into the PET backbone chain during polycondensation [10].

In recent years, montmorillonite (MMT) nanoclay has been used to modify polymers, thus providing an improvement in performance, including fire resistance. Nanoclays used as an additive to the polymer melt significantly improve the flame-retardant properties of the material, especially if they are used in synergistic systems with other non-halogenated anti-pyrenes.

The addition of Cloisite 15A to the ethylene-vinyl acetate (EVA) melt in combination with phosphorus compounds increases the flame-retardant properties of the tested blend. Observation of the charred residue revealed the formation of a dense and stable layer, which is an effective barrier against the penetration of oxygen and heat and the release of flammable gases [11]. Increased flame-retardant properties can be the result of physical and thermal cross-linking of nano-clay particles and the polymer chain and physicochemical adsorption of volatile degradation products on silicates (depending on the organic modifier of montmorillonite) [12].

There are certain technological difficulties in production of fibers from thermoplastic polymers with the addition of nanoclay using extruder method. The example might be the process of spinning and stretching of polypropylene fibers with clay addition above 1% by weight, which is technically unacceptable due to the loss of spinnability of the polymer melt [13]. The dispersion of clay nanoparticles is an important technological parameter when adding clay nanoparticles to the polymer melt. Electron microscopy (TEM) and X-ray diffraction (XRD) methods are particularly useful for assessing dispersion in a polymer matrix [14]. The presence of montmorillonite clay particles in the PET matrix affects the crystallization of the polymer, lowers the glass transition temperature and leads to the heterogeneous nucleation of the composite [15]. Further studies have shown that the addition of the clay causes the crystallization temperature to drop. This means that organophilic montmorillonite promotes PET crystal nucleation and even a small amount of organic clay is sufficient to maximize the nucleation effect [16]. In addition, surface morphology studies using the TEM method showed partial peeling of the surface structure of PET-CLAY nanocomposite film (addition of 3% C30B) [17].

An effective way of increasing flame resistance is also coating of the finished materials with special pastes containing a flame-retardant admixture. However, the resulting coatings change the surface and functional properties of processed products. It has been proven that the use of nanoclay for varnish coatings in optimal amounts of 3–5% significantly improves the flame retardancy of the impregnated product [18]. Addition of nanoclay below 1% and above 10% does not result in the desired effects. The type of nanoclay used is also important, as confirmed by flammability tests. The use of polyurethane resins with the addition of montmorillonite in form of Cloisite 30B as a coating on polyester and cotton fabrics also effectively protects the impregnated material against the harmful effects of fire [19]. The admixture of montmorillonite acts, therefore, as a flammability inhibitor.

The method of applying nanoclay in the form of an aqueous dispersion on cotton fabric was described in [20]. In this way, the limiting oxygen index (LOI) was increased by 4% for nanoclay processing, while for fabric activated by low-temperature plasma in nitrogen and then treated with nanoclay, a 5% increase in oxygen index was achieved. Such an effect of synergistic interactions was similar to the previously discussed citations.

However, there are no reports in the literature regarding the use of an analogous, aqueous dispersion of organophilized aluminosilicate for the preparation of flame-retardant polyester fibers, the possibility of flame-retardant treatment based on the application of natural zeolite was studied [21]. Therefore, in order to fill the existing gap, it was decided to carry out the relevant research, the results of which are described in this study. Commercial Cloisite 15A type montmorillonite, ground to nanometric size, was used for flame-retardant modification of PET fibers. The criterion for choosing this type of clay was the interlayer distance, which was the largest among the entire Cloisite series. The flame-retardant application was carried out using the exhaust method. This method involves the application of fiber treatment in water bath containing modifier at an appropriately high temperature (HT) and is analogous to dyeing using disperse dyes. The proposed method has been successfully tested before in the same, optimized process conditions, with the use of other flame retardants [22–24].

As a result of using the proposed method, modified PET fibers were obtained and comprehensively tested. Description of the introduction of C15A clay nanoparticles into the fiber material was based on the so-called voids approach [25]. Three main goals of this study were set: to assess the effectiveness of improving the flame-retardant properties of fibers, to analyze gaseous oxidative degradation products in terms of their toxicity in the event of a fire, and to describe the changes in the supermolecular structure of fibers resulting from the modification.

2. Materials and Methods

The materials used in this work were all commercially available technical products. PET fibers were supplied by Elana SA Toruń (Toruń, Poland). As the flame-retardant modifier—organically modified montmorillonite in the form of Cloisite®15A nanoclay, produced by Southern Clay Product Ltd. (USA), was used [26]. The finishing treatment of fibers was carried out in the laboratory dyeing device (Ahiba Turbomat—Lucerne, Switzerland) with a liquor ratio of 1:50. The conditions established were the following: temperature—130 °C, treatment time—1 h and heating rate—2.5 °C/min, respectively. The PET fibers were processed in an aqueous dispersion containing the C15A modifier and Rokacet R40 KO300G (Glyceryl Cocoate; CAS no. 68201-46-7) nonionic surfactant supplied by PCC Group (Brzeg Dolny, Poland) which was added into the bath in the amount of 0.7 g/L. C15A nanoclay dispersion was prepared using an ultrasonic bath for 15 min. After the treatment samples were washed in a solution of detergent Pretepon G (PCC Group—Brzeg Dolny, Poland) in the amount of 5 g/L. The washing time was 30 min and the temperature was 60 °C, washing was carried out in the laboratory dyeing device (Ahiba Turbomat—Lucerne, Switzerland). The effectiveness of the modifier was tested in a wide range of flame-retardant concentrations, ranging from 0 to 7.5% in relation to the fiber weight.

The examinations of the fiber flammability were carried out using the limiting oxygen index (LOI) method in accordance with PN ISO 4589.

Thermogravimetric analysis (TGA) was performed using Thermogravimetric Analyzer Q500 TA Instruments (New Castle, DE, USA). Measurements were carried out in a temperature range from 30 to 700 °C with the heating rate of 20 °/min in the air atmosphere (flow 60 mL/min). Pre-tared platinum pans were used to contain the samples, and the mass of the sample was between 20 and 30 mg. The data were evaluated by means of the Universal V4.5A (TA Instruments, (New Castle, DE, USA) software.

Evolved gas analysis (EGA) by spectrometric method (FTIR) during coupled thermogravimetric analysis (TGA) measurements were carried out using TA Instruments Q500 thermobalance (New Castle, DE, USA) with a special EGA furnace coupled with the Nicolet iS50 FTIR spectrometer (Thermo Fisher Scientific, Waltham, MA, USA). The furnace incorporates a quartz liner between the bifilar-wound heating element and the sample measurement area. This arrangement results in a small internal volume that is readily swept at normal purge rates, assuring rapid transfer and reduced dilution of decomposition off-gases. Test parameters were the following: platinum crucible, heating rate: 20 °C/min, temperature range: 30.0–750.0 °C, gas flow: air 50 mL/min. The gas flow rate through the transfer line and gas cell was kept constant. Infrared spectra over the range of 4000 to 450 cm^{-1} were collected every 15 s at a resolution of 4 cm^{-1}. The absorption bands of each recorded spectrum were simultaneously integrated over the entire spectral range. Gram-Schmidt (G-S) curves were obtained by plotting integration from each spectrum as a function of time. The average intensity of volatiles during the mass loss was determined and depicted in the G-S curves based on vector analysis. FTIR spectra analysis was performed using OMNIC Series 9.8.372 software (Thermo Fisher Scientific, Waltham, MA, USA) and a Nicolet™ FTIR Vapor Phase Spectral Library (Thermo Fisher Scientific, Waltham, MA, USA).

The wide-angle X-ray scattering (WAXS) investigations were carried out with a URD-65 Seifert (Rich. Seifert & Co. Röntgenwertk, Ahrensburg, Germany) diffractometer. CuKα radiation was used at 40 kV and 30 mA. Monochromatization of the beam was obtained by means of a nickel filter and a graphite crystal monochromator placed in the diffracted beam path. A scintillation counter was used as a detector. Investigations were performed in the range of angles from 3° to 40° in steps of 0.1°.

The small angle X-ray scattering (SAXS) experiments were performed by means of an MBraun camera, which utilizes the conventional Kratky collimation system. The front of the camera was directly mounted on the top of the tube shield of a stabilized Philips PW 1830 X-ray generator. The X-ray tube was operated at a power of 1.5 kW. CuK_α radiation was used. Scattered radiation was recorded in an acquisition time of 1200 s by means of an MBraun linear position-sensitive detector, model PSD 50 (HECUS-MBraun Graz X-Ray Systems, Graz, Austria). The detector had 1024 channels with a channel-to-channel distance of 52 μm.

Scanning electron microscopy (SEM) analyses were performed in conventional SEM mode using Jeol JSM 5500LV instrument (JEOL Ltd., Tokyo, Japan) operating at 10 kV, after coating the samples with a thin layer of gold by sputter deposition. Surfaces of samples and scorching after the process of burning of PET fibers were observed at different magnifications.

3. Results and Discussion

The presentation of the research results is divided into 3 parts corresponding to the main objectives set by the authors: assessment of the effectiveness of the introduction of flame retardant into fiber structure, analysis of gaseous products of the oxidative degradation of fibers and the characterization of the supermolecular structure of the fibers modified with Cloisite C15A.

3.1. Studies on the Effectiveness of the Flame-Retardant Modification

The assessment of the effectiveness of the proposed flame-retardant modification includes the determination of the limiting oxygen index (LOI), thermogravimetric tests carried out to confirm the results of LOI tests and microscopic observations of the charred solids obtained as a result of the controlled fiber burning.

3.1.1. Limiting Oxygen Index

The resulting flame-retardant effect of PET fibers modification was evaluated using the limiting oxygen index (LOI) method. A parameter that characterizes the method and is the lowest percentage of oxygen in the mixture with nitrogen at which the test specimen ignites and burns on its own. The measurements were performed in accordance with PN–ISO 4589 standard. The obtained results are presented in Table 1.

Table 1. Limiting oxygen index (LOI) values of studied fibers.

No.	Samples	LOI [%]
1	PET std	21.6
2	PET + 0.10% C15A	23.6
3	PET + 0.25% C15A	23.9
4	PET + 0.50% C15A	24.0
5	PET + 1.00% C15A	23.8
6	PET + 2.50% C15A	20.5
7	PET + 5.00% C15A	20.0
8	PET + 7.50% C15A	19.8

Based on the analysis of the LOI values given in Table 1, it can be concluded that the addition of C15A clay in the nanopowder form to PET fibers indeed changes the value of the limiting oxygen index. The differences in values are not large, but they are undoubtedly significant. The highest LOI value was obtained for PET fibers modified in a bath containing 0.5% C15A in relation to the fiber weight. It should be noted that a further increase in the modifier content results in a systematic reduction of LOI value. The addition of C15A in quantities of above 2.5% causes the oxygen index to decrease to the value lower than when burning PET fibers without a modifier.

3.1.2. Thermogravimetric Analysis

Thermogravimetric analyses in the air were performed in order to confirm the improvement of the flame-retardant properties of the fibers, represented by the LOI values. In the case of polymers such as PET, thermal dissociation in an oxidizing atmosphere and direct combustion of residues after thermal dissociation are almost fully separable transformations, so that the application of the TGA method gives very good results. Such an approach has already been demonstrated both for classic PET fibers [22] and for partially oriented PET fibers (POY) formed at high speeds [27]. Based on the analysis of the DTG curves, it is possible to show the shift of the exothermic peak (and its maximum) corresponding to the combustion process of the sample. When the material is modified with antipyrene, the peak shift towards higher temperatures is observed. In the observed case, the maximum temperature of the discussed effect (Figure 1) shifts from 569.7 °C for pure PET to 578.1 °C for the most favorable modification variant, that is 0.5% of C15A.

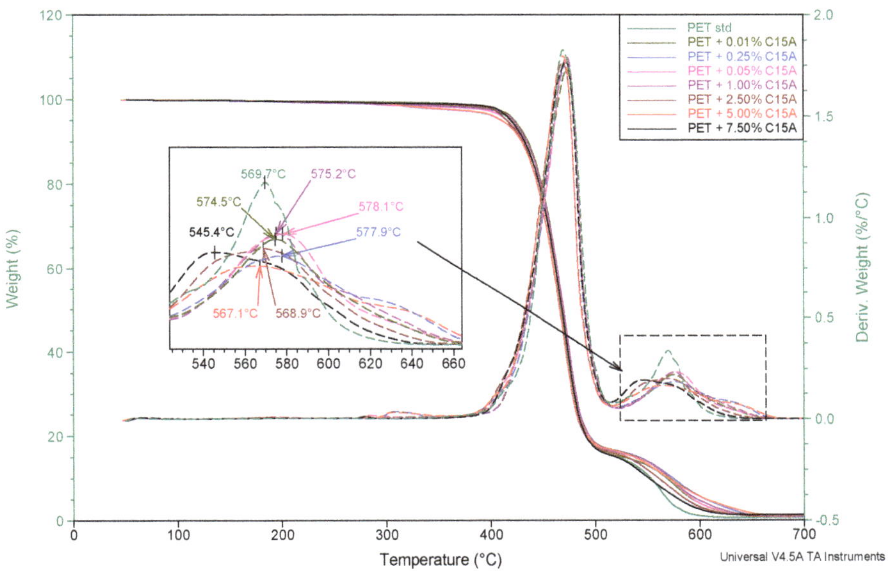

Figure 1. TG and DTG curves of polyethylene terephthalate (PET) fibers modified with Cloisite®15A in the bath at 130 °C. Heating rate 20 °/min, purge air flow 60 mL/min.

Along with the further increase of the modifier content, which was tested up to 7.5% of C15A, the opposite tendency was observed, i.e., the shift of the discussed maximum towards lower temperatures. The obtained results clearly substantiate the results of the LOI tests and indicate a real improvement of the flame retardancy of the tested PET fibers.

3.1.3. Microscopic Observations of Charred Residues.

For the optimal variant of fibers after the flame-retardant modification (0.5% C15A addition), determined on the basis of LOI and TGA tests, the controlled combustion of the sample in atmospheric air was carried out, accompanied with the analysis of the burning process and microscopic observation of the charred residues. It was found that during the burning of nanoclay-modified fibers no characteristic phenomenon of dripping was observed. Dripping of molten, burning material, which occurs in the case of raw PET can drastically contribute to the propagation of the flame zone during a fire. SEM microscopic observations of charred residues of nanoclay-modified fibers after combustion (Figure 2a,b) reveal a very clear difference with respect to the material that was not flame retardant (Figure 2c,d).

The surface of the charred residue for the sample with nanoclay is covered with a characteristic coating. The charred residues are formed in the immediate vicinity of the flame so that the said coating effectively limits the access of flammable decomposing gases released during thermal dissociation of

PET to the flame zone. Therefore, even after the direct ignition of the material, the tendency to retard fire and limit flame propagation was observed.

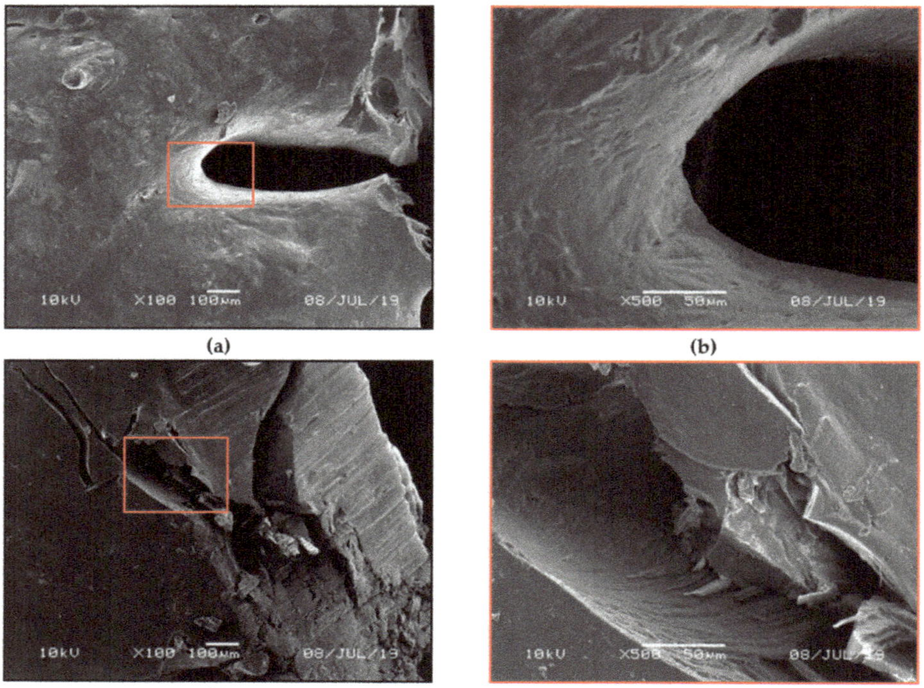

(a) (b) (c) (d)

Figure 2. SEM microphotographs of residue after combustion of PET fibers modified with Cloisite®15A (C15A) (0.5%; 130 °C) (**a**,**b**). SEM microphotographs of residue after combustion of a sample of unmodified PET fibers (**c**,**d**).

The observed phenomenon is consistent with the conclusions of many authors investigating the effect of flame retardancy after the introduction of modified montmorillonite nano-powder into various polymer matrices: PA6, PP, PP/MA [28–30], PS [31]. In our case, C15A clay dispersed in water bath was introduced into the material (matrix) of PET fibers not through physical mixing in the melt, but through penetration of the fiber structure and filling of the so-called voids in conditions of increased pressure and temperature (130 °C) for approx. 1 h.

It turns out that under the abovementioned conditions it is possible to obtain a specific structure of reduced gas permeability, described for the first time by K. Yano et al. [32] for MMT-containing polyimide nanocomposites. Since the authors of this study in the near future intend to propose a separate publication dedicated to X-ray studies of delamination of the C15A modifier and to FTIR temperature studies of the solid residue arising from the oxidative degradation process of PET fibers, the burning process inhibition mechanism of modified PET fibers will not be discussed here.

3.2. *Analysis of Volatile Products of the Oxidative Degradation of Fibers*

Another important issue is the analysis of gaseous products released during the oxidative degradation of the modified fibers during their burning process. The increase in toxicity of gases generated during decomposition of polymers in the event of a fire under the influence of specific anti-pyrenes is an extremely important problem. This phenomenon occurs especially when halogenated compounds are used to obtain flame-retardant properties of polymer materials. Despite the extremely high effectiveness of these anti-pyrenes, they have been almost eliminated from the market over the past several years. There are systematic attempts to replace them with anti-pyrenes with a completely different mechanism of flame inhibition, and the results described in this chapter also follow this trend.

3.2.1. C15A Nanoclay

The study of gases emitted during the thermal decomposition of the C15A modifier was carried out based on the analysis of the TG and DTG curves in a wide temperature range from 50 to 750 °C (Figure 3). Local maxima marked on the DTG curve determine the temperatures at which the rate of mass loss associated with the occurrence of specific transformations in the tested material is the highest. These temperatures of 257.9 °C, 313.4 °C, 600.0 °C and 672.5 °C, respectively, served as characteristic temperatures at which FTIR spectra, recorded every 15 s in the whole temperature range, were subjected to a detailed analysis aimed at identifying specific chemical compounds.

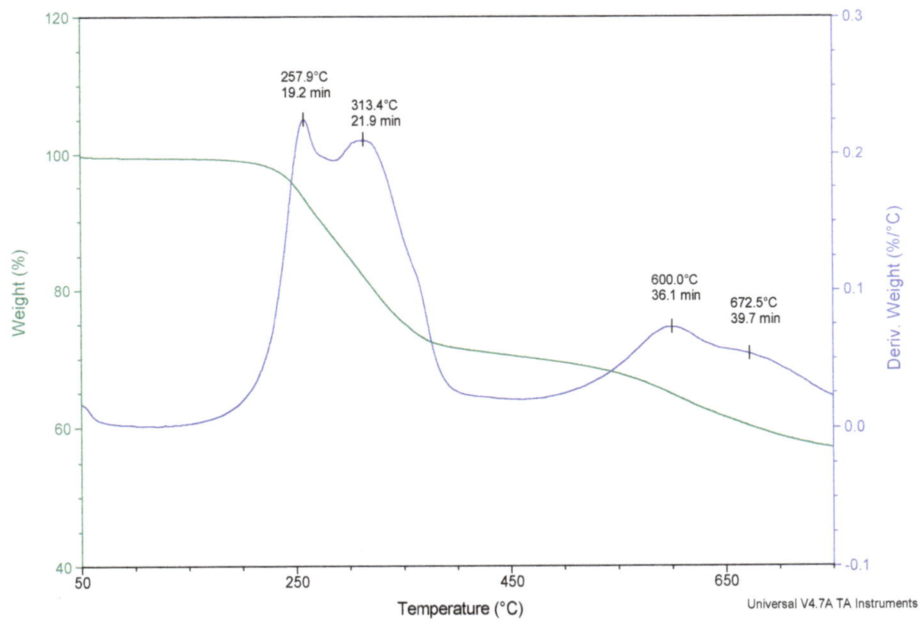

Figure 3. TG and DTG curves of C15A clay with marked FTIR analysis points.

The FTIR spectrum recorded at 257.9 °C (Figure 4) illustrates a qualitative mixture of gases released from a C15A nanoclay sample heated in the air, as the effect of the breakdown of the quaternary ammonium salts particles, introduced into the spaces between the silicate layers of aluminosilicate in the organophilization process. In the case of C15A clay, it is a sterically developed compound with 2 characteristic tails, which are essentially simple triacylglycerols containing residues of saturated fatty acids with 18 carbon atoms in the chain [26].

As a result of the thermal dissociation, a series of absorption bands appear in the recorded FTIR spectrum, among others in the ranges of 1350–1450 cm^{-1}, 1650–1700 cm^{-1} and 2700–2950 cm^{-1}, assigned respectively to the entire group of aldehydes and acids. Using the specialized Nicolet^{TM} FTIR Vapor Phase Spectral Library software, the following most characteristic compounds were identified: $C_8H_{17}CHO$ pelargonic aldehyde (nonanal) and $C_8H_{17}COOH$ pelargonic acid (nonanoic acid). Those products are formed as a result of the breakdown of quaternary ammonium salt used for organophilization of C15A clay, and more precisely the breaking of ester bonds and fragmentation of saturated fatty acid molecules, most often at about half-length of the chain. The appearance of the above compounds in the atmosphere of gases leaving the measurement cell, in the analyzed temperature range, is evidenced by the characteristic smell of geranium. It should be emphasized that these compounds do not present a toxicological hazard.

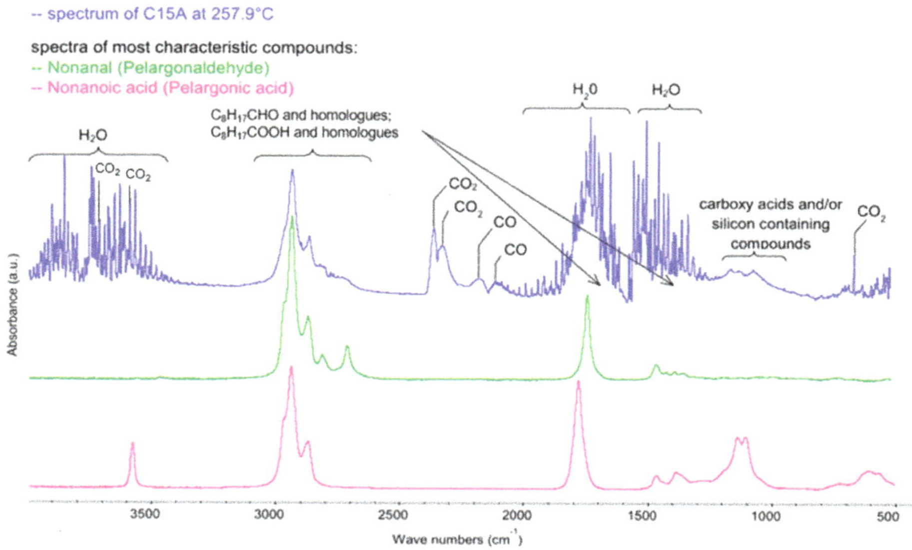

Figure 4. FTIR spectrum of gases evolved during oxidative degradation of C15A recorded at 257.9 °C (blue) with spectra of the most representative compounds from the group of aldehydes and acids—nonanal (green) and nonanoic acid (pink), respectively, present in products at the above temperature. Qualitative characteristics of the spectrum were obtained using Nicolet™ FTIR Vapor Phase Spectral Library software.

Pelargonic acid is found in the form of esters in the essential oil obtained from geranium and is a natural non-selective herbicide. In contrast, pelargonium aldehyde (nonanal) is a compound found naturally in some essential oils (including cinnamon and rose oil), and is also produced in small amounts by the human body [33] so that it is not only non-toxic but also biocompatible. In addition to the compounds mentioned above, the discussed FTIR spectrum of C15A nanoclay is predominated by water, carbon dioxide and carbon monoxide.

Analyzing the spectrum recorded at 313.4 °C, i.e., the second maximum on the DTG curve, there are no qualitative changes. The intensity of CO_2-derived bands increases, while the intensity of absorption bands dedicated to water and carbon monoxide decreases (Figure 5). The intensity of the $C_8H_{17}CHO$ and $C_8H_{17}COOH$ bands and their homologs also decreases. This situation is caused by the fact that the second maximum on the DTG curve corresponds to the same transformations as the ones discussed in the first example (spectrum recorded at 257.9 °C), i.e., the breakdown of the quaternary ammonium salt introduced between silicate stacks and the removal of water from the interlayer spaces of C15A nanoclay. However, this is a two-step process, because the cleavage of ester bonds in triacylglycerols is energetically non-homogeneous, as it depends on which carbon atom in glycerin the given bond is located at.

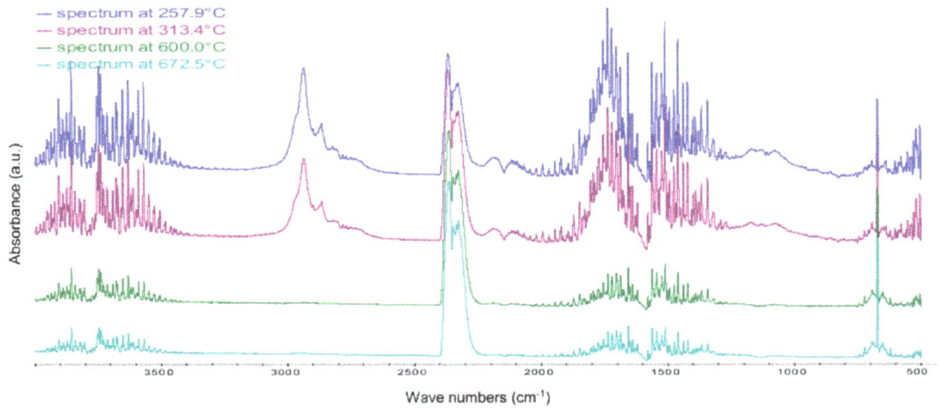

Figure 5. FTIR spectra of gases emitted during oxidative degradation of C15A recorded at: 257.9 °C (19.2 min), 313.4 °C (21.9 min), 600.0 °C (36.1 min) and 672.5 °C (39.7 min).

In the FTIR spectra of the gases emitted as a result of the oxidative degradation of the C15A, recorded at 600.0 °C and 672.5 °C (respectively third and fourth maximum on the DTG curve), only the bands associated with the presence of water and CO_2 are observed. The intensity of bands dedicated to H_2O decreases systematically with the increase of temperature, while in the case of bands dedicated to CO_2 the intensity increases slightly with the temperature (Figure 5).

3.2.2. Pure PET

Studies on the characterization of gaseous products generated as a result of the oxidative degradation of PET fibers before flame-retardant modification was done in order to settle the reference point for samples after the modification. Selected FTIR spectra were collected at temperatures corresponding to the temperatures of maxima registered on the DTG curve (Figure 6).

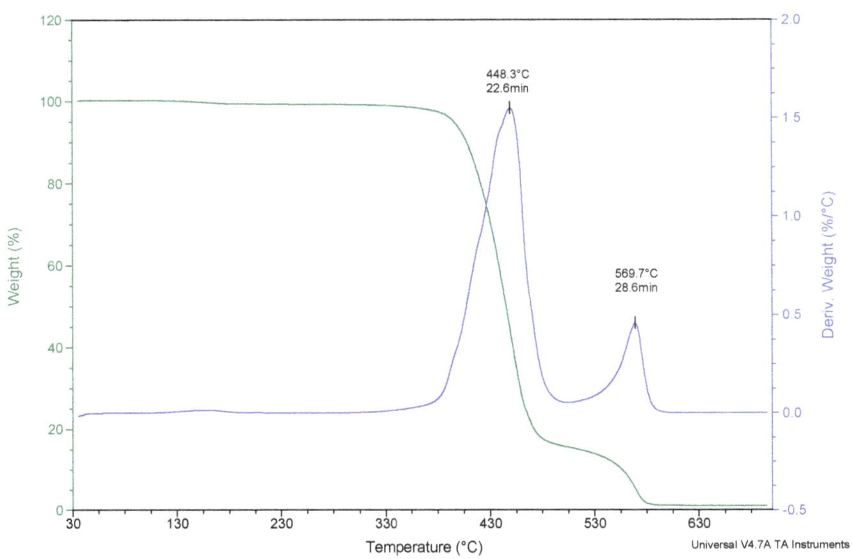

Figure 6. TG and DTG curves of pure PET with marked FTIR analysis points.

The first maximum, at 448.3 °C, corresponds to the highest mass loss rate occurring during the thermal dissociation process of the PET fiber material in the oxidizing atmosphere. In the case of pure PET, the recorded FTIR spectrum is characterized by the occurrence of a number of absorption bands corresponding to the main components of volatile oxidative degradation products, including, first of all, pairs of phthalates, benzoic acid and its homologues (main bands, in the range: 1050–1400 cm^{-1}, 1780 cm^{-1} and 3600 cm^{-1}, respectively), formaldehyde and saturated hydrocarbons (extensive low-intensity band in the range of 2640–3680 cm^{-1}), unsaturated hydrocarbons: mainly C_2H_2 (1420 cm^{-1}) and also characteristic absorption bands of water, carbon dioxide and carbon monoxide (Figure 7).

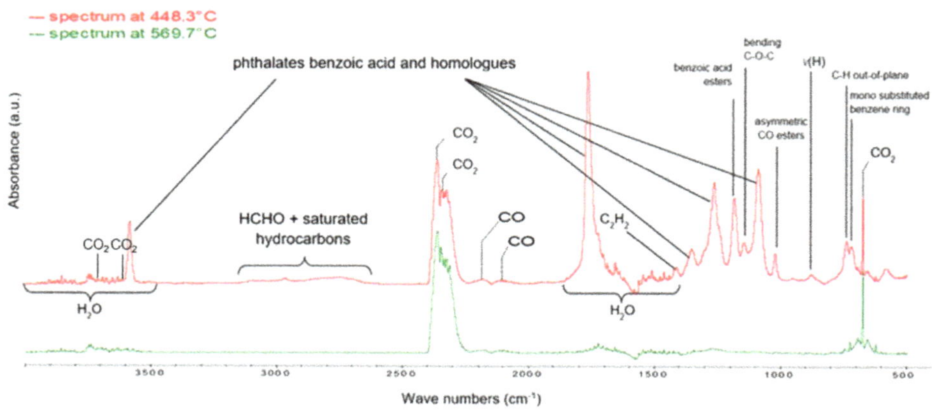

Figure 7. FTIR spectra recorded for pure PET in the range of 500–4000 cm^{-1} at temperatures characteristic of the oxidative degradation process: 448.3 and 569.7 °C, respectively, determined on the basis of DTG curves. Qualitative characteristics of the spectrum were obtained using Nicolet™ FTIR Vapor Phase Spectral Library software and [34].

The FTIR spectrum recorded at 569.7 °C corresponds to the second maximum on the DTG curve. It is associated with the highest mass-loss rate during the combustion of the solid residue after decomposition of polyethylene terephthalate. Among the gases emitted there only CO_2, small amounts of H_2O and the traces of CO could be found. Therefore, unlike the gas mixture generated during the oxidative degradation stage of PET, in the case of post-combustion of the char resulting from this degradation, the toxicity of emitted gaseous products is reduced nearly to zero.

3.2.3. PET Modified with C15A

The discussion of the results in this chapter is concluded with the analysis of gaseous products generated during the oxidative degradation of PET fibers subjected to flame-retardant treatment in a bath containing 0.5% of C15A nanoclay (based on fiber weight). For analysis, as in the previously discussed cases, selected FTIR spectra were recorded at the temperatures corresponding to the temperatures of maxima observed on the DTG curves (Figure 8). It is worth noticing that the small local maximum at 261.1 °C corresponds to the cleavage of ester bonds in the molecules of the quaternary ammonium salt present in the interlayer spaces of C15A nanoclay. Despite the relatively low content of the flame retardant used, the change in sample mass associated with the transformation of its chemical structure is reflected in the DTG curve. The beginning of thermal dissociation of C15A is almost perfectly coincident with the temperature range of the physical PET melting process. The temperature of the second maximum on the DTG curve registered for modified fiber, corresponding to the highest mass loss rate resulting from thermal dissociation of polyethylene terephthalate under the oxidative conditions, is higher by ~8 °C as compared to the results obtained for unmodified PET fibers.

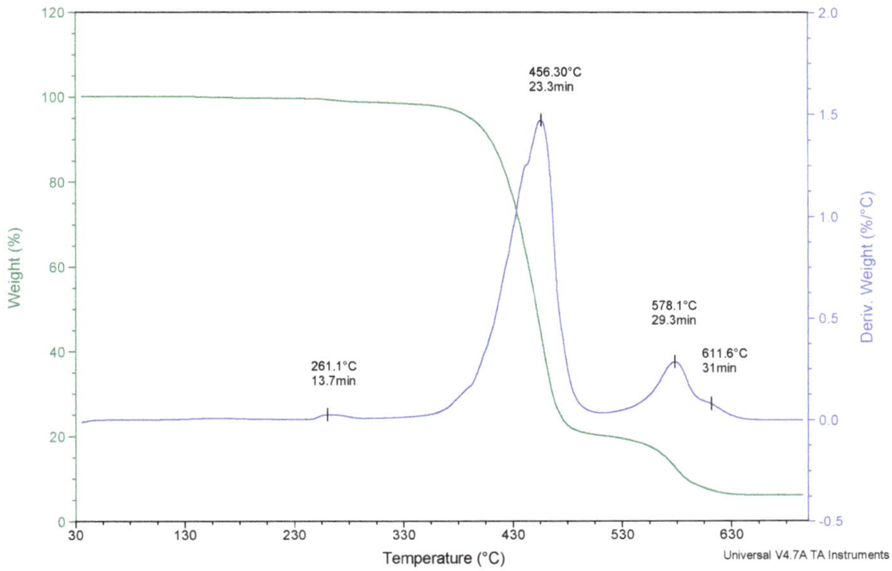

Figure 8. TG and DTG curves of PET modified with C15A (0.5%) with marked FTIR analysis points.

Thus, the previously stated theses [31,35] can be confirmed: depending on the type of polymer matrix and montmorillonite content, the temperature of nanocomposites decomposition increased from a few to nearly 50 °C. This phenomenon is explained by the limitation of the mobility of polymer chains located in the interlayer spaces of aluminosilicate. Therefore, it is assumed that in the observed samples exfoliation or at least intercalation of montmorillonite layers occurs. Despite the fact that in the studied case samples were prepared without the physical mixing of C15A nanoclay with polyethylene terephthalate in the melt, but only by exhaustion of MMT from the water bath, i.e., the introduction of modifier particles from the aqueous dispersion, as a result a modified material presenting the properties of nanocomposite was obtained, the structure of which will also be discussed based on the results of X-ray structural analysis.

Analyzing the FTIR spectra recorded for PET fiber + 0.5% of C15A (Figure 9) it should be noted that in addition to the spectrum recorded at 261.1 °C, which is analogous to the spectrum for pure C15A modifier (except for the characteristic bands corresponding to aldehydes and ketones formed after the cleavage of ester bonds, the intensity of which is very clearly reduced in this case), the other spectra are qualitatively identical to those recorded for pure PET fibers.

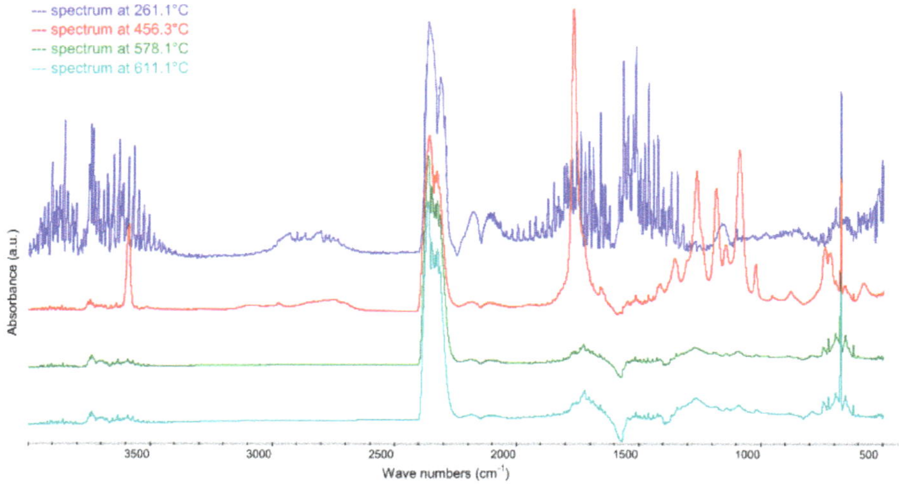

Figure 9. FTIR spectra recorded for PET modified with C15A (0.5%) in the range of 500–4000 cm^{-1} at temperatures characteristic of its oxidative degradation process: 261.1 °C (13.7 min), 456.3 °C (23.3 min), 578.1 °C (29.3 min), 611.1 °C (31.0 min), respectively, determined on the basis of DTG curves.

In order to enable the quantitative comparison of spectra aimed at determining if the gases emitted as the result of the thermal decomposition of modified fibers are toxic or not, Gram-Schmidt (G-S) curves as a function of time were generated, illustrating the intensity of gaseous product evolution in the entire analyzed temperature range: from 30 to 700 °C (Figure 10).

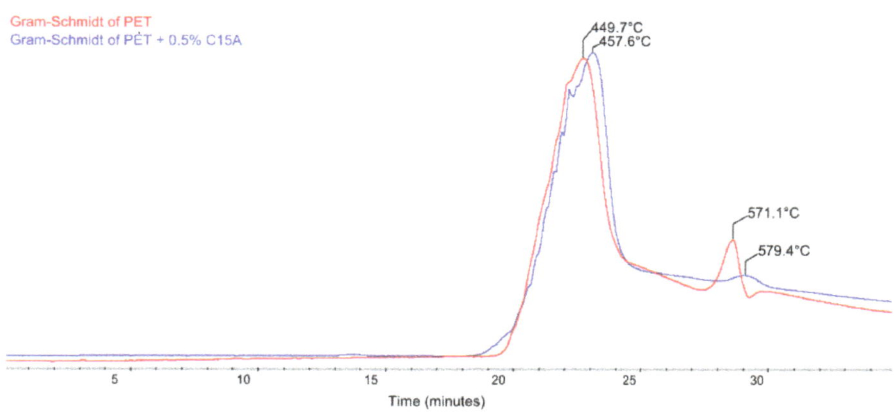

Figure 10. Gram-Schmidt curves of pure PET (red) and PET modified with 0.5% C15A (blue).

Each of the G-S curves is based on all the individual spectra. Gram-Schmidt curve is plotted by the calculated values of the infrared extinction coefficient of the evolved gases and their corresponding concentrations over the entire wavenumber range [36]. The peaks appearing on the G-S curves correspond to the maxima on the DTG curves, but are slightly shifted towards higher temperatures. In the presented case, at the heating rate of 20 °C/min, this shift is approx. 1.3–1.4 °C. This shift is a result of the delay between the TGA furnace and the detector in the FTIR. Evolved gases are swept through a transfer line which is kept at 250 °C to avoid condensation of relatively higher molecular weight gaseous products.

Since toxic gaseous products appear almost exclusively during the decomposition of the base polymer, i.e., PET, it was relevant to compare the spectra recorded at temperatures corresponding to the first maximum on the G-S curves, for pure PET (449.7 °C) and for modified PET + 0.5% C15A (457.6 °C), respectively. The overlapping of both registered spectra in the full range of wavenumbers (500–4000 cm^{-1}) revealed that, with the exception of slight differences in the absorption bands corresponding to H_2O and CO_2, in the case of bands corresponding to the presence of toxic products of PET thermal decomposition (mainly phthalates, benzoic acid homologs and formaldehyde), practically no differences in intensity were observed (Figure 11). There were also no additional bands from the emerging new (toxic) compounds, as observed for, incidentally extremely effective, halogenated flame retardants.

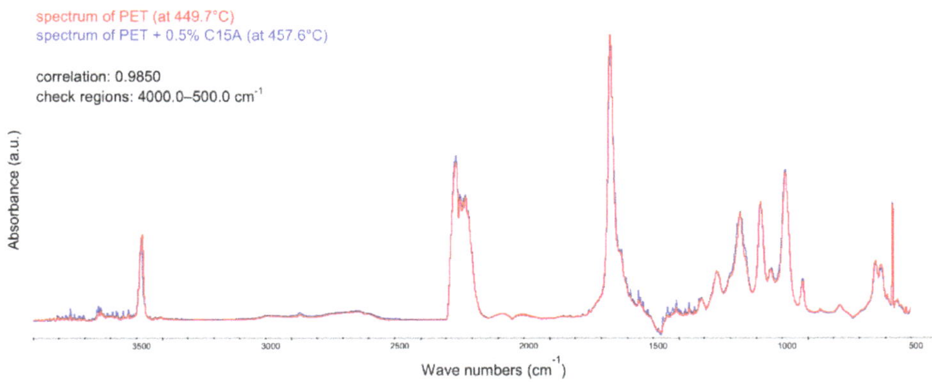

Figure 11. FTIR spectra recorded for pure PET (red) and PET modified with 0.5% C15A (blue) in the range of 500–4000 cm^{-1} at temperatures characteristic of the first step of oxidative degradation process: 449.7 and 457.6 °C, respectively, determined on the basis of Gram-Schmidt curves.

The obtained spectral correlation coefficient was 98.5%. In the case of the second maximum on the G-S curves, corresponding to the maximum evolution of gaseous products at the post-combustion stage of solid char formed after the oxidative degradation process, the value of the spectral correlation coefficient increased up to 99.3%.

3.3. Morphology and Crystallinity of Fibers

In order to determine the effect of C15A nanoclay and the conditions of its application on the supermolecular structure of flame-retardant-modified PET fibers, X-ray studies were carried out in both wide (WAXS) and small (SAXS) ranges of diffraction angles. The analysis was carried out for untreated PET fibers processed only in a water bath at 130 °C, and for two variants of the flame-retardant modification i.e., 0.5% (optimal) and 7.5% (maximum) content of C15A nanoclay in the bath.

Using the WAXS method, the basic parameters of the fiber crystal structure, such as mass crystallinity degree and the average size of crystallites, were determined. Moreover, the trend of changes in the structural parameters after the flame-retardant modification of fibers was assessed.

In Figure 12, a comparison of WAXS patterns of unmodified PET fibers (PET std) and modified fibers (PET 130 + 7.5% C15A) are presented. For C15A treated fibers, a slight decrease in the intensity of crystalline peaks can be observed, indicating a decrease in the crystallinity of these fibers. PET crystallizes in a triclinic system and according to Daubeney et al. [37], the edges of the unit cell are the single monomeric units of PET homopolymer. PET chains assume an approximately planar configuration. The chain plane is almost parallel to the (100) lattice plane.

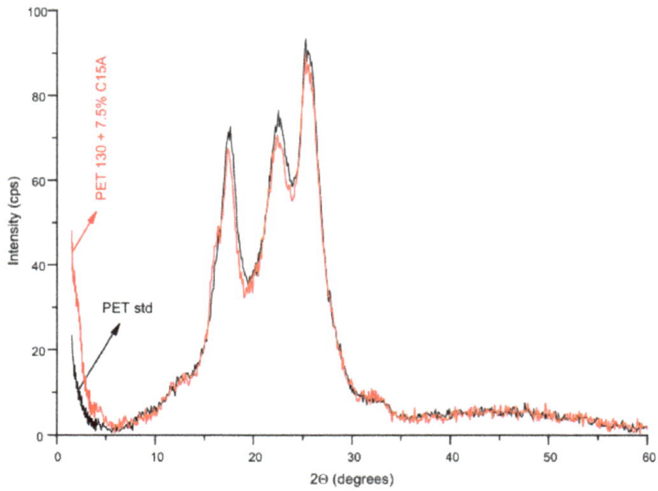

Figure 12. Examples of wide-angle X-ray scattering (WAXS) patterns of fibers subjected to analysis.

Quantitative evaluation of the crystallinity of the fibers represented by the crystallinity index was carried out based on WAXS data. For this purpose, each WAXS curve was deconvoluted into crystalline and amorphous scattering components using WaxsFit profile fitting software [38]. Each peak was modelled using a Gaussian-Cauchy peak shape. The crystallinity index was calculated as a ratio of the area under crystalline peaks to the total area of the scattering curve. Figure 13 is a representative example of a diffraction curve resolved into crystalline and amorphous scattering components using the peak fitting software.

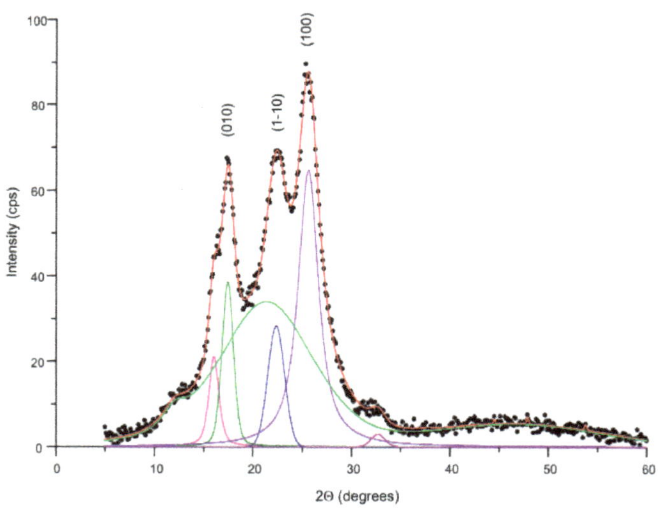

Figure 13. Deconvolution of the exemplary WAXS curve obtained for PET 130 + 7.5% C15A fibers.

In order to evaluate the variations of crystallite sizes of PET, the Scherrer equation [39] was used. The crystallite sizes were calculated in the direction perpendicular to the (010), (1-10) and (100) planes, i.e., perpendicular to the PET molecular chain axis. The results are presented in Table 2.

Table 2. Crystallinity and dimensions of crystallites obtained by means of WAXS method.

Sample	Crystallinity [%]	Dimensions of Crystallites [nm]		
		$D_{(010)}$	$D_{(1-10)}$	$D_{(100)}$
PET std	47.0	5.9	4.6	3.7
PET 130	49.2	5.5	4.2	3.7
PET 130 + 0.5% C15A	45.0	6.3	4.3	4.0
PET 130 + 7.5% C15A	43.9	6.4	4.5	3.7

Analysis of the results presented in Table 2 indicates that the treatment of fibers in a water bath without the addition of C15A, during 1 h at 130 °C, causes a more than 2% increase in the degree of crystallinity. This was to be expected because during high-temperature processing (coinciding with the temperature range of so-called cold crystallization of PET) the mobility of polymer chains increases, thus further ordering of the structure is possible. It should be noted, however, that the addition of C15A in an amount of only 0.5% (optimal modification variant) caused a more than 4% decrease in the degree of crystallinity. Increasing the amount of nanoclay in the bath to 7.5% caused a further decrease in the degree of fiber crystallinity, but not proportional to the amount of applied modifier. The determined values of the average size of crystallites do not show a specific tendency to change.

Using the WAXS method, it was additionally possible to assess the degree of dispersion of the aluminosilicate modifier in the fibers. Figure 14 shows the beginning of the diffraction curves for fibers treated with C15A and for pure C15A, covering the angle range corresponding to the characteristic montmorillonite peak, whose angular position is approximately 2.8°. According to Bragg's law, the position of this peak corresponds to the interlayer spacing of C15A equal to 3.2 nm. In PET 130 + 7.5% C15A samples, this peak shifts towards lower values of the diffraction angle, and the corresponding interlayer spacing of C15A increases to 4.7 nm. This indicates the intercalation of layers, associated with the interaction of montmorillonites with PET chains.

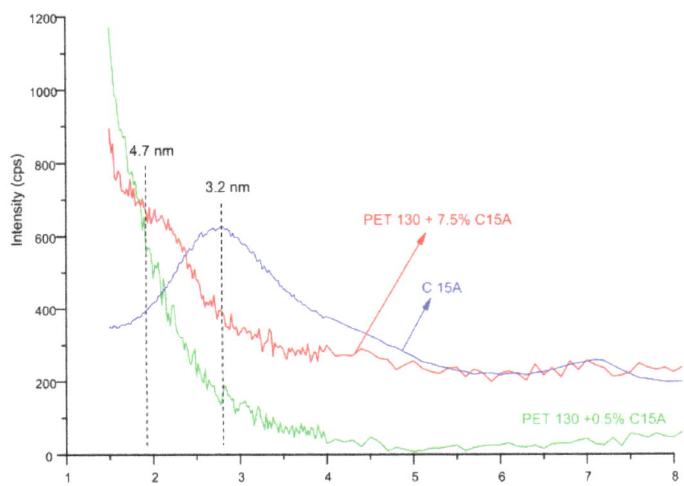

Figure 14. Comparison of the beginning of WAXS curves taken in the angular range corresponding to the characteristic montmorillonite (MMT) peak.

The characteristic montmorillonite peak disappears for the fibers with the optimal flame retardancy (PET + 0.50% of C15A). The WAXS curve of this sample, registered in the analyzed angular range is completely smooth, no local maximum or even inflection point was observed, the location of which would allow to determine the interlayer distance of montmorillonite. The reason for such a diffraction image should not be associated with the finding of the exfoliation of the modifier, but only with too low C15A concentration in the fiber. From the point of view of the WAXS resolution level, the concentration of C15A in the fiber was too low. Demonstrating C15A intercalation itself is already extremely important because it confirms that by using high-temperature treatment of PET fibers under increased pressure, it was possible to introduce a flame retardant in form of C15A to the PET fibers in the solid phase, forming a nanocomposite structure. Our separate work will be devoted to the diffraction and spectroscopic studies of the behavior of intercalated aluminosilicate under the conditions of melting and subsequent combustion of the fibers, and thus explaining the mechanism of flame retardancy.

In order to understand the distribution of nanoclay in modified fibers, SAXS research was carried out. Investigations performed independently in the direction parallel and perpendicular to the fiber axis indicated another interesting result (Figure 15). The SAXS curve taken in the direction parallel to the fiber axis does not contain the montmorillonite diffraction peak characteristic of C15A, whereas an intense and sharp peak is observed at a relatively small angle. This peak is connected with the lamellar structure of PET. On the other hand, the SAXS curve taken in the direction perpendicular to the fiber axis contains a shoulder which indicates the existence of the montmorillonite peak. These observations reveal that the lamellar structure of PET is perpendicular to the surfaces of montmorillonite layers.

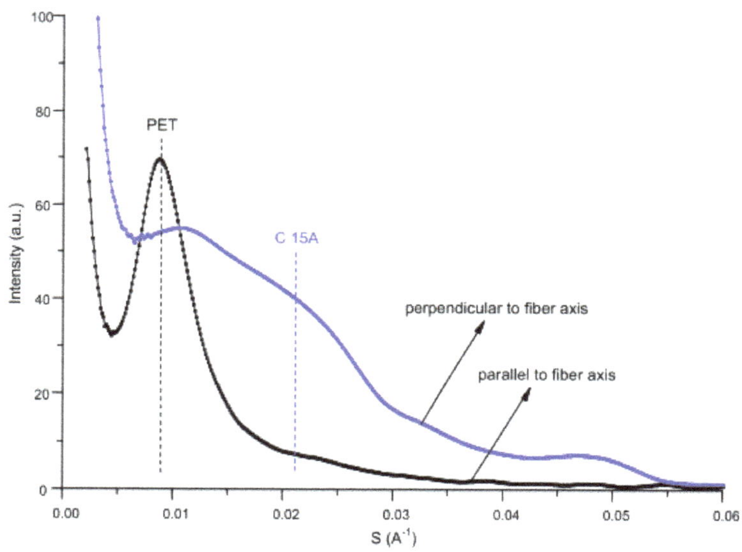

Figure 15. Comparison of small angle X-ray scattering (SAXS) curves taken in the direction parallel and perpendicular to the fiber axis for PET 130 + 7.5% C15A fibers.

4. Conclusions

The following main goals, set by the authors in the course of carrying out the described research, were achieved:

- The addition of C15A clay in the nanopowder form to PET fibers indeed changes the value of the limited oxygen index. The differences in values are not large, but they are undoubtedly significant. The highest LOI value (24.0%) was obtained for PET fibers modified in a bath containing 0.5% C15A to the fiber weight. In this case, the shift (approx. 8.5 °C)of the exothermic peak corresponding to the combustion process of the sample is observed.
- The montmorillonite in form of C15A nanomodifier was successfully introduced into the solid phase PET fiber structure during the modification process in the high-temperature bath, resulting

in improved flame retardancy. This was confirmed independently based on the results of LOI tests and TGA tests.

- The use of the TGA-FTIR combined technique has clearly demonstrated no increase in the toxicity of gaseous oxidative degradation products of modified fibers as compared to pure PET fibers, which potentially enables the use the proposed modifier instead of the forbidden halogenated flame retardants.
- The basic parameters of the fiber nanostructure were determined using diffraction methods and their nanocomposite nature was confirmed.

Author Contributions: Conceptualization, A.G. and J.F.; methodology, J.F., A.B.-S. and C.Ś.; formal analysis, J.F., C.Ś. and A.B.-S.; investigation, A.G., J.F., A.B.-S., C.Ś., M.R. and M.S.; resources, A.G., M.R. and M.S.; writing—original draft preparation, J.F., A.G. and M.R.; writing—review and editing, C.Ś. and M.R.; validation, A.G., M.R., M.S. and A.B.-S.; visualization, M.S. and M.R.; supervision, C.Ś.; project administration: J.F. All authors have read and agreed to the published version of the manuscript.

Funding: This research received no external funding.

Conflicts of Interest: The authors declare no conflict of interest.

References

1. Horrocks, A.R. Flame retardant challenges for textiles and fibres: New chemistry versus innovatory solutions. *Polym. Degrad. Stab.* **2011**, *96*, 377–392. [CrossRef]
2. Teli, M.D.; Kale, R.D. Polyester Nanocomposite Fibers with Improved Flame Retardancy and Thermal Stability. *Polym. Eng. Sci.* **2012**, *52*, 1148–1154. [CrossRef]
3. Didane, N.; Giraud, S.; Devaux, E.; Lemort, G.; Capon, G. Thermal and fire resistance of fibrous materials made by PET containing flame retardant agents. *Polym. Degrad. Stab.* **2012**, *97*, 2545–2551. [CrossRef]
4. Yang, S.-C.; Kim, J.P. Flame Retardant Polyesters. III. Fibers. *J. Appl. Polym. Sci.* **2008**, *108*, 2297–2300. [CrossRef]
5. Carosio, F.; Laufer, G.; Alongi, J.; Camino, G.; Grunlan, J.C. Layer-by-layer assembly of silica-based flame retardant thin film on PET fabric. *Polym. Degrad. Stab.* **2011**, *96*, 745–750. [CrossRef]
6. Carosio, F.; Di Blasio, A.; Cuttica, F.; Alongi, J.; Frache, A.; Malucelli, G. Flame Retardancy of Polyester Fabrics Treated by Spray-Assisted Layer-by-Layer Silica Architectures. *Ind. Eng. Chem. Res.* **2013**, *52*, 9544–9550. [CrossRef]
7. Alongi, J.; Ciobanu, M.; Tata, J.; Carosio, F.; Malucelli, G. Thermal Stability and Flame Retardancy of Polyester, Cotton, and Relative Blend Textile Fabrics Subjected to Sol–Gel Treatments. *J. Appl. Polym. Sci.* **2011**, *119*, 1961–1969. [CrossRef]
8. Kim, Y.-H.; Jang, J.; Song, K.-G.; Lee, E.-S.; Ko, S.-W. Durable Flame-Retardant Treatment of Polyethylene Terephthalate (PET) and PET/Cotton Blend Using Dichlorotribromophenyl Phosphate as New Flame Retardant for Polyester. *J. Appl. Polym. Sci.* **2001**, *81*, 793–799. [CrossRef]
9. Chen, D.-Q.; Wang, Y.-Z.; Hu, X.-P.; Wang, D.-Y.; Qu, M.-H.; Yang, B. Flame-retardant and anti-dripping effects of a novel char-forming flame retardant for the treatment of poly(ethylene terephthalate) fabrics. *Polym. Degrad. Stab.* **2005**, *88*, 349–356. [CrossRef]
10. Guo, D.-M.; Fu, T.; Ruan, C.; Wang, X.-L.; Chen, L.; Wang, Y.-Z. A new approach to improving flame retardancy, smoke suppression and anti-dripping of PET: Via arylene-ether units rearrangement reactions at high temperature. *Polymer* **2015**, *77*, 21–31. [CrossRef]
11. Mochane, M.J.; Luyt, A.S. Synergistic effect of expanded graphite, diammonium phosphate and Cloisite 15A on flame retardant properties of EVA and EVA/wax phase-change blends. *J. Mater. Sci.* **2015**, *50*, 3485–3494. [CrossRef]
12. Nazare, S.; Kandola, B.K.; Horrocks, A.R. Flame-retardant unsaturated polyester resin incorporating nanoclays. *Polym. Adv. Technol.* **2006**, *17*, 294–303. [CrossRef]
13. Rault, F.; Campagne, C.; Rochery, M.; Giraud, S.; Devaux, E. Polypropylene multifilament yarn filled with clay and/or graphite: Study of a potential synergy. *J. Polym. Sci. Part B Polym. Phys.* **2010**, *48*, 1185–1195. [CrossRef]

14. Eckel, D.F.; Balogh, M.P.; Fasulo, P.D.; Rodgers, W.R. Assessing Organo-Clay Dispersion in Polymer Nanocomposites. *J. Appl. Polym. Sci.* **2004**, *93*, 1110–1117. [CrossRef]
15. Gashti, M.P.; Moradian, S. Effect of Nanoclay Type on Dyeability of Polyethylene Terephthalate/Clay Nanocomposites. *J. Appl. Polym. Sci.* **2012**, *125*, 4109–4120. [CrossRef]
16. Gurmendi, U.; Eguiazabal, J.I.; Nazabal, J. Structure and Properties of Nanocomposites with a Poly(ethylene terephthalate) Matrix. *Macromol. Mater. Eng.* **2007**, *292*, 169–175. [CrossRef]
17. Ghasemi, H.; Carreau, P.J.; Kamal, M.R.; Tabatabaei, S.H. Properties of PET/Clay Nanocomposite Films. *Polym. Eng. Sci.* **2012**, *52*, 420–430. [CrossRef]
18. Chuang, C.S.; Tsai, K.C.; Yang, T.H.; Ko, C.H.; Wang, M.K. Effects of adding organo-clays for acrylic-based intumescent coating on fire-retardancy of painted thin plywood. *Appl. Clay Sci.* **2011**, *53*, 709–715. [CrossRef]
19. Devaux, E.; Rochery, M.; Bourbigot, S. Polyurethane/Clay and Polyurethane/POSS Nanocomposites as Flame Retarded Coating for Polyester and Cotton Fabrics. *Fire Mater.* **2002**, *26*, 149–154. [CrossRef]
20. Shahidi, S.; Ghoranneviss, M. Effect of Plasma Pretreatment Followed by Nanoclay Loading on Flame Retardant Properties of Cotton Fabric. *J. Fusion Energy* **2014**, *33*, 88–95. [CrossRef]
21. Grancarić, A.M.; Tarbuk, A. EDA Modified PET Fabric Treated with Activated Natural Zeolite Nanoparticles. *Mater. Technol. Adv. Perform. Mater.* **2009**, *1*, 58–63. [CrossRef]
22. Fabia, J.; Gawłowski, A.; Graczyk, T.; Ślusarczyk, C.Z. Changes of crystalline structure of poly(ethylene terephthalate) fibers in flame retardant finishing process. *Polimery* **2014**, *7–8*, 557–561. [CrossRef]
23. Gawłowski, A.; Fabia, J.; Graczyk, T.; Janicki, J. Method for Obtaining Flame Retardant Polyester and Polyamide Fibers Using an Aqueous Solution of Sodium Silicate. Patent Application No P.429004, 22 February 2019.
24. Gawłowski, A.; Fabia, J.; Graczyk, T.; Ślusarczyk, C.; Janicki, J.; Sarna, E. Study of PET fibres modified with phosphorus-silicon retardants. *J. Therm. Anal. Calorim.* **2016**, *125*, 1327–1334. [CrossRef]
25. Toda, T.; Yoshida, H.; Fukunishi, K. Amorphous structure changes in poly(ethylene terephthalate) induced by annealing under dry and wet conditions and its dye uptake properties. *Polymer* **1997**, *38*, 5463–5469. [CrossRef]
26. Mokhtar, N.M.; Lau, W.J.; Ismail, A.F.; Ng, B.C. Physicochemical study of polyvinylidene fluoride-Cloisite15A® composite membranes for membrane distillation application. *RSC Adv.* **2014**, *4*, 63367–63379. [CrossRef]
27. Gawłowski, A.; Fabia, J.; Ślusarczyk, C.; Graczyk, T.; Pielesz, A. Flame retardant modification of partially oriented PET fibers—Structural conditions of applications. *Polimery* **2017**, *11–12*, 848–854. [CrossRef]
28. Gilman, J.W. Flammability and thermal stability studies of polymer layered-silicate (clay) nanocomposites. *Appl. Clay Sci.* **2000**, *15*, 31–49. [CrossRef]
29. Bourbigot, S.; Bras, M.L.; Dabrowski, F.; Gilman, J.W.; Kashiwagi, T. PA-6 clay nanocomposite hybrid as char forming agent in intumescen formulations. *Fire Mater.* **2000**, *24*, 201–208. [CrossRef]
30. Gilman, J.W.; Jackson, C.L.; Morgan, A.B.; Harris, R., Jr. Flammability properties of polymer-layered-silicate nanocomposites. Polypropylene and polystyrene nanocomposites. *Chem. Mater.* **2000**, *12*, 1866–1873. [CrossRef]
31. Zhu, J.; Morgan, A.B.; Lamelas, F.J.; Wilkie, C.A. Fire properties of polystyrene-clay nanocomposites. *Chem. Mater.* **2001**, *13*, 3774–3780. [CrossRef]
32. Yano, K.; Usuki, A.; Okada, A.; Kurauchi, T.; Kamigaito, O. Synthesis and properties of polyimide-clay hybrid. *J. Polym. Sci. Part A Polym. Chem.* **1993**, *31*, 2493–2498. [CrossRef]
33. Kohlpaintner, C. Aldehydes, Aliphatic and Araliphatic. In *Ullmann's Encyclopedia of Industrial Chemistry;* Wiley-VCH: Weinheim, Germany, 2003; pp. 11–12. ISBN 978-3-527-30385-4.
34. Sovová, K.; Ferus, M.; Matulková, I.; Španěl, P.; Dryahina, K.; Dvořák, O.; Civiš, S. A study of thermal decomposition and combustion products of disposable polyethylene terephthalate (PET) plastic using high resolution Fourier transform infrared spectroscopy, selected ion flow tube mass spectrometry and gas chromatography mass spectrometry. *Mol. Phys.* **2008**, *9–10*, 1205–1214. [CrossRef]
35. Sur, G.S.; Sun, H.L.; Lyu, S.G.; Mark, J.E. Synthesis, structure, mechanical properties, and thermal stability of some polysulfone/organoclay nanocomposites. *Polymer* **2001**, *42*, 9783–9789. [CrossRef]
36. Feng, J.; Hao, J.; Du, J.; Yang, R. Using TGA/FTIR TGA/MS and Cone Calorimetry to Understand Thermal Degradation and Flame Retardancy Mechanism of Polycarbonate Filled with Solid Bisphenol A Bis(Diphenyl Phosphate) and Montmorillonite. *Polym. Degrad. Stab.* **2012**, *4*, 605–614. [CrossRef]

37. Daubeney, R.P.; Bunn, C.W.; Brown, C.J. The crystal structure of polyethylene terephthalate. *Proc. R. Soc. Lond.* **1954**, *226A*, 531–542.
38. Rabiej, M. Application of the particle swarm optimization method for the analysis of wide-angle X-ray diffraction curves of semicrystalline polymers. *J. Appl. Crystallogr.* **2017**, *50*, 221–230. [CrossRef]
39. Alexander, L.E. *X-Ray Diffraction Methods in Polymer Science*; Wiley: New York, NY, USA, 1969.

© 2020 by the authors. Licensee MDPI, Basel, Switzerland. This article is an open access article distributed under the terms and conditions of the Creative Commons Attribution (CC BY) license (http://creativecommons.org/licenses/by/4.0/).

polymers

MDPI

Article

Acoustic, Mechanical and Thermal Properties of Green Composites Reinforced with Natural Fibers Waste

Tufail Hassan [1], Hafsa Jamshaid [1,2,*], Rajesh Mishra [3,*], Muhammad Qamar Khan [1], Michal Petru [4,5], Jan Novak [5], Rostislav Choteborsky [3] and Monika Hromasova [3]

1 Faculty of Textile Engineering, National Textile University, Faisalabad 37610, Pakistan; tufailhassan12@gmail.com (T.H.); qamarkhan154@gmail.com (M.Q.K.)
2 Protective Textile Group, National Textile University, Faisalabad 37610, Pakistan
3 Faculty of Engineering, Czech University of Life Sciences Prague, Kamýcká 129, 165 00 Praha-Suchdol, Czech Republic; choteborsky@tf.czu.cz (R.C.); hromasova@tf.czu.cz (M.H.)
4 Department of Machinery Construction, Institute for Nanomaterials, Advanced Technology and Innovation, Technical University of Liberec, Studentska 2, 46117 Liberec, Czech Republic; michal.petru@tul.cz
5 Faculty of Mechanical Engineering, Technical University of Liberec, Studentska 2, 46117 Liberec, Czech Republic; jan.novak1@tul.cz
* Correspondence: hrntu@hotmail.com (H.J.); rkmily@gmail.com (R.M.)

Received: 7 February 2020; Accepted: 11 March 2020; Published: 13 March 2020

Abstract: The use of acoustic panels is one of the most important methods for sound insulation in buildings. Moreover, it has become increasingly important to use green/natural origin materials in this area to reduce environmental impact. This study focuses on the investigation of acoustic, mechanical and thermal properties of natural fiber waste reinforced green epoxy composites. Three different types of fiber wastes were used, e.g., cotton, coconut and sugarcane with epoxy as the resin. Different fiber volume fractions, i.e., 10%, 15% and 20% for each fiber were used with a composite thickness of 3 mm. The sound absorption coefficient, impact strength, flexural strength, thermal conductivity, diffusivity, coefficient of thermal expansion and thermogravimetric properties of all samples were investigated. It has been found that by increasing the fiber content, the sound absorption coefficient also increases. The coconut fiber-based composites show a higher sound absorption coefficient than in the other fiber-reinforced composites. The impact and flexural strength of the cotton fiber-reinforced composite samples are higher than in other samples. The coefficient of thermal expansion of the cotton fiber-based composite is also higher than the other composites. Thermogravimetric analysis revealed that all the natural fiber-reinforced composites can sustain till 300 °C with a minor weight loss. The natural fiber-based composites can be used in building interiors, automotive body parts and household furniture. Such composite development is an ecofriendly approach to the acoustic world.

Keywords: sound absorption coefficient; impact strength; flexural strength; coefficient of thermal expansion; bio-composite

1. Introduction

The word noise is derived from the Latin word "nausea" which means the sensation of discomfort. Noise is the unwanted sound, it may be loud, distracting or annoying. Noise pollution refers to the unwanted sound waves in the environment produced by humans that becomes a threat to the health of both human and animals. Noise has become the third largest source of pollution which causes diverse environmental effects on the economy and human health [1]. Advancement in technology and lifestyle has, in many ways, caused an increase in air, soil, water and noise pollution [2]. Sound absorption takes place when sound waves strike the surface of any porous material. Some of the sound is reflected back while some waves or energy are absorbed by the material [3]. Vehicular transportation is one of the major sources for producing noise pollution. Various types of materials are used to prevent noise pollution like wall cladding, acoustic barriers and acoustic ceilings which cover a wide range of applications [4]. Verma et al. described that composites, ceramics and plastics are vastly used for acoustic panels as compared to other materials. Traditionally, carbon and glass fiber-reinforced composites were used for acoustic absorbance. The advantage of synthetic fiber-reinforced composite is its high strength and mechanical properties which are more suitable to use in structural applications. The main problems associated with synthetic fiber-reinforced composites are the environmental aspects, e.g., production process, application and afterlife disposal. It is harmful to the environment because it is not biodegradable and made from nonrenewable resources [5].

The most important thing is to protect our environment from pollution, and it can be achieved without compromising the performance and quality of the product. The solution is to use biodegradable materials which are obtained from natural and renewable sources. Due to environmental concern, plant fiber-reinforced composites are receiving greater attention of researchers and industrialists because they are biodegradable, combustible and lightweight [6].

Traditionally, noise is controlled by non-biodegradable and expensive materials such as polymer fibers, glass wool, fabric fillers and polymer foams. Mineral fibers like mineral wool and glass fibers are widely used for the manufacturing of soundproofing materials in buildings and industries but both of them are harmful to the environment and risky to human health [7]. In the 1970s, public health regulations banned the use of carcinogenic materials, which enabled the transition of soundproofing materials from asbestos to synthetic fibers. These fibers are non-biodegradable, cause heavy pollution and have a high carbon footprint which is harmful to the environment. Along with synthetic fibers, wood is the most desirable candidate for industries to manufacture sound-absorbing materials with relatively good physical and mechanical properties. Due to its diverse characteristics, wood has a huge demand for various industries, which causes shortages in the supply of wood. A large number of trees are being cut annually for the manufacturing of acoustic panels causing unrecoverable damage to our ecosystem. In such scenarios, different industries, especially acoustic material industries, must begin to search for a suitable and environmentally friendly substitute that can fulfill the demand. Currently, the most commonly used materials for sound absorption are fiber-reinforced polymer composites, e.g., glass fibers composites. In addition, polystyrene, poly(vinyl chloride), polyurethane and poly(acrylic ester)s are some polymers that are widely used by the acoustic industries. These materials are either expensive or hazardous during normal use [8–10]. Sound absorption performance can be divided into several classes based on the sound absorption coefficients as shown in Table 1.

Table 1. Classes of sound absorption coefficient [11].

Sound Absorption Classes	Sound Absorption Coefficient Range
A	0.90–1.00
B	0.80–0.85
C	0.60–0.75
D	0.30–0.55
E	0.15–0.25
F	0.00–0.10

Recently, natural fiber-reinforced composites have received great attention from researchers and industrialists as a replacement of synthetic fiber-reinforced composite. They have relatively good mechanical and physical characteristics that can be used in various applications. Natural fibers are bio-degradable, nonabrasive, nonhazardous, lightweight and renewable materials. The most commonly used natural fibers as a reinforcement in composites are hemp, flax, luffa, banana, jute, sisal, sugarcane, ramie, betel nut, etc. [12]. Various studies show that natural fibers, e.g., tea leaf, rice-straw, coconut and kenaf fibers have a very good capability of sound absorption and are widely used in industries for manufacturing of sound-absorbing panels [13–17].

A lot of work has been done on natural fiber-reinforced composites so far by different researchers. Chen et al. investigated the sound absorption coefficient of ramie fiber-reinforced polylactic acid composites. They found that the sound absorption coefficient is ranged between 0.089 and 0.353 in a frequency range of 250–1600 Hz [18]. Wang established in his research that the sound absorption properties of rock wool are found to be similar to that of glass fibers [19].

Khusairy et al. investigated the acoustic properties of banana fiber-reinforced epoxy composites in the frequency range of 500–6000 Hz. The sound absorption coefficient was found to be 0.11. They further described that by changing fiber diameter, flow resistivity changes hence it causes a significant change in sound absorption coefficient [20]. Chen et al. described that the sound absorption coefficient of ramie fiber-reinforced polylactic acid is found to be 0.121 in the frequency range between 500 and 6000 Hz [18]. Bratu et al. used different waste materials, e.g., steelworks slag, fiberglass waste, wood waste, waste ash from the burning shells of oil seeds from plants as reinforcement in formaldehyde composite. They found excellent results of sound absorption coefficient (SAC) 0.8–0.9 in the frequency range of 400–3200 Hz [21]. Yang et al. investigated ramie, flax and jute reinforced epoxy composites. SAC of these fibers is found to be between 0.6 and 0.65 at a frequency range of 2000 Hz. They further described that multiscale surface structure and hollow lumen of natural fiber is mainly responsible for high SAC [1]. Abdullah et al. investigated sugarcane bagasse fiber and banana fiber-reinforced polyester resin composites. They found that the banana fiber composite has SAC 0.6835 while bagasse fiber has SAC 0.6338 at 4000 Hz frequency. They also prepared hybrid composites from these fibers and found that SAC is greater than the individual fiber component [22]. Jayamani et al. investigated the acoustic properties of rice-straw/polypropylene and kenaf/urea formaldehyde composites. They found that the SAC of rice-straw/polypropylene composite is 0.008 and kenaf/urea composite is 0.065, respectively, at 2000 Hz frequency [23].

Moretti et al. in 2016 investigated the acoustic and thermal properties of mineral fiber basalt in order to use it in the application of building and construction. The sound reverberation method was used for checking the acoustic properties. They found that with increasing frequency of incident sound waves, sound absorption properties also tend to improve. Further, they found that basalt fibers exhibit relatively good sound absorption and thermal insulation properties, hence they are best-suited materials for panels in building and construction [24].

In 2015, Jayamani et al. investigated the SAC of sisal fiber-reinforced polylactic acid composites. At a frequency of 2000 Hz, they show SAC of 0.085. They further found that by increasing fiber content, SAC also increases [25]. The same researchers, in another work, investigated the acoustic behavior of different natural fiber-reinforced epoxy composites. They used coconut/coir, sugarcane bagasse and kenaf fibers as reinforcement and found that the SAC of the coconut/coir reinforced composite is 0.078, sugarcane bagasse reinforcement is 0.075 and kenaf reinforcement 0.078, respectively, at 6000 Hz frequency. Coconut/coir reinforced composite shows higher SAC than the sugarcane bagasse composite because coconut fiber is more porous as compared to sugarcane bagasse which causes increased SAC [26].

Ricciardi et al. designed the acoustic panels manufactured from recycled materials like paper waste and textile fiber waste as reinforcement and bonded them with glue, making samples of 2–20 mm thickness. They measured SAC in frequency ranges of 100–6400 Hz through the transfer function impedance tube. They found that by increasing the thickness of samples, SAC also increases [27].

Jiang et al. investigated the seven-hole hollow polyester fibers (SHPF) reinforced chlorinated polyethylene (CPE) composites. It has been observed that for 3 mm thickness of a sample, SAC was 0.695 at frequency 2500 Hz. Further by increasing the fiber content SAC also increases [28]. Zhang et al. investigated different fibers, e.g., flax, carbon, glass, ramie and jute for reinforcement of epoxy (Bisphenol-A) composites. These fibers are used in the form of plain-woven fabric in the manufacturing of composites. It was found that the flax, ramie and jute reinforced composites have higher SAC than carbon and glass reinforced composite for a wider frequency range up to 10,000 Hz [29]. Zhang et al. investigated the acoustic properties of flax and basala wood reinforced composites at 250–10,000 Hz range. They found that the SAC of the of flax reinforced composites is 0.96 and that of the basala wood reinforced composite is 0.58 [30].

The use of wood and synthetic fiber/polymers composites for sound insulation applications is not environmentally friendly. Abundant cutting of trees is done for wood, whereas synthetic fibers/polymer composites have a non-degradability issue. Therefore, there is an urgent need to work in this area and develop relatively economic, biodegradable and environmentally friendly sound absorption material. It will greatly help the saving of trees and will reduce the use of non-biodegradable materials for acoustic applications. The overall impact will be to protect our environment from several pollution hazards including noise.

2. Materials and Methods

2.1. Materials

Three different types of fiber wastes, e.g., cotton fly, coconut/coir husk and sugarcane waste were used in the manufacturing of composite samples. Cotton fly was collected from a yarn manufacturing industry, coconut husk was purchased from a local market and sugarcane fibers were extracted from bagasse. The cross-sectional images of the fibers were taken by microscopy and shown in Figure 1.

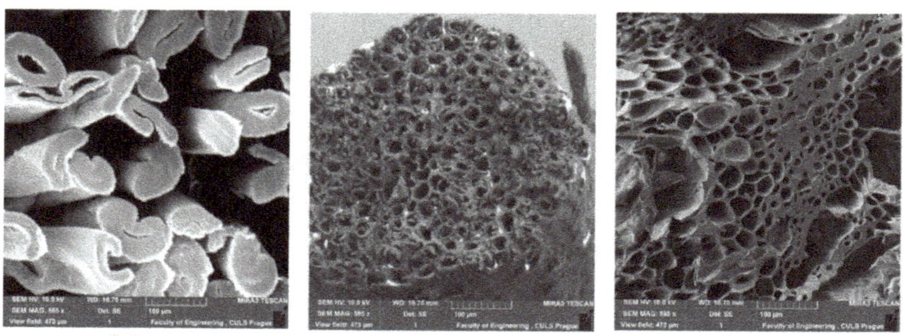

Cotton Coconut Sugarcane

Figure 1. Microscopic images of fiber cross-sections.

The properties and chemical composition of these fibers are given in Tables 2 and 3, respectively.

Table 2. Fiber properties.

Fiber Type	Fiber Fineness (Tex)	SD (±)	Fiber Length (mm)	Force at Break (cN)	SD (±)	Elongation at Rupture (mm)	SD (±)	Tenacity (cN/Tex)	SD (±)
Sugarcane	36	1.12	3–7	330.6	14.22	1.32	0.04	9.2	0.27
Coconut	32	0.83	3–9	360.6	14.43	4.52	0.13	11.2	0.56
Cotton Fly	0.19	0.007	12–15	2.70	0.11	3.5	0.09	14.2	0.66

SD: Standard Deviation.

Table 3. Chemical composition of fibers.

Fiber Type	Cellulose (%)	Hemicellulose (%)	Lignin (%)	Pectin (%)	Wax (%)
Sugarcane	28.3–50	20–36.3	21.2–24	N/A	0.9
Coconut	19.9–55	11.9–15.4	32.7–53.3	4.7–7	N/A
Cotton	82.7	5.7	28.2	5.7	0.6

Green epoxy resin CHS-G530 (new commercial name EnviPOXY®530) was purchased from company SPOLCHEMIE in Ústí nad Labem, Czech Republic. It has low molecular weight and contains low crude oil content and high renewable content. Chemically it contains 4,4′-Isopropylidenediphenol, oligomeric reaction products with 1-chloro-2-3-epoxypropane. It is called "green" due to obtaining 1-chloro-2-3-epoxypropane (epichlorohydrine) from glycerine originating from waste generated at biodiesel production. Its properties are given in Table 4.

Table 4. Properties of green epoxy resins.

Properties	Value
Physical State	Liquid at 20 °C
Color	Yellowish to Yellow
Boiling Point	270–280 °C (at very low pressure)
Density (g/cm^3)	1.16 at 20 °C
Water Solubility (g/L)	6–9 at 20 °C
Viscosity (Poise)	8–10 at 25 °C
Solubility	Soluble in Acetone
Storage Temperature	5–25 °C
Epoxide Index (mol/kg)	5.4–5.7
Mass Equivalent of Epoxide, EEW (g/mol)	176–186
Color (Hz, G*)	Max. 100
Hydrolyzable Chlorine Content (%)	Max. 0.03
Non-Volatile Substances (2 h/140 °C)	Min. 99.5%

2.2. *Methods*

2.2.1. Fiber Preparation

In the case of sugarcane, the waste of sugarcane bagasse was collected from a juice extractor. Joints were removed and treatment of the bagasse was done by adding 4 g/L sodium hydroxide in water. The bagasse was immersed in NaOH solution for three hours. The fibers were separated from treated sugarcane manually and cut into a short length of approximately 3–7 mm. Coconut fibers were opened and then cut into short lengths ranging from 3 to 9 mm. Cotton fly having a length of 12–15 mm fiber was used for the manufacturing of composite.

2.2.2. Samples Preparation

In order to develop composite samples, a mold of 20 cm × 20 cm (length × width) and 3 mm thickness was prepared. Fibers were uniformly distributed within the boundaries of the mold. In order to ensure uniform fiber distribution, the spreading was done layer by layer so that the averaging effect will minimize the variation of mass. This method was especially beneficial at low fiber content. No visible voids were found in the prepared samples. A mixture of epoxy and hardener (CHS-HARDENER P11) was prepared as per manufacturer (SPOLCHEMIE) guidelines with a ratio of 100:32 and stirred well for uniform mixing. The mixture was dispensed over the fibers in the mold very carefully to ensure uniform distribution of epoxy throughout the samples. A Teflon sheet was placed on both sides of the material in order to keep them in hydraulic press machine model TDF-110 under 100 bar pressure for 30 min at 80 °C. It was followed by curing of the sample at room temperature. In total, nine types of samples were developed by repeating the same procedure with fiber content 10%, 15% and 20% for all three types of fibers. The samples were cut in the size for different characterization as per standard requirement (Figure 2).

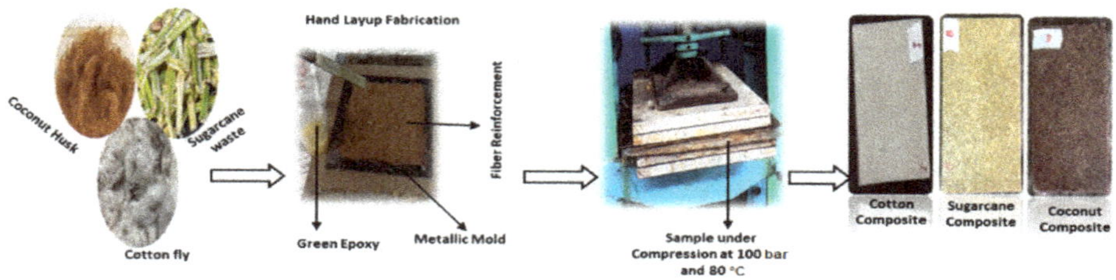

Figure 2. Illustration scheme for fabrication of composite samples.

2.3. Characterization

2.3.1. Acoustic Testing

In order to measure the sound absorption coefficient (SAC), two microphone impedance tube (Brüel & Kjær, model 4206, Nærum, Denmark) with a diameter of 100 mm was used in a frequency range 0 Hz to 1600 Hz as per ASTM E1050-08 standard. The samples were cut with a 140 mm diameter. Ten measurements were carried out and the average was reported. Graphs were plotted for SAC with respect to fiber content and fiber type.

2.3.2. Impact Test

To investigate the impact properties of the samples, Charpy impact tester from Zwick/Roell HIT 50P, Ulm, Germany were used according to ISO-179-1 standard. Samples were cut into size 80 mm × 10 mm for testing. Ten measurements were carried out and the average was reported.

The thickness and width of the samples were measured by Vernier caliper before the test. Specimens were placed on the specific slot and the pendulum was allowed to impact with 50 J energy in order to hit and break the specimen. The absorbed energy was recorded, and impact strength was calculated using Equation (1)

$$\partial cu = WB/bh \times 103, \tag{1}$$

where WB, is the energy at break, in joules; b, is the width of the specimens in millimeters; h, is thickness of the specimens in millimeters.

2.3.3. Flexural Strength Testing

In order to investigate the flexural strength of composites, the samples were cut into size 13 mm × 120 mm and the test was performed by universal testing machine Zwick/Roell Z100 by ASTM D 7264 standard. The gauge length of 80 mm, extension rate of 1 mm/min and load of 5 kN was used. The flexural strength was calculated by the Equation (2)

$$\sigma = 3PL/2bh^2, \tag{2}$$

where P represents load; L represents gauge length; b represent width; h represents thickness. Ten measurements were carried out and the average was reported.

2.3.4. Light Microscopy

The light microscope MIC-209 was used to capture images of impact broken samples with 20% fiber content from all three types of composites. The images were taken to check the mode of fiber failure, i.e., slippage or rupture of fibers in composites during impact testing.

2.3.5. Thermal Properties

C-Therm thermal conductivity analyzer (TCi) was used to measure the thermal properties like conductivity, resistance and diffusivity of the samples. The C-Therm (Tci) thermal conductivity analyzer allows determining accurate values for thermal conductivity and thermal effusivity of materials without extensive sample preparation or damage to the sample.

This highly accurate technique is based on the transient plane source (TPS) method. The primary difference between the traditional and modified transient plane source techniques is that the modified method offers a single-side interface compared to the double-sided interface requirements of the traditional version.

The modified transient plane source (MTPS) technique has many advantages in comparison to other available testing methods, e.g., guarded hot plate, hot wire or hot probe. The non-invasive nature of the C-Therm TCi's MTPS sensors allows testing of materials of any size in-situ or in laboratories without destruction of the specimen. Moreover, testing can be done in seconds with consistent and accurate results.

The TCi consists of a sensor, power control device and computer software. A spiral-type heating source is located at the center of the sensor where heat is generated. The generated heat enters the material through the sensor during which a voltage drop occurs rapidly at the heating source. The thermal conductivity is calculated through the voltage drop data. Before conducting the measurements, all samples were conditioned at standard atmospheric conditions (25 ± 2 °C, 65% ± 2% RH) for 24 h. The average of ten measurements for each sample was taken and the mean values of the thermal properties were calculated. The tested data were statistically analyzed.

2.3.6. Linear Coefficient of Thermal Expansion

The increase or decrease in the length of any material by changing temperature is represented by the linear coefficient of thermal expansion (LCTE). Measurement of LCTE is very important in order to know the dimensional changes of materials at different temperatures and one can decide the specific end-use of a material in terms of temperature condition. Dilatometer DIL 2010STD (Orton, Westerville, OH, USA) was used to measure the LCTE of specimens with high sensitivity using the ASTM E831 standard. The setup consists of three main components:

- Sample holder and push rod;
- Furnace or Chamber;
- Linearly Variable Differential Transformer (LVDT) for measuring changes in dimensions.

The sample holder and push rod support the thermocouple and the sample. Different varieties of the sample holder are available like silica or alumina etc. Furnace provides heat to the specimens in vacuum. LVDT is used as a measuring head which measures the dimensional changes of the specimen with respect to temperature. Samples were cut into size 2.54 cm × 1 cm. A temperature range of 40–120 °C was used during the test with a 5 °C/min increasing rate.

2.3.7. Thermogravimetric Analysis

To investigate the thermal stability of the samples, NETZSCH TG 209F1 Libra instrument TGA having crucible weight 139 g (NETZSCH-Gerätebau GmbH, Selb, Germany) and sample weight 2 g was used. Weight loss with increasing temperature was measured in a temperature range of 25–300 °C with an increasing rate of 10 °C/min.

3. Results and Discussion

3.1. Acoustic Properties

3.1.1. Effect of Fiber Type on Sound Absorption Coefficient (SAC)

The effect of different types of fiber reinforcement in composites on the sound absorption coefficient was studied. The main aim of the current research is to develop green composite materials from natural fiber waste which is impregnated with green epoxy resin. The targeted products are to be used in lightweight construction, indoor panels and automotive components. Thus, the mechanical, thermal and stability at higher temperature are essential properties which are investigated.

In all the above applications, there is an essential requirement of sound absorption. Therefore, a composite panel with some acoustic performance is an added advantage. It may be pointed out that the prepared composite panels have some sound insulation as compared to the pure green epoxy panel which shows absolutely no absorption at all. The increase of SAC with the increase of fiber volume fraction is demonstrated.

The results revealed that the coconut fiber-reinforced composites show the highest SAC followed by sugarcane fiber-reinforced composites and cotton fiber-reinforced composites, respectively, as shown in Table 5 and Figure 3d.

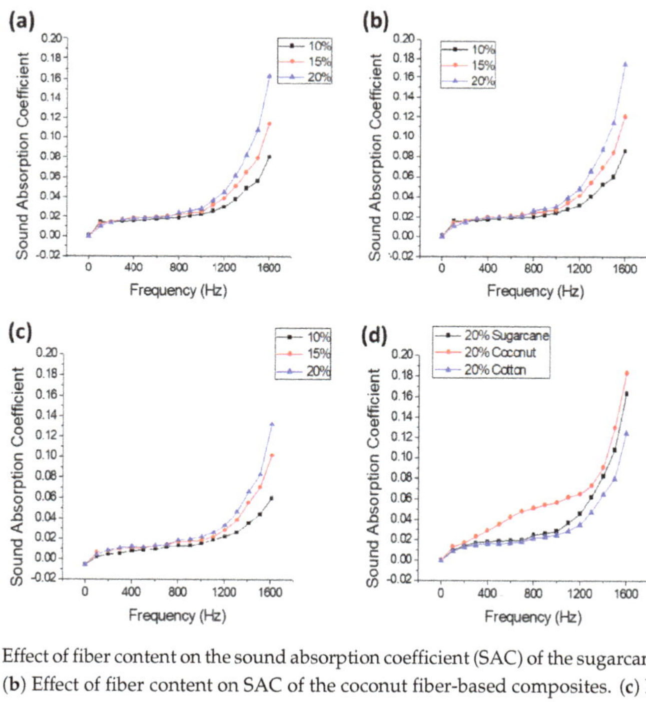

Figure 3. (**a**) Effect of fiber content on the sound absorption coefficient (SAC) of the sugarcane fiber-based composites. (**b**) Effect of fiber content on SAC of the coconut fiber-based composites. (**c**) Effect of fiber content on SAC of the cotton fiber-based composites. (**d**) Effect of fiber type on acoustic properties of composites with 20% fiber content.

The sugarcane fiber-reinforced composites have a higher SAC than the cotton fiber-reinforced composites; this is because the SAC of the fibers mainly depends upon their porosity. Large hollow lumen structure of fibers has the ability to absorb sound more efficiently than small lumen structures. It must be noted that the overall hollow lumen of the coconut fiber is larger than of sugarcane and cotton. Sugarcane fibers have a larger cumulative lumen diameter than cotton lumen diameter. Previously, researchers investigated and established that larger lumen structure has the ability to absorb more sound than small lumen structure. Overall, the SAC of materials under the present study is relatively lower and therefore, they can be used as good sound blockers rather than absorbers.

Table 5. Sound absorption coefficient of samples at 1600 Hz frequency.

Reinforcement Type	Fiber Volume Fraction (%)	Fiber Weight Fraction (%)	SAC at 1600 (Hz)	SD (±)
Pure Green Epoxy	0	0	0.000	0.000
Sugarcane	10	4.75	0.081	0.002
	15	7.13	0.115	0.004
	20	9.50	0.163	0.005
Coconut	10	6.12	0.090	0.002
	15	9.18	0.130	0.005
	20	12.24	0.183	0.006
Cotton	10	14.25	0.059	0.001
	15	21.38	0.097	0.004
	20	28.50	0.125	0.006

SD: Standard Deviation.

3.1.2. Effect of Fiber Content on Sound Absorption Coefficient

The effect of fiber content on the sound absorption coefficient was studied in the frequency range of 0–1600 Hz. A systematic increase in SAC is observed with an increase of fiber content as shown in Table 5 and Figure 3a–c. It is due to the fact that higher fiber volume fraction allows the sound waves to pass through a more tortuous path causing higher friction, thus higher energy loss takes place. Furthermore, natural fibers have a porous surface and internal structure with a hollow lumen which helps in the dissipation of sound energy [31].

3.2. Mechanical Properties

3.2.1. Effect of Fiber Type on Impact Strength of Composites

The ability of the material to resist fracture when load is applied at high speed is known as impact strength. In composites with epoxy resin, the curing process enables the formation of a complex 3D network which increases the mechanical performance substantially. The impact properties of the composites with cotton, coconut and sugarcane fibers have been investigated. There is a substantial improvement of impact strength as compared to that of pure green epoxy. The results revealed that the cotton fiber-reinforced composites have relatively higher impact strength than the sugarcane and coconut fiber-reinforced composites as shown in Table 6 and Figure 4.

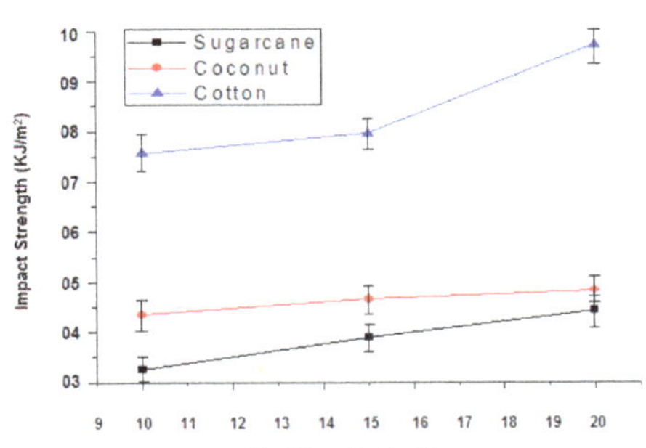

Figure 4. Effect of fiber type and fiber content on the impact properties of the sugarcane, coconut and cotton fiber-reinforced composites.

Table 6. Impact strength of samples.

Reinforcement Type	Fiber Volume Fraction (%)	Fiber Weight Fraction (%)	Impact Strength (KJ/m^2)	SD (±)
Pure Green Epoxy	0	0	2.75	0.12
Sugarcane	10	4.75	3.25	0.15
	15	7.13	3.89	0.17
	20	9.50	4.43	0.18
Coconut	10	6.12	4.35	0.18
	15	9.18	4.66	0.17
	20	12.24	4.83	0.22
Cotton	10	14.25	7.57	0.28
	15	21.38	7.96	0.37
	20	28.50	9.73	0.38

SD: Standard Deviation.

It should be noted that tenacity of the cotton fiber is higher than the sugarcane and coconut fiber as shown in Table 2. Previously researchers found that high tenacity and long fiber length leads to an increase in impact strength [32]. Further, the coconut fiber-reinforced composites have higher impact strength than the sugarcane fiber-based composites because the tenacity of the coconut fiber is greater than that of the sugarcane fiber.

3.2.2. Effect of Fiber Content on Impact Strength of Composites

Results revealed that when increasing fiber content, impact strength also increases (Figure 5). It has been observed that 10%, 15% and 20% of the sugarcane fiber content shows an impact strength of 3.25 kJ/m^2, 3.89 kJ/m^2 and 4.43 kJ/m^2, respectively. Overall, a 36% increase in impact strength is observed with the increase of fiber content from 10% to 20%. Similarly, an 11% increase in impact strength is observed, when the coconut fiber content is increased from 10% to 20%. In the case of the cotton fiber-reinforced composites, there is a 28% increase in impact strength when the fiber content increases from 10% to 20%. In a fiber-reinforced composite system, the reinforcement is the main constituent which is responsible to bear the impact load. Similar observations have been made by several other researchers [33,34].

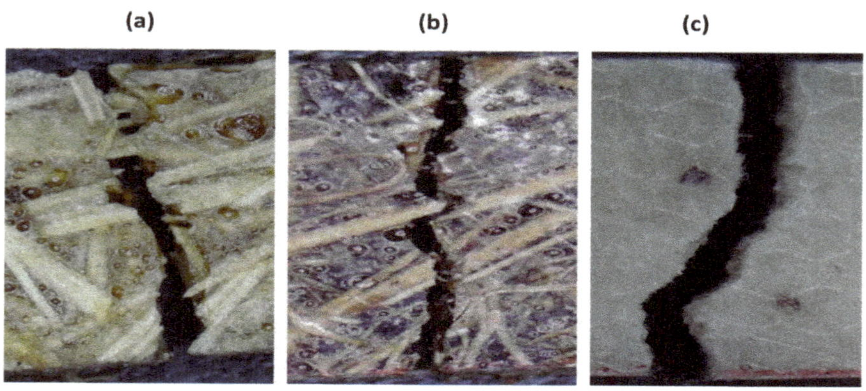

Figure 5. Microscopic images of impact broken samples: (**a**) sugarcane (**b**) coconut (**c**) cotton fiber-reinforced composites.

3.2.3. Effect of Fiber Type on Flexural Strength of Composites

The ability of a material to resist bending deformation under load is called flexural strength. It mainly depends upon the type of reinforcement and matrix. The results of the study revealed that the flexural strength of the cotton fiber-reinforced composite is higher than the sugarcane fiber and coconut fiber composites as shown in Table 7 and Figure 6. It can be justified based on the fact that the tenacity of the cotton fibers is higher than the coconut and sugarcane fibers. It is also well known that bending rigidity is directly proportional to the tensile modulus. Researchers have previously reported that by increasing the tenacity of the fiber, the flexural strength of the composites can be increased [35,36].

Table 7. Flexural properties of samples.

Reinforcement Type	Fiber Volume Fraction (%)	Fiber Weight Fraction (%)	Flexural Strength (MPa)	SD (±)
Pure Green Epoxy	0	0	15.4	0.76
Sugarcane	10	4.75	16.6	0.77
	15	7.13	19.3	0.79
	20	9.50	29.7	1.03
Coconut	10	6.12	29.2	1.12
	15	9.18	35.6	1.34
	20	12.24	39.9	1.50
Cotton	10	14.25	71.9	2.98
	15	21.38	77.6	3.25
	20	28.50	81.7	4.03

SD: Standard Deviation.

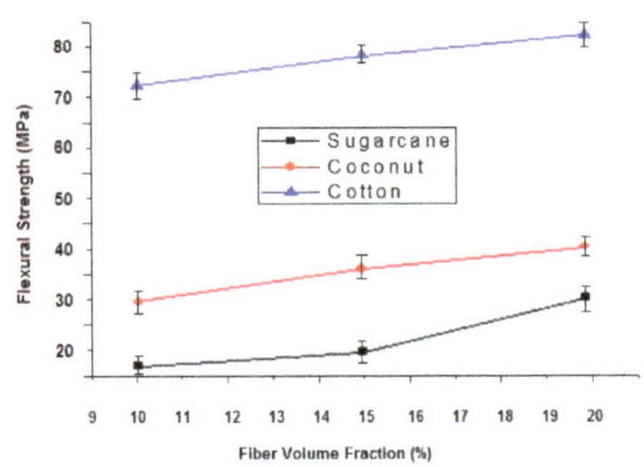

Figure 6. Effect of fiber type and fiber content on flexural strength of the sugarcane, coconut and cotton fiber-reinforced composites.

3.2.4. Effect of Fiber Content on Flexural Strength of Composites

It has been observed that when increasing fiber content, flexural strength also increases. The flexural rigidity has a direct proportionality with the modulus of elasticity or tenacity. The fiber reinforcement is the major load-bearing element in a composite system. By increasing the fiber volume fraction (FVF) from 10% to 20%, there is a substantial increase in flexural strength. FVF can be further increased by impregnation using the infusion method. As the set up used in the current investigation was a relatively simpler one, there was a restriction to maximum FVF and preparing void-free samples. Previously, researchers have found a significant increase in flexural strength by increasing fiber content in composites [37,38].

3.3. Thermal Properties

3.3.1. Effect of Fiber Type and Fiber Content on Thermal Conductivity, Resistance and Diffusivity

The thermal properties evaluated by the C-Therm thermal conductivity analyzer (TCi) are given in Table 8.

Table 8. Thermal properties of samples.

Reinforcement Type	Fiber Volume Fraction (%)	Fiber Weight Fraction (%)	Thermal Conductivity ($W \cdot m^{-1} \cdot K^{-1}$)	SD (±)	Thermal Diffusivity (10^{-6} $m^2 \cdot s^{-1}$)	SD (±)	Thermal Resistance ($K \cdot m^2 \cdot W^{-1}$)	SD (±)
Pure Green Epoxy	0	0	0.435	0.02	0.262	0.012	0.0069	0.00021
Sugarcane	10	4.75	0.414	0.03	0.275	0.012	0.0072	0.00026
	15	7.13	0.378	0.01	0.277	0.013	0.0079	0.00033
	20	9.50	0.342	0.02	0.287	0.015	0.0088	0.00037
Coconut	10	6.12	0.373	0.01	0.265	0.011	0.0080	0.00032
	15	9.18	0.312	0.01	0.269	0.012	0.0096	0.00048
	20	12.24	0.303	0.01	0.274	0.012	0.0120	0.00007
Cotton	10	14.25	0.421	0.02	0.266	0.013	0.0071	0.00029
	15	21.38	0.386	0.02	0.270	0.014	0.0078	0.00042
	20	28.50	0.373	0.01	0.276	0.013	0.0083	0.00038

SD: Standard Deviation.

Thermal Conductivity

Thermal conductivity, λ, is a measure of the rate at which heat is transferred through the unit area of the sample across unit thickness under a specified temperature gradient and thus is defined by the relation below.

$$\lambda\left[Wm^{-1}K^{-1}\right] = \frac{Q}{Ft\frac{\Delta T}{h}}, \tag{3}$$

$\lambda\left(Wm^{-1}K^{-1}\right) = \frac{Q}{F\tau\frac{\Delta T}{h}}$ where Q is the amount of conducted heat, F is the area through which heat is conducted, t is the time of heat conduction, ΔT is the drop of temperature and h is the sample thickness.

The thermal conductivity is a material intrinsic property and dependent on its porosity (volume of air content). The higher the density, the higher the conductivity. The pure epoxy resin shows a higher conductivity as compared with the fiber-reinforced composites. It is due to the fact that all fibers have a significant volume of air entrapped inside their structure. The comparison of the thermal conductivity of the reinforced composites is shown in Figure 7.

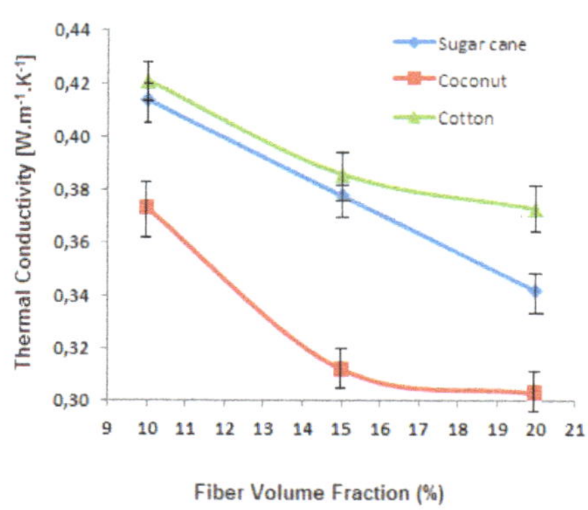

Figure 7. Effect of fiber type and fiber content on thermal conductivity of the sugarcane, coconut and cotton fiber-reinforced composites.

It is observed that thermal conductivity decreases with increasing content of the fiber reinforcement. It is because of the fact that the fibers have lower conductivity themselves in comparison with the matrix (green epoxy resin and hardener). Among the three types of fibers, cotton fiber-reinforced composites show higher conductivity due to higher fiber density and relatively lower porosity of the cotton fibers as compared to coconut and sugarcane. Further, the porosity in coconut fibers is much higher due to the overall lumen area which is responsible for the lowest thermal conductivity.

Thermal Diffusivity

Thermal diffusivity describes the rate of heat spread through a material. The fiber-reinforced composite panels show improved diffusivity in comparison to pure green epoxy resin. The comparison of thermal diffusivity for the reinforced composites is shown in Figure 8.

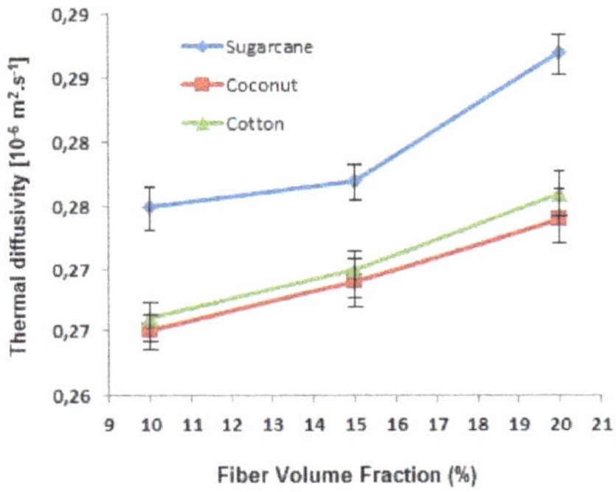

Figure 8. Effect of fiber type and fiber content on thermal diffusivity of the sugarcane, coconut and cotton fiber-reinforced composites.

It is found that the composites improve diffusivity due to the fibrous reinforcement. A higher volume fraction of fibers leads to an increase in thermal diffusivity. Among the fiber types investigated, the sugarcane-based composite panels diffuse heat more significantly in comparison with coconut and cotton-based composites. This may be attributed to the chemical composition of sugarcane which is rich in hemicellulose and lignin.

Thermal Resistance

Thermal resistance is defined as the ratio of the temperature difference between the two faces of a material to the rate of heat flow per unit area. Thermal resistance determines the heat insulation property of a material. The higher the thermal resistance, the lower is the heat loss. The thermal resistance, R, is connected with the thermal conductivity, λ, and the thickness, h, as follows.

$$R\left[m^2K^{-1}W^{-1}\right] = \frac{h}{\lambda}. \quad (4)$$

Thus, thermal resistance is critically dependent on thickness and thermal conductivity. As all the composite panels have almost the same thickness, the resistance is inversely proportional to the conductivity. The fiber-reinforced composites have higher thermal resistance as compared to pure epoxy resin. The comparison of thermal resistance for the reinforced composite panels is shown in Figure 9.

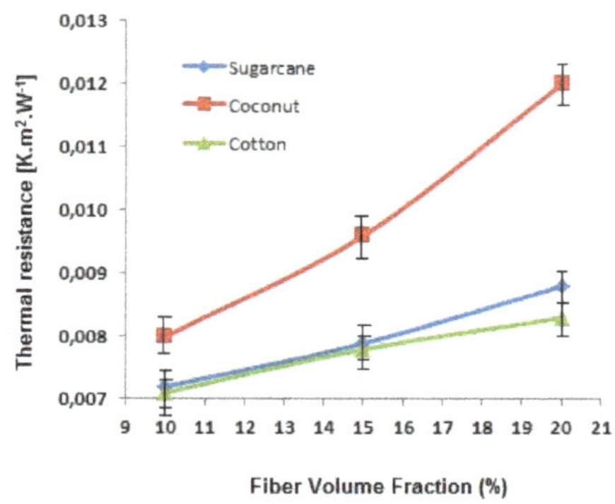

Figure 9. Effect of fiber type and fiber content on thermal resistance of the sugarcane, coconut and cotton fiber-reinforced composites.

It is observed that thermal resistance increases linearly when increasing the fraction of the reinforcing fiber. Among the three types of fibers used, coconut fiber has a relatively higher porosity and thus the lowest thermal conductivity. This leads to the highest thermal resistance among the fiber types investigated.

3.3.2. Effect of Fiber type on Coefficient of Thermal Expansion in Composites

Thermal expansion is basically the property of a material to change its length, area, volume or shape due to an increase in temperature. By increasing temperature, the average kinetic energy of the molecules increases, and the molecular vibration also increases. The relative expansion of material divided by the temperature is known as the coefficient of thermal expansion. It has been observed that the thermal expansion coefficient of the cotton fiber-reinforced composite is higher than both the sugarcane and coconut fiber composite as shown in Table 9 and Figure 10. The coconut fiber composite has a higher coefficient of thermal expansion (CTE) than the sugarcane fiber composite. It might be based on the chemical composition of the constituent fibers. The percentage of cellulose content in cotton is higher than the coconut and sugarcane fibers as shown in Table 3. Cellulose inherently has a linear structure and could thermally expand to a greater extent than hemicellulose and lignin. The structure of lignin contains aromatic rings which increase thermal stability. Researchers have investigated the thermal degradation of cellulose, hemicellulose and lignin of sugarcane bagasse. It is found that the thermal stability of lignin is higher than of cellulose and hemicellulose. Furthermore, the thermal stability of hemicellulose is higher than of cellulose [39].

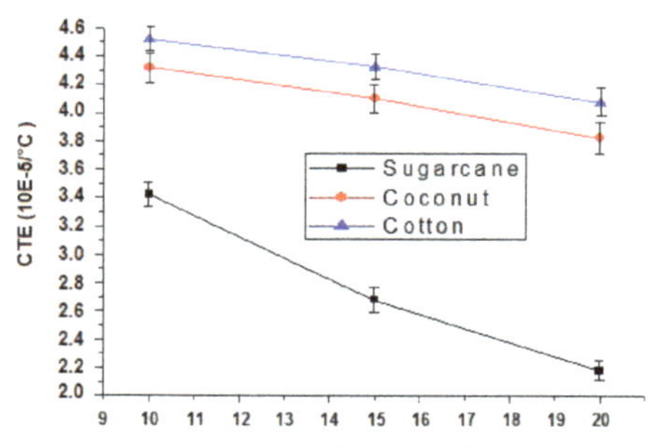

Figure 10. Effect of fiber type and content on thermal expansion coefficient of composites.

Table 9. Coefficient of thermal expansion of developed samples.

Reinforcement Type	Fiber Volume Fraction (%)	Fiber Weight Fraction (%)	CTE (10^{-5}/°C)	SD (±)
Pure Green Epoxy	0	0	6.342	0.32
Sugarcane	10	4.75	3.422	0.16
	15	7.13	2.681	0.14
	20	9.50	2.182	0.12
Coconut	10	6.12	4.323	0.24
	15	9.18	4.114	0.16
	20	12.24	3.832	0.18
Cotton	10	14.25	4.519	0.28
	15	21.38	4.333	0.28
	20	28.50	4.084	0.18

SD: Standard Deviation.

3.3.3. Effect of Fiber Content on Coefficient of Thermal Expansion of Composites

It has been observed that, in all cases, increasing the fiber content decreases the coefficient of thermal expansion as shown in Table 9 and Figure 10. It is due to the fact that the thermal expansion of epoxy (matrix) is much higher than fibers (reinforcement) therefore by increasing fiber content (FVF), results in decreasing in CTE [40,41]. By lowering the coefficient of thermal expansion, the composite materials will deform to a relatively smaller extent in high-temperature applications.

3.3.4. Thermogravimetric Analysis (TGA)

The weight loss of composite as a function of increasing temperature was measured by TGA. Untreated fiber contains hemicellulose and lignin thus it can store water. It has been observed that weight loss starts from 80 to 115 °C which indicates dehydration of fibers as shown in Figure 11. The results revealed that the cotton, coconut and sugarcane fiber-reinforced composites show initial degradation and weight loss of 1.5%, 2.8% and 2.3%, respectively. There is no significant weight loss in the samples at a temperature range of 115 to 270 °C. Significant weight loss is observed at 270–300 °C, which corresponds to the degradation of the matrix. Furthermore, the selected reinforcement fibers contain cellulose and it starts degradation around 270 °C [35]. The coconut, sugarcane and cotton fiber-reinforced composites show a relatively smaller overall weight loss of 7%, 5.8% and 3%, respectively, at 300 °C. It must be pointed out that the fiber-reinforced composites show lower weight loss as compared to the pure green epoxy which shows a 9.8% weight loss. The thermal stability of the fiber-reinforced composites is sensitive to the weight content of reinforcing fibers, rather than their volume content. At the same volume fraction, the weight content is dependent on the density of constituent fibers. As the cotton fiber has a higher density as compared to coconut and sugarcane, the weight fraction is substantially higher and the effect is clearly reflected in TGA results. The weight loss is inversely proportional to the cellulose content in these fiber types. The non-cellulosic constituents degrade at relatively lower temperatures resulting in higher weight loss in coconut and sugarcane fiber-reinforced composites. However, cotton fiber-reinforced composites can survive for a relatively higher temperature range due to a high content of cellulose which is thermally more stable [40,41].

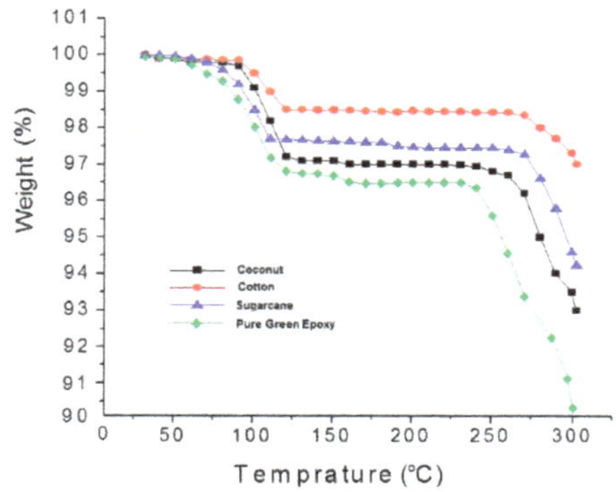

Figure 11. TGA analysis of the cotton, coconut and sugarcane reinforced epoxy composite with a 15% fiber volume fraction (FVF).

4. Conclusions

The effect of different cellulosic fibrous waste and their content (FVF) on acoustic, mechanical and thermal properties of green/bio epoxy composites has been investigated. The results reveal that an increase in fiber content tends to increase the sound absorption coefficient. Compared to pure epoxy resin, the SAC of the fiber-reinforced composites is higher. As most of the natural cellulosic fibers are highly porous in nature, they have much higher SAC compared to the matrix phase, therefore SAC increases with increasing fiber content. Among the samples investigated, coconut/coir fiber-reinforced composites show the highest sound absorption coefficient followed by sugarcane and cotton fiber-reinforced composites for 10%, 15% and 20% fiber content based on overall porosity of the constituent fibers. Fiber porosity has a direct relation with sound absorption.

Impact strength, as well as flexural strength, increases by increasing fiber content in all the cellulosic fiber types because the fiber/reinforcement phase is the main load-bearing constituent in a composite system. Among the samples investigated, cotton fiber-reinforced composites have the highest impact and flexural strength followed by coconut and sugarcane fiber composites. It is due to higher tenacity of the cotton fiber as compared to both coconut fiber and sugarcane fiber.

An increase in fiber content decreases the thermal conductivity and thus increases thermal resistance. It is because of the fact that the fibers have lower conductivity themselves in comparison with the matrix (green epoxy resin and hardener). Among the three types of fibers, cotton fiber-reinforced composites show higher conductivity due to higher fiber density and relatively lower porosity of the cotton fibers as compared to coconut and sugarcane. Further, the porosity in coconut fibers is much higher due to the overall lumen area which is responsible for the lowest thermal conductivity. It is found that the composites improve thermal diffusivity due to the fibrous reinforcement. A higher volume fraction of fibers leads to an increase in thermal diffusivity. Among the fiber types investigated, the sugarcane-based composite panels diffuse heat more significantly in comparison with coconut and cotton-based composites. This may be attributed to the chemical composition of sugarcane which is rich in hemicellulose and lignin.

The coefficient of thermal expansion (CTE) decreases with an increase in fiber content because the CTE of epoxy resin is much higher than the reinforcing cellulosic fibers. Furthermore, cotton fiber-reinforced composites show higher CTE compared to coconut and sugarcane fibers due to high cellulose content. Cellulose is thermally more sensitive than hemicellulose and lignin. Thermogravimetric analysis reveals that the composite with 15% of the coconut, sugarcane and cotton fiber content show 7%, 5.8% and 3% weight loss, respectively, at 300 °C, while pure green epoxy shows a weight loss of 9.8%.

Based on the performance analysis of the samples in the current study, coconut fiber-reinforced composites are the most suitable materials among all the three types investigated as far as acoustic performance is concerned. On the other hand, higher tenacity of the cotton fiber enables it to be the best reinforcement when mechanical performance, e.g., impact strength and flexural strength are desired. Further, the higher cellulose content in the case of cotton proves to be thermally more stable (minimum weight loss) as compared to the coconut and sugarcane fiber-based counterparts, especially at elevated temperatures. However, cotton tends to expand more in the linear direction. The composites will change dimension but can survive hot conditions. The possible applications of proposed composites are in building construction, indoor panels, and automotive body parts with some sound insulation. For more effective and efficient sound insulation, the composite panels themselves are not sufficient and additional porous fibrous layers can be added on top of composite panels. They can be used as a separator and as a possible replacement of pure/virgin wood in household furniture, etc.

Author Contributions: Conceptualization, H.J.; Data curation, M.Q.K.; Formal analysis, H.J. and R.M.; Funding acquisition, M.P.; Investigation, T.H., M.Q.K., M.P., J.N., R.C. and M.H.; Methodology, T.H., R.M., M.Q.K. and J.N.; Project administration, H.J. and M.P.; Resources, H.J., R.C. and M.H.; Supervision, H.J. and R.M.; Validation, J.N., R.C. and M.H.; Visualization, T.H., M.Q.K. and R.C.; Writing—original draft, T.H. and R.M.; Writing—review & editing, H.J., R.M., M.P., R.C. and M.H. All authors have read and agreed to the published version of the manuscript.

Acknowledgments: The result was obtained through the financial support of the Ministry of Education, Youth and Sports of the Czech Republic and the European Union (European Structural and Investment Funds—Operational Programme Research, Development and Education) in the frames of the project "Modular platform for autonomous chassis of specialized electric vehicles for freight and equipment transportation", Reg. No. CZ.02.1.01/0.0/0.0/16_025/0007293.

Conflicts of Interest: The authors declare no conflicts of interest.

References

1. Yang, W.; Li, Y. Sound absorption performance of natural fibers and their composites. *Sci. China Technol. Sci.* **2012**, *55*, 2278–2283. [CrossRef]
2. Yilmaz, N.D.; Powell, N.B.; Banks-Lee, P.; Michielsen, S. Hemp-fiber based nonwoven composites: Effects of alkalization on sound absorption performance. *Fibers Polym.* **2012**, *13*, 915–922. [CrossRef]
3. Verma, D.; Gope, P. Coir fiber reinforcement and application in polymer composites: A review. *J. Mater. Environ. Sci.* **2013**, *4*, 263–276.
4. Bell, L.; Bell, D. *Industrial Noise Control: Fundamentals and Applications*, 2nd ed.; CRC Press: Boca Raton, FL, USA, 1994; Volume 1.
5. Sargianis, J.; Kim, H.; Andres, E.; Suhr, J. Sound and vibration damping characteristics in natural material based sandwich composites. *Compos. Struct.* **2013**, *96*, 538–544. [CrossRef]
6. Mohanty, A.; Misra, M.; Drzal, L. *Natural Fibers, Biopolymers, and Biocomposites*; CRC Press: Boca Raton, FL, USA, 2005; Volume 1.
7. Nagendra, M.; Sowjanya, M.; Kumar, S. *Development of Noise Absorbing Composite Materials Using Agro Waste Products*; Project Reference # 40S-BE-1595; Jain University: Bengaluru, India, 2000; Volume 1, pp. 1–6.
8. Zulkifli, R.; Nor, M.M.; Tahir, M.M.; Ismail, A.R.; Nuawi, M.Z. Acoustic properties of multi-layer coir fibres sound absorption panel. *J. Appl. Sci.* **2006**, *8*, 3709–3714. [CrossRef]
9. Zhou, H.; Li, B.; Huang, G. Sound absorption characteristics of polymer microparticle. *J. Appl. Polym. Sci.* **2006**, *101*, 2675–2679. [CrossRef]
10. Jayamani, E. Sound Absorption and Impedance Study of Lignocellulosic Fibre Based Composites for Acoustical Applications. Ph.D. Thesis, University of Malaysia Sarawak, Kota Samarahan, Sarawak, Malaysia, 2015.
11. ASTM C423—17. *Standard Test Method for Sound Absorption and Sound Absorption Coefficients by the Reverberation Room Method*; ASTM: Washington, DC, USA, 2017.
12. Joserph, P.; Joseph, K.; Thomas, S. Effect of processing variables on the mechanical properties of sisal-fiber-reinforced polypropylene composites. *Compos. Sci. Technol.* **1999**, *59*, 1625–1640. [CrossRef]
13. Danihelová, A.; Němec, M.; Gergel, T.; Gejdoš, M.; Gordanová, J.; Sčensný, P. Usage of recycled technical textiles as thermal insulation and an acoustic absorber. *Sustainability* **2019**, *11*, 2968. [CrossRef]
14. Němec, M.; Igaz, R.; Gergel, T.; Danihelová, A.; Ondrejka, V.; Krišťák, L.; Gejdoš, M.; Kminiak, R. Acoustic and thermophysical properties of insulation materials based on wood wool. *Akustika* **2019**, *33*, 115–123.

15. Roy, S.; Bhowmik, S.; Davim, J.P.; Kumar, K. Estimation of mechanical and tribological properties of epoxy-based green composites. In *Green Approaches to Biocomposite Materials Science and Engineering*; IGI Global Publishers: Hershey, PA, USA, 2016; Volume 1.
16. Jadhav, A.C.; Pandit, P.; Gayatri, T.N.; Chavan, P.P.; Jadhav, N.C. *Production of Green Composites from Various Sustainable Raw Materials*; Green Composites; Springer: Singapore, 2018; pp. 1–24.
17. Seddeq, H. Factors influencing acoustic performance of sound absorptive materials. *Aust. J. Basic Appl. Sci.* **2009**, *3*, 4610–4617.
18. Chen, D.; Li, J.; Ren, J. Study on sound absorption property of ramie fiber-reinforced poly (L-lactic acid) composites: Morphology and properties. *Compos. Part A Appl. Sci. Manuf.* **2010**, *41*, 1012–1018. [CrossRef]
19. Wang, C.; Torng, J. Experimental study of the absorption characteristics of some porous fibrous materials. *Appl. Acoust.* **2001**, *62*, 447–459. [CrossRef]
20. Khusairy, M.; Jayamani, E.; Heng, S.; Hamdan, S.; Kakar, A. An experimental and simulation studies on sound absorption coefficients of banana fibers and their reinforced composites. *Nano Hybrids Compos.* **2017**, *12*, 9–20.
21. Bratu, M.; Vasile, O.; Dumitrescu, O. Sound-absorbing properties of composite materials reinforced with various wastes. *Environ. Eng. Manag.* **2011**, *10*, 1046–1051. [CrossRef]
22. Abdullah, F.; Azharia, A. Sound absorption cofficient of natural fibers hybrid reinforced polyester composites. *J. Teknol.* **2015**, *76*, 31–36.
23. Jayamani, E.; Hamdan, S.; Heng, S.; Rahman, R. Sound absorption property of agricultural lignocellulosic residue fiber-reinforced polymer matrix composites. *Appl. Mech. Mater.* **2014**, *663*, 464–468. [CrossRef]
24. Moretti, E.; Belloni, E.; Agosti, F. Innovative mineral fiber insulation panels for buildings: Thermal and acoustic characterization. *Appl. Energy* **2016**, *169*, 421–432. [CrossRef]
25. Jayamani, E.; Hamdan, S.; Rahman, M.R.; Bakri, M.K.B.; Kakar, A. An investigation of sound absorption coefficient on sisal fiber poly lactic acid bio-composites. *J. Appl. Polym. Sci.* **2015**, *132*. [CrossRef]
26. Jayamani, E.; Heng, S.K.; bin Bakri, M.K.; Hamdan, S. Comparative study of sound absorption coefficients of coir/kenaf /sugarcane bagasse fiber-reinforced epoxy composites. *Key Eng. Mater.* **2017**, *730*, 48–53. [CrossRef]
27. Ricciardi, P.; Belloni, E.; Cotana, F. Innovative panels with recycled materials: Thermal and acoustic performance and life cycle assessment. *Appl. Energy* **2014**, *134*, 150–162. [CrossRef]
28. Jiang, S.; Xu, Y.Y.; Zhang, H.P.; White, C.B.; Yan, X. Seven-hole hollow polyester fibers as reinforcement in sound absorption chlorinated polyethylene composites. *Appl. Acoust.* **2012**, *73*, 243–247. [CrossRef]
29. Zhang, S.P.; Li, Y.; Zheng, Z.Y. Effect of physiochemical structure on energy absorption properties of plant fibers reinforced composites: Dielectric, thermal insulation, and sound absorption properties. *Compos. Commun.* **2018**, *10*, 163–167. [CrossRef]
30. Zhang, J.; Shen, Y.; Jiang, B.; Li, Y. Sound absorption characterization of natural materials and sandwich structure composites. *Aerospace* **2018**, *5*, 75. [CrossRef]
31. Jayamani, E.; Hamdan, S.; Bakri, M.K.B.; Kok Heng, S.; Rahman, M.R.; Kakar, A. Analysis of natural fiber polymer composites: Effects of alkaline treatment on sound absorption. *J. Reinf. Plast. Compos.* **2016**, *35*, 703–711. [CrossRef]
32. Garkhail, S.; Heijenrath, R.; Peijs, T. Mechanical properties of natural-fibre-mat-reinforced thermoplastics based on flax fibres and polypropylene. *Appl. Compos. Mater.* **2000**, *7*, 351–372. [CrossRef]
33. Rana, A.K.; Mandal, A.; Mitra, B.C.; Jacobson, R.; Rowell, R.; Banerjee, A.N. Short jute fiber-reinforced polypropylene composites: Effect of compatibilizer. *J. Appl. Polym. Sci.* **1998**, *69*, 329–338. [CrossRef]
34. Mohanty, S.; Nayak, S.; Verma, S.; Tripathy, S. Effect of MAPP as a coupling agent on the performance of jute–PP composites. *J. Reinf. Plast. Compos.* **2004**, *23*, 625–637. [CrossRef]
35. Alavudeen, A.; Rajini, N.; Karthikeyan, S.; Thiruchitrambalam, M.; Venkateshwaren, N. Mechanical properties of banana/kenaf fiber-reinforced hybrid polyester composites: Effect of woven fabric and random orientation. *Mater. Des.* **2015**, *66*, 246–257. [CrossRef]
36. Abdellaoui, H.; Bensalah, H.; Echaabi, J.; Bouhfid, R.; Qaiss, A. Fabrication, characterization and modelling of laminated composites based on woven jute fibres reinforced epoxy resin. *Mater. Des.* **2015**, *68*, 104–113. [CrossRef]
37. Kaur, E.; Kalra, E. Effect on tensile and flexural properties of coconut fiber-reinforced with fresh plus recycled high density polyethylene based natural fiber composites. *Int. J. Eng. Pure Appl. Sci.* **2016**, *2*, 191–195.

38. Pan, N. Theoretical determination of the optimal fiber volume fraction and fiber-matrix property compatibility of short fiber composites. *Polym. Compos.* **1993**, *14*, 85–93. [CrossRef]
39. Ouensanga, A.; Picard, C. Thermal degradation of sugar cane bagasse. *Thermochim. Acta* **1988**, *5*, 89–97. [CrossRef]
40. Nogi, M.; Ifuku, S.; Abe, K.; Handa, K.; Nakagaito, A.N.; Yano, H. Fiber-content dependency of the optical transparency and thermal expansion of bacterial nanofiber-reinforced composites. *Appl. Phys. Lett.* **2009**, *5*, 210–215. [CrossRef]
41. Song, Y.S.; Lee, J.T.; Ji, D.S.; Kim, M.W.; Lee, S.H.; Youn, J.R. Viscoelastic and thermal behavior of woven hemp fiber-reinforced poly(lactic acid) composites. *Compos. Part B Appl. Eng.* **2012**, *43*, 856–860. [CrossRef]

© 2020 by the authors. Licensee MDPI, Basel, Switzerland. This article is an open access article distributed under the terms and conditions of the Creative Commons Attribution (CC BY) license (http://creativecommons.org/licenses/by/4.0/).

polymers

MDPI

Article

Valorization of Chicken Feet By-Product of the Poultry Industry: High Qualities of Gelatin and Biofilm from Extraction of Collagen

José C. C. Santana [1,*], Roberta B. Gardim [2], Poliana F. Almeida [1,3], Giovanna B. Borini [2], Ada P. B. Quispe [4], Segundo A. V. Llanos [4], Jorge A. Heredia [5], Stella Zamuner [2], Felix M. C. Gamarra [6,*], Thiago M. B. Farias [7], Linda L. Ho [1] and Fernando T. Berssaneti [1]

1 Department of Production Engineering, Polytechnic School of University of São Paulo, Av. Prof. Luciano Gualberto, 1380-Butantã, São Paulo 05508-010, Brazil; poliana.almeida@svc.ifmt.edu.br (P.F.A.); lindalee@usp.br (L.L.H.); fernando.berssaneti@usp.br (F.T.B.)
2 Industrial Engineering Post Graduation Program, Nine July University, Vergueiro Avenue, 235/249, Liberdade, São Paulo-SP 01504-000, Brazil; roberta.bgardim@hotmail.com (R.B.G.); giovanna_borini@hotmail.com (G.B.B.); stella.rz@uni9.pro.br (S.Z.)
3 Federal Institute of Mato Grosso, São Vicente Campus, São Vicente da Serra, 78106-000 MT, Brazil
4 Chemical Engineering Department, National University of Pedro Ruiz Gallo. Calle Juan XXIII 391, Lambayeque 14013, Peru; abarturen@unprg.edu.pe (A.P.B.Q.); svasquezll@unprg.edu.pe (S.A.V.L.)
5 Business School, Universidad del Pacífico, Calle Sanchez Cerro 2141 Jesús Maria, Lima 11, Peru; ja.herediap@up.edu.pe
6 University of Brasília, Gama Campus, St. Leste Projeção A—Gama Leste, Brasília-DF 72444-240, Brazil
7 Federal University of São Paulo, Baixada Santista Campus, Santos, São Paulo 11015-020, Brazil; thiago.michel@unifesp.br
* Correspondence: jccurvelo@yahoo.com.br (J.C.C.S.); fmartings@gmail.com (F.M.C.G.)

Received: 11 December 2019; Accepted: 19 February 2020; Published: 2 March 2020

Abstract: In this research, products with high quality were obtained from natural sources. The sensorial qualities, chemical characterization, and physical properties of gelatin extracted from chicken feet were compared with commercial gelatins. The extraction process was performed using acetic acid on a concentration ranging from 0.318% to 3.682%, processing time between 1.0 h and 8.4 h and extraction temperature between 43.3 °C and 76.8 °C. After the end of each assay, the yield was measured. Results showed that, under the best conditions, the collagen extraction yield was above 8%, and comprised 78.525 g/100 g of protein. Collagen analyzed by ICP-MS was composed of 99.44% of macro-minerals that are of great importance to human health. ATR-FTIR analysis showed that approximately 70.90% of the total protein from chicken feet is collagen, whereas, in commercial gelatin, only 30.31% is collagen. When comparing chicken gelatin with commercial gelatin, most sensory attributes were similar and chicken gelatin gained acceptance by more than 80% of the consumers. Additionally, the collagen films obtained from chicken feet and swine showed water absorption, odors, and texture characteristics similar to commercial material, such as latex and celofane. Consequently, due to its similarity to human skin, it is possible to apply it as a biocurative.

Keywords: chemical quality; sensorial quality; chicken feet; collagen; health care product; gelatin; biofilms

1. Introduction

The demand and requirements of consumers and the international regulations for food products is frequently inclined towards healthy products with improved quality. These demands include foods with high nutritional value (e.g., fiber, proteins, and micronutrients), and foods with excellent sensory aspects, fewer calories, and with low sugar, salt, fat, and other undesirable characteristics. This scenario has stimulated research exploring new sources of food and utilizing by-products or wastes. For these reasons, there is interest in these new food products called functional foods [1–3].

In this context, collagen and gelatin extracted from natural sources have been demonstrated in many applications as food additives, and have recently been explored in the pharmaceutical and medical fields. Usually, the most popular commercial product is gelatin from mammals (pigs and cattle), which is sometimes subjected to greater restrictions and skepticism among consumers, owing to socio-cultural and health concerns [3–6]. However, owing to its nutritional value, the demand for collagen and gelatin from porcine skin has been increasing progressively since 1990 [7,8]. In addition, studies have recently explored other natural sources for collagen and gelatin, including duck feet, fish, and poultry [8,9].

Thus, collagen and gelatin from natural sources like chicken feet merit special attention because, although chicken feet are considered waste material in various countries, they contain important nutrients and have essential health beneficial properties. In Brazil, chicken feet are regarded as waste from the poultry industry because Brazilian consumers do not have a habit of using chicken by-products like chicken feet and carcasses [3].

The world's largest poultry producer is the USA. Brazil overtook China as the second largest producer of chicken meat, reaching more than 10 million tons in 2009 and reached more than 13 million in the last year [10]. However, this places a very low commercial price on by-products such as chicken feet, feathers, skin, and bones. If collagen is determined to be component of these by-products, the production of artificial skin, bio-curatives, gelatin, jelly, and spreads from chicken feet could become alternate economically viable options [3,11].

The main products of the poultry industry include live animals, eggs, and various cuts of meat. Some by-products are also commonly produced from the carcass, meat bran, bone meal, and feathers in order to add value to products [12–15]. In this regard, studies have been conducted to improve the quality and add value to poultry products, such as by introducing new techniques to prevent the microbial contamination of poultry meat [13,16–21]. Other methods include improvements in the animal's diet to improve meat quality and carcass yield [22–25]. The use of ultraviolet, infrared, and gamma irradiation techniques and computer vision techniques have also been used for quality control of chicken breasts [12,21,26,27]. Ultimately, this has led to the development of new products such as jelly and chocolate spread from chicken feet collagen [3,11].

Most commercial gelatins are derived from mammalian sources, mainly pigs and cattle, but, due to sociocultural restrictions (Islam and Judaism) and the frequent occurrence of diseases such as bovine spongiform encephalopathy, which cause problems for human health, the use of products derived from mammals for processing of functional foods, cosmetics, and pharmaceuticals ends up being limited. Thus, studies have reported extraction of gelatin from chicken feet and have focused on the extraction method [8,28–30]. However, to the best of our knowledge, there are no previous studies that have investigated the quality of chicken feet collagen and gelatin as well and determined the presence of macronutrients and micronutrients in these products using analytical techniques. Thus, as chicken feet are waste from the meat industry, the extraction of their collagen and applying in the production of gelatins and biofilms will add value to this waste and will solve the socio-cultural problems associated with other by-products from the meat industry [3–5,11]. Therefore, this work focused on the extraction and chemical characterization of chicken feet collagen using some techniques such as spectroscopy-ATR-FTIR, ICP-MS, and others, and by comparing the sensorial qualities of the gelatin products with commercial gelatin.

2. Material and Methods

The reagents and equipment used in the preparation of biotherapeutic films were supplied by Nine July University (UNINOVE), acquired through accredited and qualified suppliers, while chicken tarsi were purchased from CEAGESP (Company of Warehouses and General Warehouses of São Paulo, São Paulo, Brazil). Collagen extraction from chicken tarsi was based on the work of [3].

2.1. Collagen Extraction Procedure

Warehouse Company and General Stores Company of Sao Paulo (CEAGESP), Brazil provided the chicken feet. Chicken feet were washed, the nails were removed, and the feet were again washed with cold water to remove any residual dirt or debris. Finally, the chicken feet were chopped, weighed, and placed in contact with acetic acid solution under the conditions used in the factorial design [3–5,31]. The assays followed a 2^3 factorial design using acetic acid concentration, $[A_c]$ (%, m/m), processing time, t (h), and extraction temperature, T (°C), as factors that influence the extraction yield (y_{exp}). The coding of the variables followed Equations (1a)–(1c):

$$x_1 = [A_c] - 2 \tag{1a}$$

$$x_2 = \frac{t-5}{2} \tag{1b}$$

$$x_3 = \frac{T-60}{10} \tag{1c}$$

The extraction systems comprised beakers to a total volume of 1.0 L, kept under constant agitation (magnetic stirring) at the planned temperature. Each factorial design assay used 100.0 g of chicken feet. The assays were performed in triplicate. Extraction yield was calculated based on the initial mass of chicken feet, using Equation (2):

$$y_{exp} = 100 \times \frac{M_{collagen}}{M_{chicken\ feet}} \tag{2}$$

After execution of the tests, a model was obtained by the least squares method and its adjustment was verified using the analysis of variance (ANOVA). Optimization was performed using the response surface methodology (RSM) in Software Statistica 6.0 for Windows®, São Paulo, Brazil, based on the concept proposed by [32–35]. The extracted material was distributed in Petri dishes and placed in a vacuum oven at 55 °C for 12 h. The dry material was then ground to obtain a powder and was characterized according to the standards presented in Association of Official Analytical Chemists, AOAC [36].

2.2. Gelatin Preparation

The techniques used for good practices and food security were based on Resolution RDC n° 216/2004 of the National Health Surveillance Agency (ANVISA) on 15 September 2004 [32,35]. Requirements for analysis of chemical and microbiological composition in food are presented in this law. The research was approved by the ethics committee of the institution, and the group of evaluators was composed of undergraduate students from the UNINOVE chosen at random in university living environments, but with similar distribution between genders.

A commercial gelatin (cowhide) acquired from the Brazilian market was used as a comparative sample. In water heated to 60 °C, chicken feet collagen powder was mixed with sugar, dye, and artificial flavor, and then distributed in small plastic cups that were refrigerated for gel formation, following the preparation procedure of the manufacturer, Dr. Oetker®. Chicken gelatin powder, protein, sugar content, colorants, and other compositional components of the gelatin formulations [11] are shown in Table 1, for commercial gelatin as well. The flavors used in the gelatin formulations were grape and pineapple because to these flavors are the most consumed in Brazil.

The acceptability of gelatin samples was assessed using sensory affective tests by comparing with the sensorial qualities of a commercial gelatin. Twenty-five milliliters of gelatin samples was served to 50 consumers in codified plastic cups covered with a thin layer of plastic film, using a monadic presentation and a 9-point hedonic scale. The consumers also registered their consumption intent for each sample (yes or no). Sensorial characteristics such as appearance, aroma, flavor, texture, and overall aspects of the gelatins were evaluated.

Table 1. Composition of gelatins produced in this study.

Contents (%)	Chicken Gelatin	Commercial Gelatin
Protein	70.9	18
Sugar	28.4	72
Citric acid	0.2	ND *
Salt	0.01	0.3
Colorant	0.1	ND
Flavor	0.4	ND
Microorganism (UFC/100 mL)	0	0
Pathogenic M. (UFC/100 mL)	0	0

* ND = not disclosed by manufacturer. UFC: colony forming unit

The form used to translate the sensorial responses of consumers to a numeric value in a hedonic scale was as follows: 1—I disliked extremely, 2—I disliked very much, 3—I disliked moderately, 4—I disliked slightly, 5—I perceived no difference (neither like nor dislike), 6—I liked slightly, 7—I liked moderately, 8—I liked very much, and 9—I liked extremely [32,37–39]. Based on the frequency of responses, sensorial data were compared using Student's *t*-test according to Equations (3) and (4) [38,40,41]. Gelatins of the same flavors were compared in pairs on the attributes: appearance, aroma, flavor and texture:

$$S_G = \sqrt{\frac{\sum_{i=1}^{n_1}(x_i - \overline{x_1})^2 + \sum_{j=1}^{n_2}(x_j - \overline{x_2})^2 + \sum_{k=1}^{n_3}(x_k - x_3)^2}{N_1 + N_2 + N_3 + \ldots - N_t}} \quad (3)$$

$$t = \frac{x_1 - \overline{x_2}}{S_G} \quad (4)$$

where N_1, N_2, N_3, and N_t are the sample numbers of each test.

2.3. Determination of Percent Composition

(a) Moisture: The moisture content of gelatin and chicken feet was determined by drying at 105 °C for 8 h, according to the standard method of the *Association of Official Analytical Chemists*, (A.O.A.C.) [36] and as described by [31] and [42]. The results are expressed as the percent weight loss during the drying process.

(b) Protein: The protein content in gelatin and chicken feet was determined by the Kjeldahl method [36]. The correction factor used was 5.55 for gelatin and 6.25 for chicken feet.

(c) Lipids: Determination of the lipid contents in gelatin and chicken feet samples was performed by the Soxhlet method. This uses direct extraction of fat using an organic solvent (ether). After 4 h, the ether was evaporated in an oven at 100 °C for 1 h; the residue was then cooled in a desiccator to measure the mass of the extracted oil [36,43,44].

(d) Ash: The total ash content of previously dried samples was determined by calcining at 500–600 °C for 4 h, following the methodology described by [43] and by A.O.A.C. [36].

2.4. Determination of Mineral Contents by the ICP-MS Technique

Gelatin mineral analysis was determined by inductively coupled mass spectroscopy (ICP-MS) following the method from [45,46]. Samples with diameter smaller than 0.08 m were homogenized (ultrasound strainer) and digested in a mixture of 5:2 nitric and hydrofluoric acid in a microwave oven (100 plus DGT model—Provecto Analytic) and analyzed in a quadrupole ICP-MS system (PerkinElmer ELAN 6100).

The equipment was operated under standard conditions with cell collision technology (CCT); using a collision cell with 8% (*v/v*) He in H_2, on an automatic mode. The configuration of the instrument uses a mixed flame torch and Peltier inlet with a 1.5 mm quartz nozzle and an increased sensitivity with an Xi cone interface (plasma display). A parallel nebulizer (Miramist Burgener) was used for high solid content. To improve precision, an aspiration rate of less than 0.4 mL/min was selected. The elements determined were: Mn, Fe, Al, K, Ca, Ti, Mg, Na, Cr, V, Ni, Zn, Pb, Li, Cu, La, Ce, Th, U, Sr, and Be. Previously, calibration curves for each analyzed element were prepared with a robust ICP-MS analytical setting and gravimetric standard solutions from the dilution of each derived element stock solution [47,48].

2.5. FTIR Procedure

FTIR spectra of collagen samples were recorded using a horizontal ATR, through a Nicolet iS5 FTIR spectrometer equipped with an ATR/iD3 with an argon horizontal cell (Thermo Fisher Scientific®, EUA) at 16 °C. The spectra in the range of 400–4000 cm^{-1} were rationed and the automatic signals gained were collected in 32 scans at a resolution of 4 cm^{-1} against a background spectrum recorded from a clean empty cell at 16 °C [41,45,49–51].

2.6. Determination of Gel Strength

For this, a texture analyzer TA-XT2, Stable Micro System (Surrey, UK) was used for determination of gel strength. The gelatins prepared at different concentrations were conditioned and cooled at 25 °C, and poured in standard bloom jars. According with [52], the common gelatin concentration used to analyze gel strength of commercial gelatin in the gelatin industry is 6.67% (*w/w*); this value was used to prepare the commercial gelatin sample (cowhide). It was then refrigerated at 5–7 °C for 16 h prior to gel strength measurement. The operating conditions used were a cross-head speed of 0.5 mm/s, a load cell of 5 kg, and a flat bottom plunger of 0.5 cm in diameter. The bloom value (g) is obtained after the plunger penetrates into the gel to a depth of 4 mm. At this depth, the maximum force reading was obtained and translated as the bloom strength of the gel [53].

2.7. Chicken Collagen Film Production

The research works [4,54] were considered for the production of the collagen films from chicken. Consequently, the biofilms were produced in different types of plasticizers (propylene glycol and glycerine), as formulated in Table 2.

Table 2. Formulation of bio-therapeutic film of collagen from chicken.

Formulation I (with Glycerin)	Formulation II (with Propylene Glycol)
Collagen from chicken 10%	Collagen from chicken 10%
Glycerin 5%	Propylene glycol 5%
Methylparaben 0.1%	Methylparaben 0.1%
Distilled water 100 mL	Distilled water 100 mL

For the formulation, 10 g of collagen from chicken were weighed and stored. After that, 0.1 g of methylparaben and 5 g of glycerin (Formulation I) or propylene glycol (Formulation II) were weighed, and were transferred for a glass beaker. Seventy milliliters of distilled water were immediately added and homogenized, followed by a water bath until the solubilization of methylparaben was completed. Collagen from chicken was added posteriorly to homogenize until fully solubilized. The solution was filtered to remove impurities present. The pH range verified should be between 5 and 6. The volume of the solution was completed with distilled water until 100 mL and fractionated in acrylic petri dishes, followed by infrared drying at 70 °C.

2.8. Production of Biotherapeutic Collagen Films from Swine

The production of biotherapeutic collagen films from swine was in accordance with the research by [4,54]. For this purpose, 150 bloom of gelatin (gel forming capacity) will be used in different types of plasticizers as shown in Table 3.

Table 3. Formulation of biotherapeutic with collagen from chicken.

Formulation III (with Glycerin)	Formulation IV (with Propyleneglycol)
Gelatin from swine 10%	Gelatin from swine 10%
Glicerin 5%	Propyleneglicol 5%
Methilparaben 0.1%	Methilparaben 0.1%
Distilled water 100 mL	Distilled water 100 mL

Ten grams of collagen from chicken were weighed and stored. Immediately, 0.1 g methylparaben and 5 g glycerin (formulation III) or propylene glycol (formulation VI) were weighed, and transferred to a glass beaker, with approximately 70 mL of distilled water added, homogenized and followed by being put in a water bath until complete solubilization of methylparaben. After that, collagen from chicken was added to homogenize until complete solubilization. The solution was filtered to remove impurities present. The pH was remained between 5 and 6. The volume of the solution was completed with distilled water up to 100 mL and fractionated in acrylic petri dishes following the infrared dryer at 70 °C. After the formulation of films, these were characterized.

2.8.1. Swelling Index Determination

The swelling test allows the prospect of degradation that is related to the degree of hydration of the system to be checked and determined in advance. This test is important to verify if the studied material presents structural stability during the period necessary for the formation of the new regeneration tissue. Free films of collagen from chicken feet were cut into 2.5 × 2 cm pieces and placed in glass petri dishes and left in the desiccator with silica gel for 24 h. After the elapsed time, they were removed from the desiccator and weighed, where these values were adopted as the weight of an initial zero time (*Ws*). Free films were left in 20 mL of 0.9% NaCl solution according to the established times: 1, 10, 30, and 60 min. At the end of each stipulated time, swollen films were weighed and their values recorded (*Wi*). Its calculation is done using Equation (5) [55,56]. The swelling index (*Si*) was determined in triplicate:

$$Si(\%) = 100 \times \left(\frac{Wi - Ws}{Ws}\right) \tag{5}$$

where:

- *Wi* represents the membrane mass after swelling at times 1, 10, 30, and 60 min;
- *Ws* represents the dry membrane mass at time zero.

The data were collected in the swelling test (Ii%), and a comparative graph was generated for each sample.

2.8.2. Determination of Water Vapor Transmission (*WVT*)

The objective of the *WVT* test was to verify the permeabiization of biotherapeutic collagen films from chicken feet. *WVT* tests the biofilms were made in triplicate with propylene glycol and glycerin, and according to method B E96-66 from *American Society for Testing Materials/EUA* (ASTM) [56].

The biotherapeutic films were placed in glass Petri dishes containing 40% saturated NaCl solution per 72 h into desiccator. After the established time, they were prepared properly for the water vapor transmission rate (*WVT*) tests. In each cup (film of 10 cm area is weighed), 10 mL water was added.

Then, the films to be investigated were individually fixed to the edge of the clip-film domes. The kits (cup + distiller water + film) were weighed (time zero) and placed in a desiccated silica gel desiccator. Samples were weighed at 24 h, 48 h, 72 h, and 96 h. For each interval, the values were recorded to calculate the water vapor transmission rate of these films. In addition, for the samples preparation, a control kit was prepared, which served as a comparative basis. *WVT* was calculated using Equation (6) [55]:

$$TVW\left(\frac{g}{hcm^2}\right) = 24{\cdot}g{\cdot}t{\cdot}a \tag{6}$$

where:

- "g" is mass loss,
- "t" the time in hour,
- "a" area of film (10 cm^2).

The values obtained were shown in Figure 10 with the time intervals.

Additionally, the swelling index test and the water vapor transmission test were made in the collagen films developed from swine (frequently more used), in order to compare with the collagen films from chicken feet.

3. Results and Discussion

3.1. Extraction and Optimization of the Chicken Collagen Process

Table 4 shows the factorial design used in this experiment and the results of the extractions of collagen (Y), in the experimental setting (Y_{exp}) and those calculated by the model (Y_{calc}), according to each assay performed. It is noted that the extraction varied from 1.7% to 8.5% of collagen in the initial mass of the chicken feet. After execution of each assay, the collagen powder was obtained after drying at 50 °C for 12 h in a vacuum drier.

Table 4. Experimental design and the results of the collagen extraction process.

Assay	x_1	x_2	x_3	$[A_c]$ (%)	t (h)	T (°C)	Y_{exp} (%)	Y_{calc} (%)
1	−1	−1	−1	1	3	50	4.4384	4.2493
2	1	−1	−1	3	3	50	5.1980	5.1505
3	−1	1	−1	1	7	50	1.6962	1.8785
4	1	1	−1	3	7	50	5.1659	5.4897
5	−1	−1	1	1	3	70	8.5713	8.3821
6	1	−1	1	3	3	70	8.1620	8.1145
7	−1	1	1	1	7	70	2.8835	3.0657
8	1	1	1	3	7	70	6.1218	6.4457
9	0	0	0	2	5	60	6.2332	6.0421
10	0	0	0	2	5	60	6.2700	6.0421
11	0	0	0	2	5	60	5.5907	6.0421
12	1.682	0	0	3.682	5	60	7.6714	7.4079
13	−1.682	0	0	0.318	5	60	4.1287	4.2016
14	0	1.682	0	2	8.364	60	2.5070	1.9702
15	0	−1.682	0	2	1.034	60	5.4152	5.7617
16	0	0	1.682	2	5	76.82	6.1483	6.0530
17	0	0	−1.682	2	5	43.18	7.0203	6.9252

where: $[A_c]$ = acetic acid concentration, t is time, T = temperature, and Y = extraction yield.

The yields found in this work are similar to those reported by Lim et al. [57], which ranged from 1.72% to 5.33% depending on the extraction method used. Hao et al. [58] in a study with sturgeon skin pretreated with $Ca(OH)_2$ obtained a gelatin yield of 2.40% to 3.52%. Jamilah and Hervinder [59] reported gelatin yields of 5.39% and 7.81% upon collagen extraction from red and black tilapia, respectively. Chew et al. [60] found a gelatin yield of 7.25% from the fin after the extraction process.

Therefore, the results presented in trials 5 and 6 were greater than all the results reported previously. However, in a study on ray skin (*Raja kenojei*) with 6 h-extraction, Ref. [61] identified a gelatin yield of 17.48% and Ref. [54] obtained a gelatin yield of 12% during collagen extraction from giant squid.

The analysis of variance was employed for determining significant variables, as shown in Table 5. The regression equations were submitted to the F-test for the coefficient of determination R^2 and explaining the variances at 95% of the confidence level. According to [31–35], the first F-test (F_{calc}/F_{tab}) must be more than one for it to be significant, the second F-test must be less than 1 for it to be predictive, and R^2 explaining the variance must be next at 1.0 and 100, respectively. Table 5 lists the significant parameters and statistical test results of the models.

Table 5. Evaluation of model fit by variance analysis (ANOVA) at a 95% confidence level.

Source	Square Sum	Degree Freedom	Square Mean	F_{calc}
Regression	57.832	11	5.257	
Residual	1.146	5	0.229	22.947
Fitting fault	0.854	3	0.285	
Error	0.292	2	0.146	1.950
Total	58.977	16		
% explaining variance				98.059
% maximum explanable variance				99.505
Coefficient of determination (R^2)				0.9806

$F_{tab1\,(95\%.\ 11,\ 5)} = 4.070$ and $F_{tab2\ (95\%.\ 3,\ 2)} = 19.160$.

The calculated value for the F_1 test was 22.947, being 4.5 times larger than the tabulated F_1 (4.070), while the calculated F_2 was 1.950, being 10 times smaller than the tabulated F_2 (19.160); this indicates that the model is statistically significant and is adjusted to the experimental data. Another parameter that corroborated for the analysis was the coefficient of determination (R^2), which was equal to 0.9806 and is close to 1.0, as indicated by [31–35]. In this manner, it can be affirmed that the model is adjusted and can be used to predict the value of the yield of collagen extraction from chicken feet, for the conditions presented in this work.

Equation (7) is the best fitting model to predict the extraction yield (y) value on the influence of acetic acid (x_1), processing time (x_2), and temperature (x_3):

$$\begin{aligned} y \;=\; & 6.0421 + 0.9531x_1 - 1.1271x_2 + 1.9282x_3 - 0.0839x_1^2 - 0.7692x_2^2 + 0.1580x_3^3 \ldots + \\ & \ldots - 0.7732x_3^3 + 0.7947x_1x_2 - 0.1750x_1x_3 - 0.6192x_2x_3 + 0.1172x_1x_2x_3 \end{aligned} \tag{7}$$

Figures 1–3 are the response surfaces obtained in the Statistica 6.0® software and are used in the optimization of collagen extraction from chicken feet. As shown in Figure 3, the highest yields are found when operated with the highest concentrations of acetic acid and with the shortest processing times.

Figure 2 shows that, if the temperature and acetic acid concentration are at their highest values, the yield is at a maximum.

Figure 3 shows that, in the shortest processing times and the highest temperature values, the yield tends to reach the maximum value. To find the best conditions, one must observe the level curves (lines) on the response surface, which tend to the highest values of extraction yield (in red).

Thus, it can be concluded that, by using acetic acid concentrations between 3.000% and 3.682%, operating times between 1 h and 3 h, and temperatures between 70 °C and 76.82 °C yielding greater than 8.0% collagen can be obtained.

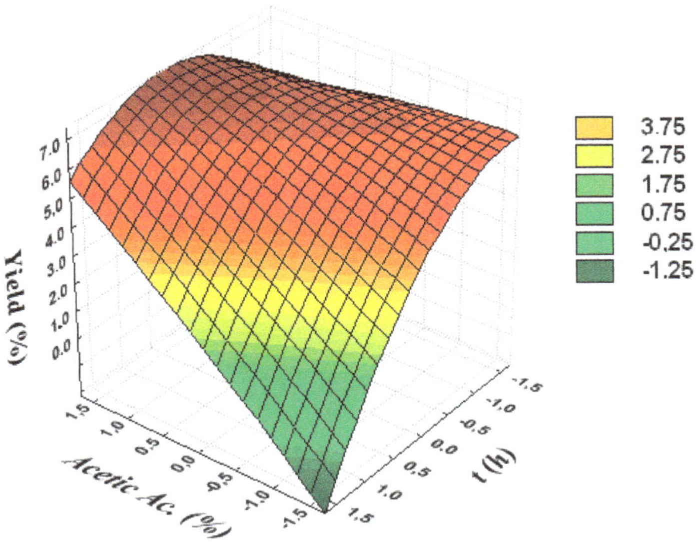

Figure 1. Response surface to show the mutual influence of acetic acid concentration and processing time on extraction yield.

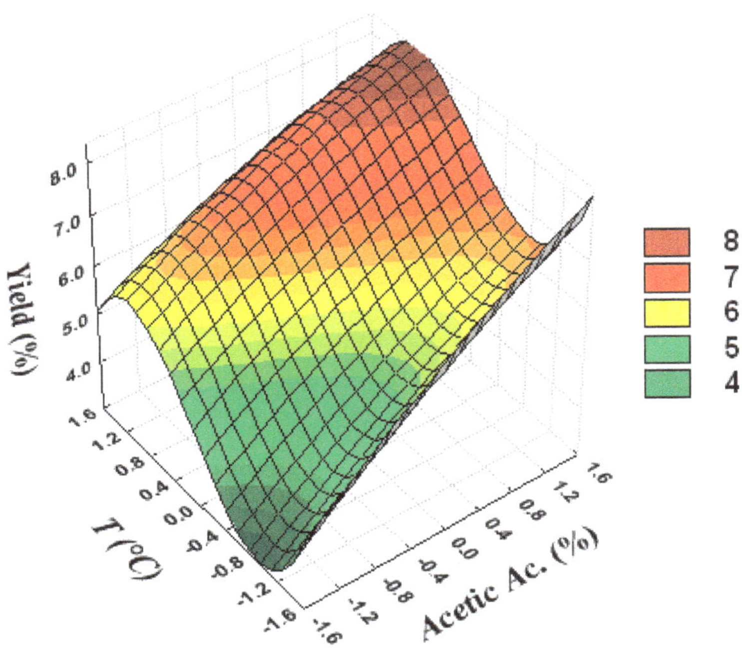

Figure 2. Response surface to show the mutual influence of temperature and acetic acid concentration on extraction yield.

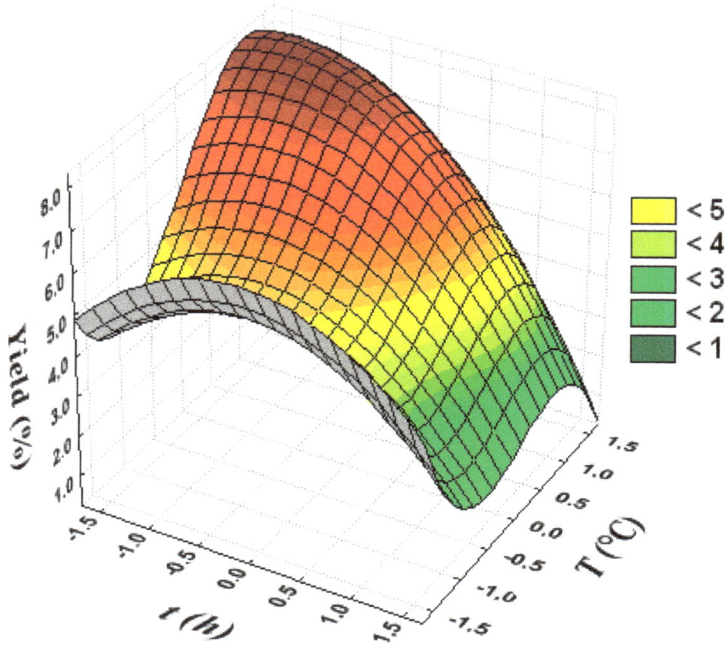

Figure 3. Response surface to show the mutual influence of processing time and temperature on extraction yield.

3.2. Chemical and Bloom Analysis

Table 6 shows the chemical composition of chicken feet used for the extraction of collagen, with respect to the protein, lipid, ash, and water content. The powdered collagen composition was 9.7 g, 4.8 g, 6.9 g, and 78.5 g/100 g for moisture, ash, lipids, and proteins, respectively. Cliché et al. [62] also presented the composition of chicken feet, which had the following crude protein, ash, fat, and moisture content values of 17.42%, 12.04%, 5.98%, and 62.05%, respectively, which is similar to that found in our work.

Table 6. Chemical composition of chicken feet and their powdered collagen.

Composition *	Chicken Feet (g/100 g)	Powdered Collagen (g/100 g)
Moisture	60.009	9.749
Ashes	9.943	4.807
Lipids	12.875	6.919
Proteins	17.173	78.525

* Average of three batches under optimal extraction conditions.

The results of collagen gel strength from chicken feet are presented in Table 7. Upon comparing the gel strength of chicken feet gelatin to commercial gelatin at 6.67% collagen, this chicken feet gelatin has roughly 45% greater gel strength than that of commercial gelatin.

Consequently, Figure 4 shows the variation in gel strength with the composition of the collagen in the gelatin. As expected, gel strength increases with the collagen composition in gelatin and, in this case, showed a linear increase. Using a linear equation obtained from Table 7 data, it was observed that, to obtain a gel strength similar to that of commercial gelatin, one requires only 5% collagen concentration from chicken feet gelatin. This equates to a reduction of about 24% (*w/w*) in the final gelatin composition, and thus makes chicken feet gelatin more economically feasible than commercial gelatin.

Table 7. Gel strength analysis of collagen from chicken feet.

Gelatin	Collagen Contents (*w*/*w*%)	Gel Strength (kPa) *
Chicken feet	1.00	19.87
	2.00	44.65
	3.33	96.47
	4.00	119.1
	6.67	294.79
	10.0	466.87
Commercial	6.67	204.05

* Average of three batches under optimal extraction conditions.

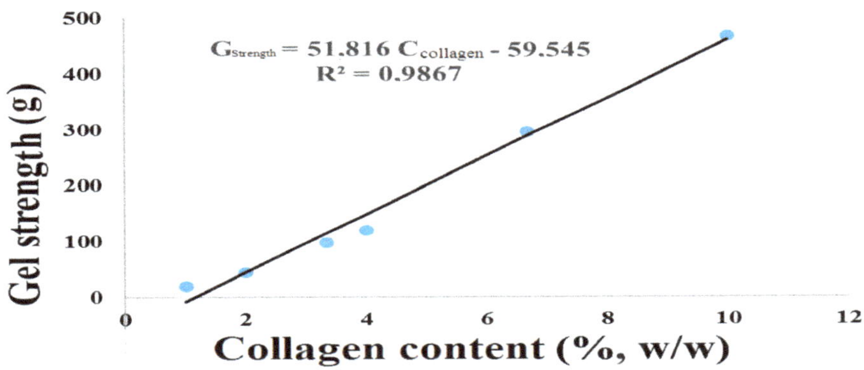

Figure 4. Variation of gel strength with the collagen content in gelatin.

In a study on fish gelatin extraction with or without treatment by transglutaminase enzymes in a hydrolysis process, Norziah et al. [53] produced gelatin with low gel strength, varying between 70 g and 100 g, in addition to a commercial gelatin from halal bovine with a gel strength of 336.2 g. All gelatins presented in this study had low gel strength; however, in the same work, the authors cite a commercial fish gelatin with gel strength (435.9 g) approaching that found in this study. Thus, it is possible to suggest that the gel strength of chicken feet gelatin found in this work is superior or equal to that reported in previous studies.

3.3. Collagen Composition from Analysis of the FTIR

Recently, Ref. [27] used infrared spectroscopy to control the quality of intact chicken breast fillets by predicting the major component scores of qualities. In addition, Ref. [41] reports that this technique can be used to monitor the shelf life of products and the origin of agricultural products. Thus, in order to contribute to the quality control of chicken products, we used the FTIR technique in chicken jelly and commercial gelatin samples to verify the collagen content of both.

Figure 5a shows the FTIR spectra of collagen extracted from chicken feet on a thermal bath of 4% acetic acid solution at 60 °C for 4 h and Figure 5b shows the FTIR spectra of collagen from commercial gelatin. Each spectrum is an average of 32 scans of three batches on the optimal extraction conditions presented in the optimization section. FTIR spectra of the collagen extracted from chicken feet showed major peaks in the amide region. Specifically, chicken collagen showed a vibration peak at the wave numbers of 1652.01 cm^{-1} for amide I, of 1539.87 cm^{-1} for amide II, of 1241.29 cm^{-1} for amide III, of 2923.72 cm^{-1} for amide B, and of 3399.56 cm^{-1} for amide A. The FTIR spectra of commercial gelatin showed Amide II at 1556.53 cm^{-1}, amide I at 1651.32 cm^{-1}, Amide B at 2921.49 cm^{-1}, and amide A in a range of 3391.84–3467.09 cm^{-1}, and did not show any amide III peaks. However, a high protein content of low molecular weight has been found [51,63].

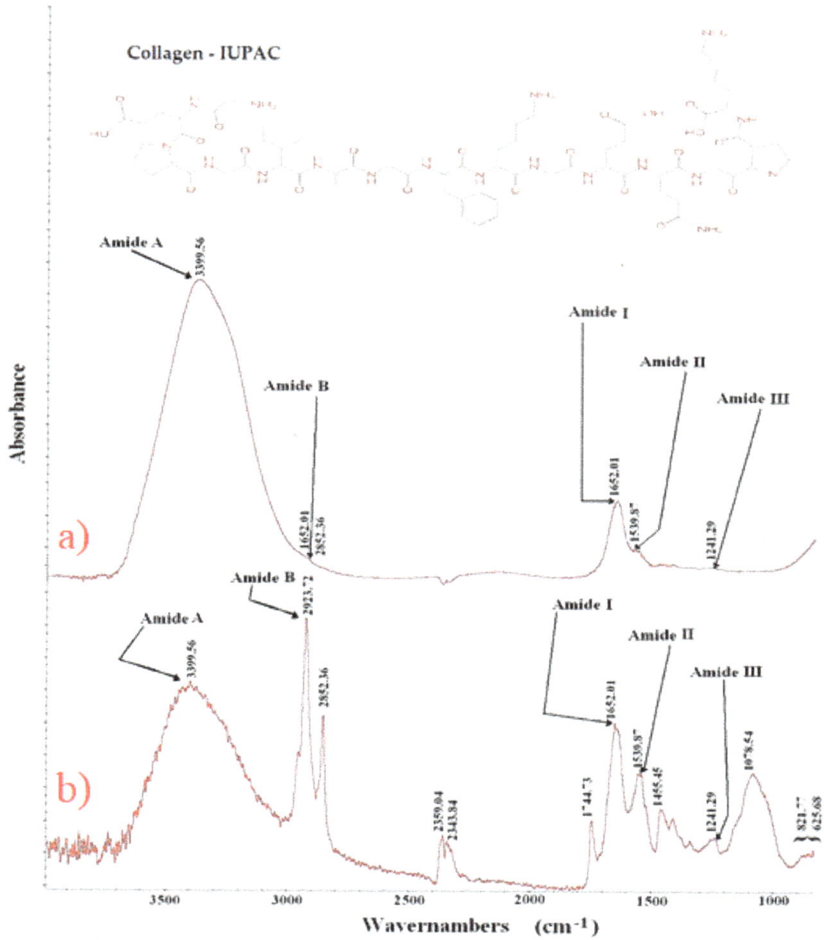

Figure 5. FTIR spectra of gelatins. (**a**) collagen from chicken feet, (**b**) collagen from commercial gelatin.

The low molecular weight peptides formed during extraction for long times were more likely to form covalent cross-links during the freeze-drying process [50,64–66]. This affected the collagen content from commercial gelatin and reduced one of the main qualities expected in gelatins. The process used in this work did not have the same effect on chicken gelatin. Results showed that collagen composition from chicken feet was greater than 70.90% while that from cowhide gelatin (commercial) was only 35%.

3.4. Mineral Composition

The chemical composition of ash was determined by mass spectroscopy (ICP-MS) and the results are reported in Table 8. Macro-minerals (Na, K, Ca, Mg, P, and S) that are of great importance to health human are presented and make up most of the elements (99.44% ash content) in the collagen composition.

Table 8. Results of ICP-MS analysis on powdered collagen from chicken feet.

Elements	Ash Content (μg/g) *	Total in Collagen (mg/100 g)
Li	0.050	2.10^{-4}
Be	0.090	4.10^{-4}
B	0.050	2.10^{-4}
Na	5938	28.55
Mg	1834	8.817
Al	10.44	0.050
Si	22.53	0.108
P	168.1	0.808
S	1505	7.236
K	853.9	4.105
Ca	1692	8.136
Ti	2.060	0.010
V	0.010	5.10^{-5}
Cr	<dl	-
Mn	0.100	5.10^{-4}
Fe	3.210	0.015
Co	<dl.	-
Ni	0.150	7.10^{-4}
Cu	1.350	0.006
Zn	<dl	-
Br	5.700	0.027
Sr	3.810	0.018
Ag	1.300	0.006
Cd	<dl	-
Sn	0.140	6.10^{-4}
Ba	1.970	0.009
Pb	14.54	0.070

<d.l. = below detection limit. * average of 6 analyses of three batches under optimal extraction conditions.

As seen in Table 8, there are high levels of Na, at about 6 mg/g, resulting in 28.546 mg of Na per 100 g of powdered collagen. In the descending of quantity in 100 g of powdered collagen, Na is followed by Mg, Ca, S, and K, with contents of 8.817 mg, 8.136 mg, 7.236 mg, and 4.105 mg, respectively. Phosphorus is an important element for energy changes in cells via adenosine triphosphate transfer/adenosine diphosphate (ATP/ADP), and shows a significant amount in collagen from chicken feet.

Other important micro-minerals to human health such as Cu, Fe, and Mn are present in trace amounts (0.04% of ash). Similarly, the presence of toxic elements such as Pb, Ba, Al, Li, and Be is in trace amounts, which does not compromise this food (0.22% of ash content) [32,67], and the elements are all derived from chicken feet.

However, the commercial (powder) gelatin used in this work as a comparison has 108 mg of sodium per 5.7 g of powder, which is equivalent to 1847 mg per 100 g. Thus, the collagen obtained in this work presents low sodium content, which is in conformity with the new Brazilian laws that oblige companies to reduce the levels of sodium in their food products.

Using mass spectroscopy, Haug et al. [5] have found about 5.08 mg, 0.29 mg, 0.02 mg, and 7.13 mg of Na, Ca, K, and other heavy metals in each 1 g of fish gelatin. These values are 43.50% of the total ash content. Their macro-mineral percentage is lower than that of chicken feet gelatin, indicating that our gelatin is nutritionally superior to fish gelatin with respect to minerals.

Under the conditions used in this work, the quality of collagen powder allows production of gelatins that can be used to produce various cosmetics, facial masks, or even artificial tissues for wound healing in humans [68]. Thus, these products can be more easily marketed because of a good representation of chemical, nutritional, and sensorial qualities in their labels. This can possibly bring overall health improvement to humans, and being superior to commercial products, can facilitate greater acceptance by consumers [39].

3.5. Sensorial Qualities

Table 9 shows the sensorial comparison between chicken and commercial gelatin for two flavors. The table show that the consumer acceptances were higher than 80% for all samples tested in this study. The Student's *t*-test showed no significant differences between the pineapple flavored gelatins for all sensorial qualities studied. However, for grape-flavored gelatins differences were observed for aroma, flavor, and overall vision; in all cases, commercial gelatins presented higher means than chicken collagen gelatins, mainly with the grape-flavored gelatin. However, since the values attributed to chicken collagen gelatin presented an average between 6.3 and 7.7 on a hedonic scale, they can be considered as accepted by consumers, of which 83% confirmed that they would consume this product.

Table 9. Sensory comparison of chicken and commercial gelatins at a 95% confidence level.

Flavour	Sample	Consumption Intention (%)	Attribute Evaluation *				
			Aroma	Appearance	Flavor	Texture	Overall
Pineapple	Commercial	94%	7.0 ± 1.1^{a}	7.1 ± 1.8^{b}	7.1 ± 1.6^{c}	6.7 ± 1.7^{d}	7.0 ± 1.6^{e}
	Experimental	83%	7.1 ± 1.9^{a}	7.5 ± 2.4^{b}	6.3 ± 2.8^{c}	6.5 ± 2.4^{d}	6.8 ± 2.4^{e}
Grape	Commercial	88%	8.3 ± 0.9^{f}	8.1 ± 0.7^{h}	8.3 ± 0.6^{i}	7.9 ± 0.9^{k}	8.2 ± 0.8^{l}
	Experimental	83%	7.5 ± 1.6^{g}	7.7 ± 1.5^{h}	6.9 ± 2.5^{j}	6.3 ± 2.4^{k}	7.1 ± 2.0^{m}

* similar letter indicates that there is no difference between the samples; $t_{\text{tabled, 95\%}} = 1.66$.

The high values of commercial gelatin may be because of the high levels of sugar in its composition. The highest averages were obtained by commercial gelatin because of 72% sugar in its composition. Chicken gelatin had lower levels of sugar (28%) and high collagen content (71%), which gives it greater value with respect to health and nutrition for the consumers compared to commercial gelatin. Additionally, it can be more readily consumed by people with diabetes, owing to its low sugar content. Chicken gelatin may thus be considered as the best gelatin, since its sensory qualities were the best as assessed by consumers, with an acceptance average varying between 6.3 and 7.5 times on the hedonic scale for all sensorial attributes.

Figure 6 shows that the majority of gelatins presented similar values for each attribute, which were close to seven points in a hedonic scale, thus indicating the similarity between chicken collagen (experimental) and commercial gelatins. However, grape-flavored commercial gelatin was significantly different among the others, since its average was above eight points on the hedonic scale

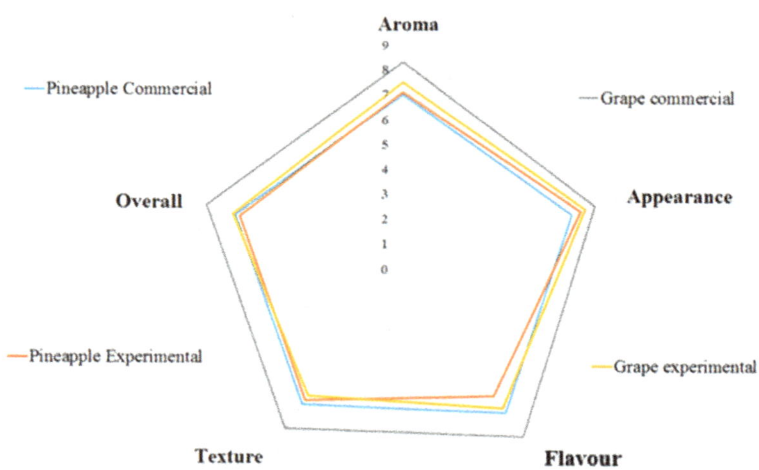

Figure 6. Response to sensorial qualities of gelatins.

All attributes of chicken collagen gelatin were satisfactorily evaluated by consumers, since their average is close to seven points on the hedonic scale, indicating that this product has an excellent sensorial quality.

3.6. Biotherapeutic Collagen Films Properties

Table 10 shows that some organoleptic properties collected of biotherapeutic collagen from chicken feet. The films were completed with propylene glycol, and films made with glycerine showed pH, color, and odor, similar characteristics except the texture.

Table 10. Comparative organoleptic properties of biotherapeutic films obtained from chicken feet using propylene glycol and glycerin.

Sample	pH	Color	Texture	Odor
Biotherapeutic colllagen films from chicken feet with glycerine	5.72	Dark yellow	Malleable, soft, slightly sticky touch but, with soft touch	Characteristic
Biotherapeutic colllagen films from chicken feet with propylene glycol	5.78	Dark yellow	Malleable, soft, slightly sticky touch but, with rough touch	Characteristic

The biotherapeutic collagen film sample as shown in Figure 7. From this, it can be observed that films A (prepared with glycerin) are softer and than film B (prepared with propylene glycol). The texture found in Figure 7A showed similarities with texture of the latex used to make disposable gloves and condoms. However, the biotherapeutic collagen films prepared with propylene glycol (see Figure 7B) proved to be more rustic and rougher, having a similar appearance with cellophane material.

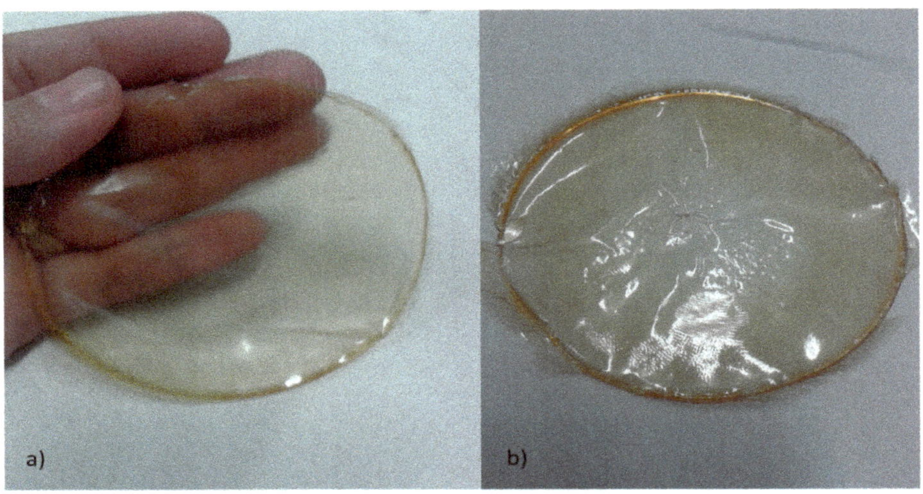

Figure 7. Film sample obtained from collagen of chicken feet with: (**a**) glycerin and (**b**) propylene glycol.

Additionally, for comparative interpretation, biotherapeutic collagen films from swine were produced using one sample with propylene glycol and another sample with glycerin, as are shown in Figure 8. From this, it can be observed that the results were similar with those films produced from chicken feet, but the films from swine were odorless and lighter in color. In addition, comparative organoleptic characteristics of biotherapeutic collagen films from swine are shown in Table 11.

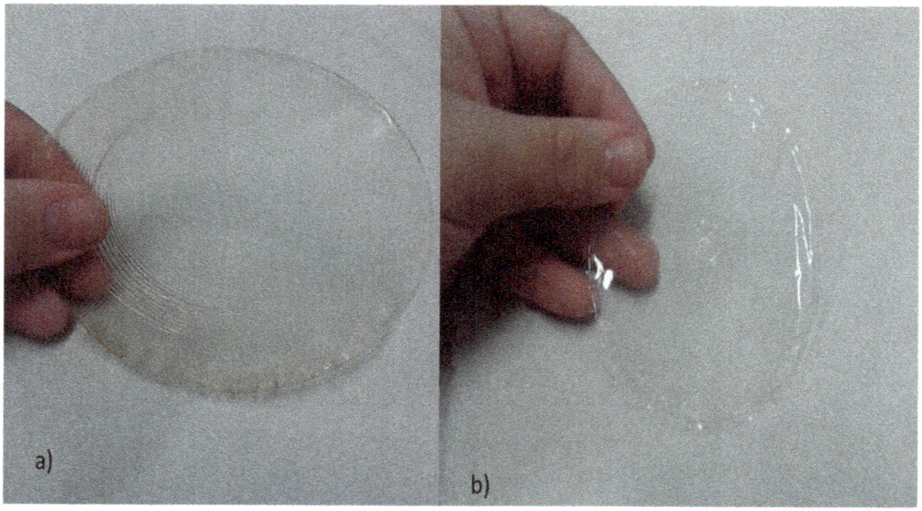

Figure 8. Films sample obtained from collagen of swine skin with: (**a**) glycerin and (**b**) propylene glycol.

Table 11. Comparative organoleptic characteristic of biotherapeutic collagen films from swine.

Sample	pH	Color	Texture	Odor
Biotherapeutic colllagen films from swine with glycerine	5.63	Light yellow Almost transparent	Malleable, soft, dry and soft touch	Odorless
Biotherapeutic colllagen films from swine with propylene glycol	5.71	Light yellow Almost transparent	Malleable, soft, dry and rough touch	Odorless

3.7. *Quality of Biotherapeutic Collagen Films from Chicken and Swine*

Swelling test allows checking and determining, in advance, the prospect of degradation, which is related to the degree of hydration of the system. These tests are important to verify the qualities and if the material has structural stability during the period necessary for the formation of the new regeneration tissue [52]. Right after 1 min, both films' formulations were swollen to 99% of weight, and this remained unchanged until 60 min. This showed that there was a low degradation of the films. Films swell very fast. Comparing Figure 9a,b, it is noted that there is no difference between the collagen films of chicken feet and that of pig collagen. The swelling of the films is explained by its water absorption power, which is determined by the presence of hydrophilic groups and the high degree of crosslinking of the collagen microspheres [4,5].

Water vapor transmission rate (*WVT*) tests were performed on biofilms produced from collagen of chicken feet and swine collagen, both with the plasticizers: glycerin and propylene glycol. Results are shown in Figure 10a,b. Regardless of the type of collagen and plasticizer used, all biofilms have similar results, with a satisfactory percentage of water vapor transmission, which allows its use as a biotherapeutic, since both allow similar transpiration to human skin.

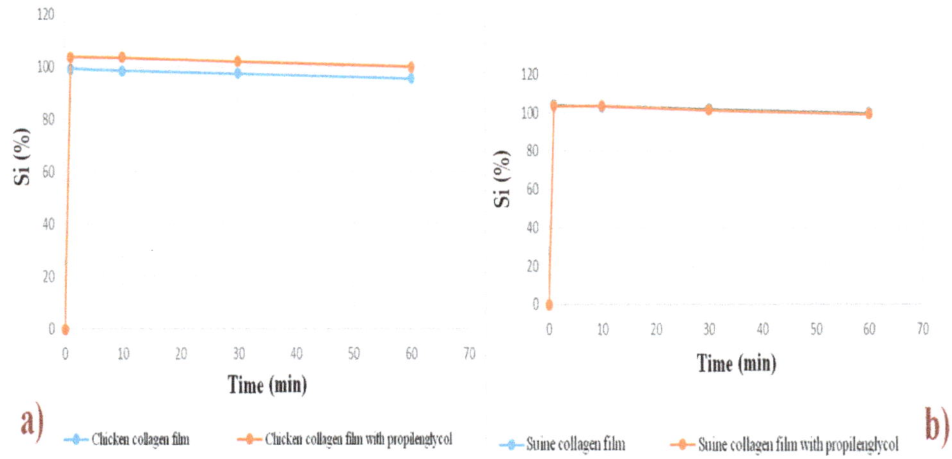

Figure 9. Polymer degradation process, based in swelling index curves. (**a**) chicken collagen samples and (**b**) swine collagen samples.

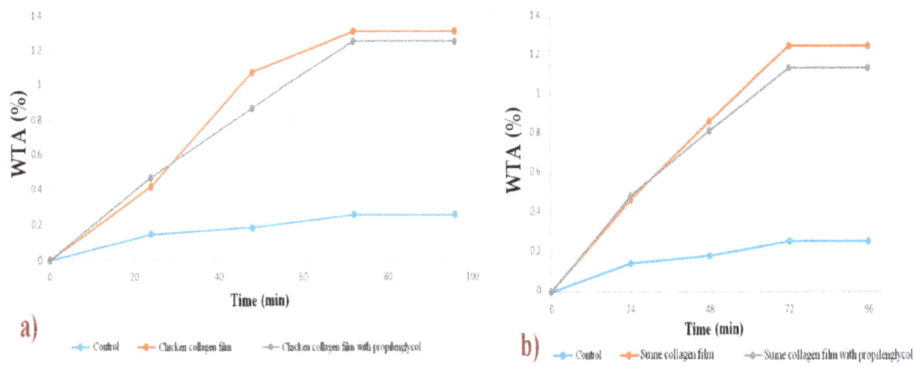

Figure 10. Water vapor transmission rate (*WVT*) test for films sample obtained from (**a**) chicken collagen samples and (**b**) swine collagen.

4. Conclusions

In this work, the best extraction condition for collagen from chicken feet was found at 3.000%–3.682% acetic acid concentration, with variation of time between 1–3 h, and variation of temperature between 70 °C–76.82 °C. The gelatin analyzed by ICP-MS was composed of 99.44% Na, Ca, K, Mg, P, and S, which are macro-minerals of great importance to human health. Furthermore, FTIR characterization has shown that the collagen composition of chicken feet gelatins is two times greater than that of commercial cowhide gelatin. Therefore, the gelatin obtained is nutritionally richer than the marketed gelatin. In addition, the sensorial qualities of this gelatin were similar to commercial gelatin, and more than 80% of surveyed panelists would consume the chicken gelatin. In addition, the collagen biofilms were obtained from chicken feet using propylene glycol, and glycerol as plasticizers showed satisfactory results for swelling index (with 99% absorption capacity) and *WVT* test (without significant change), and had similar behavior to collagen films made from swine when they were compared. In addition, the collagen films made from chicken feet displayed texture characteristics similar to the appearance of latex and celofane commercial products. Additionally, these collagen films had various similar characteristics to the collagen film made from swine. However, more complementary research is necessary to validate their applications. This work shows interesting results and adds important contributions such as the conversion of chicken feet, by-products in the poultry industry with low value in Brazil, to new material with high quality, such as the gelatin and films. Additionally, the data obtained provide important scientific information for food, heath, pharmaceuticals, and other areas. Therefore, chicken feet merit special attention because they appear to be a good alternative source for material and high-quality products.

Author Contributions: All authors contributed according with the fallowing distribution: Conceptualization, R.B.G.; J.C.C.S.; P.F.A.; T.M.B.F. and G.B.B.; Methodology, P.F.A.; S.Z. and J.C.C.S.; Formal Analysis, R.B.G.; P.F.A.; T.M.B.F.; G.B.B.;and A.P.B.Q.; Resources, A.P.B.Q.; S.A.V.L.; J.A.H.; F.M.C.G.; T.M.B.F.; F.T.B. and J.C.C.S.; Writing—Original Draft Preparation, F.T.B.; G.B.B.; J.C.C.S.; S.Z. and F.M.C.G.; Writing—Review & Editing; A.P.B.Q.; S.A.V.L.; J.A.H.; F.M.C.G.; F.T.B.; and L.L.H.; Supervision, L.L.H.; F.T.B. and J.C.C.S.; Project Administration, S.Z.; F.M.C.G.; F.T.B., S.A.V.L.; J.A.H.; and J.C.C.S. All authors have read and agreed to the published version of the manuscript.

Funding: CNPq funded number 305987/2018-6 and CAPES financial funded number 01.

Acknowledgments: The authors thank CNPq, CAPES, and FAPIC/CNPq (UNINOVE) for their financial support.

Conflicts of Interest: All authors declare have no conflict of interest, and we warrant that this article is the authors´ original work.

References

1. Liang, Q.F.; Wang, L.; He, Y.Q.; Wang, Z.B.; Xu, J.M.; Ma, H.L. Hydrolysis kinetics and antioxidant activity of collagen under simulated gastrointestinal digestion. *J. Funct. Foods.* **2014**, *11*, 493–499. [CrossRef]
2. Lentz, K.A. Current Methods for Predicting Human Food Effect - Mini-Review. *AAPS J.* **2008**, *10*, 282–288. [CrossRef]
3. Almeida, P.F.; Araújo, M.G.O.; Santana, J.C.C. Collagen extraction from chicken feet for jelly production. *Acta Sci. Technol.* **2012**, *34*, 345–351. [CrossRef]
4. Gómez-Estaca, J.; Montero, P.; Fernández-Mantín, F.; Gómez-Guillén, M.C. Physico-chemical and film-forming properties of bovine-hide and tuna-skin gelatin: A comparative study. *J. Food Eng.* **2009**, *90*, 480–486. [CrossRef]
5. Haug, I.J.; Draget, K.I.; Smidsrod, O. Physical and rheological properties of fish gelatin compared to mammalian gelatin. *Food Hydrocoll.* **2004**, *18*, 203–213. [CrossRef]
6. Karim, A.A.; Bhat, R. Extraction and characterization of gelatin from Atlantic salmon (*Salmo salar*) skin. *Food Hydrocoll.* **2009**, *23*, 563–576. [CrossRef]
7. Sheela, A.K. Gelatin Market-Global Industry Analysis, Size, Share, Growth, Trends and Forecast, 2012–2018. Available online: http://www.transparencymarketresearch.com/gelatin.html (accessed on 12 October 2019).
8. Widyasari, R.; Rawdkuen, S. Extraction and characterization of gelatin from chicken feet by acid and ultrasound assisted extraction. *Food Appl. Biosci. J.* **2014**, *2*, 83–95.
9. Huda, N.; Seow, E.K.; Normawati, M.N.; Nik Aisyah, N.M. Preliminary study on physicochemical properties of duck feet collagen. *Int. J. Poult. Sci.* **2013**, *12*, 615–621. [CrossRef]
10. ABPA—Brazilian Association of Animal Protein. Summary on the Poultry Sector. Available online: http://abpa-br.com.br/setores/avicultura/publicacoes/relatorios-anuais/2018 (accessed on 15 November 2019).
11. Almeida, P.F.; Lannes, S.C.S. Effects of chicken by-product gelatin on the physicochemical properties and texture of chocolate spread. *J. Texture Stud.* **2016**, *48*, 392–402. [CrossRef]
12. Ji, B.; Wang, J.; Liu, W. Color-based automatic quality control for roasting chicken. *Comput. Electron. Agric.* **2016**, *123*, 49–56. [CrossRef]
13. Kim, J.H.; Han, G.P.; Shin, J.E.; Kil, D.H. Effect of dietary calcium concentrations in phytase-containing diets on growth performance, bone mineralization, litter quality, and footpad dermatitis score in broiler chickens. *Anim. Feed Sci. Technol.* **2017**, *229*, 13–18. [CrossRef]
14. Promket, D.; Ruangwittayanusorn, K.; Somchan, T. The study of carcass yields and meat quality in crossbred native chicken (Chee). *Agric. Agric. Sci. Procedia* **2016**, *11*, 84–89. [CrossRef]

15. Zhao, M.M.; Gong, D.Q.; Gao, T.; Zhang, L.; Li, J.L.; Lv, P.A.; Yu, L.L.; Gao, F.; Zhou, G.H. In ovo feeding of creatine pyruvate increases hatching weight, growth performance, and muscle growth but has no effect on meat quality in broiler chickens. *Livest. Sci.* **2017**, *206*, 59–64. [CrossRef]
16. Argyri, A.A.; Papadopoulou, O.S.; Nisiotou, A.; Tassou, C.C.; Chorianopoulos, N. Effect of high-pressure processing on the survival of Salmonella Enteritidis and shelf-life of chicken fillets. *Food Microbiol.* **2018**, *70*, 55–64. [CrossRef] [PubMed]
17. Noori, S.; Zeynali, F.; Almasi, H. Antimicrobial and antioxidant efficiency of nanoemulsion-based edible coating containing ginger (Zingiber officinale) essential oil and its effect on safety and quality attributes of chicken breast fillets. *Food Control.* **2018**, *84*, 312–320. [CrossRef]
18. Silva, F.; Domingues, F.C.; Nerín, C. Control microbial growth on fresh chicken meat using pinosylvin inclusion complexes based packaging absorbent pads. *LWT Food Sci. Technol.* **2018**, *89*, 148–154. [CrossRef]
19. Sharma, H.; Mendiratta, S.K.; Agrawal, R.K.; Gurunathan, K.; Kumar, S.; Singh, T.P. Use of various essential oils as bio preservatives and their effect on the quality of vacuum packaged fresh chicken sausages under frozen conditions. *LWT Food Sci. Technol.* **2017**, *81*, 118–127. [CrossRef]
20. Yang, S.; Sadekuzzaman, M.; Ha, S.D. Reduction of Listeria monocytogenes on chicken breasts by combined treatment with UV-C light and bacteriophage List Shield. *LWT Food Sci. Technol.* **2017**, *86*, 193–200. [CrossRef]
21. Zouaghi, F.; Cantalejo, M.J. Study of modified atmosphere packaging on the quality of ozonated freeze-dried chicken meat. *Meat Sci.* **2016**, *119*, 123–131. [CrossRef]
22. Bahadori, Z.; Esmaielzadeh, L.; Karimi-Torshizi, M.A.; Seidavi, A.; Olivares, J.; Rojas, A.; Salem, A.Z.M.; Khusro, A.; López, S. The effect of earthworm (Eisenia foetida) meal with vermi-humus on growth performance, hematology, immunity, intestinal microbiota, carcass characteristics, and meat quality of broiler chickens. *Livest. Sci.* **2017**, *202*, 74–81. [CrossRef]
23. Begli, H.E.; Torshizi, R.V.; Masoudi, A.A.; Ehsani, A.; Jensen, J. Relationship between residual feed intake and carcass composition, meat quality and size of small intestine in a population of F2 chickens. *Livest. Sci.* **2017**, *205*, 10–15. [CrossRef]
24. Li, X.-K.; Wang, J.-Z.; Wang, C.-Q.; Zhang, C.-H.; Li, X.; Tang, C.H.; Wei, X.-L. Effect of dietary phosphorus levels on meat quality and lipid metabolism in broiler chickens. *Food Chem.* **2016**, *205*, 289–296. [CrossRef] [PubMed]
25. Sun, Y.Y.; Tang, S.; Chen, Y.; Li, D.L.; Bi, Y.L.; Hua, D.K.; Chen, C.; Luo, Q.Y.; Yang, L.; Chen, J.L. Effects of light regimen and nutrient density on growth performance, carcass traits, meat quality, and health of slow-growing broiler chickens. *Livest. Sci.* **2017**, *198*, 201–208. [CrossRef]
26. Chen, Q.; Cao, M.; Chen, H.; Gao, P.; Fu, Y.; Liu, M.; Wang, Y.; Huang, M. Effects of gamma irradiation on microbial safety and quality of stir fry chicken dices with hot chili during storage. *Radiat. Phys. Chem.* **2016**, *127*, 122–126. [CrossRef]
27. Yang, Y.; Zhuang, H.; Yoon, S.C.; Wang, W.; Jiang, H.; Jia, B. Rapid classification of intact chicken breast fillets by predicting principal component score of quality traits with visible/near-Infrared spectroscopy. *Food Chem.* **2018**, *244*, 184–189. [CrossRef]
28. Rahman, N.; Jamalulail, S. Extraction, physicochemical characterizations and sensory quality of chicken feet gelatin. *Borneo Sci.* **2012**, *30*, 1–13.
29. Sarbon, M.N.; Badii, F.; Howell, N.K. Preparation and characterization of chicken skin gelatin as an alternative to mammalian gelatin. *Food Hydrocoll.* **2013**, *30*, 143–151. [CrossRef]
30. Zhou, C.; Li, Y.; Yu, X.; Yang, H.; Ma, H.; ElGasim, A.; Yagoub, A.; Cheng, Y.; Hu, J.; Naa, P.; et al. Extraction and characterization of chicken feet soluble collagen. *LWT-Food Sci. Technol.* **2016**, *74*, 145–153. [CrossRef]
31. Irwandi, J.; Faridayanti, S.; Mohamed ES, M.; Hamzah, M.S.; Torla, H.H.; Che Man, Y.B. Extraction and characterization of gelatin from different marine fish species. *Int. Food Res. J.* **2009**, *16*, 381–390.
32. Almeida, S.S.; Alves, W.A.L.; Araújo, S.A.; Santana, J.C.C.; Narain, N.; Souza, R.R. Use of simulated annealing in standardization and optimization of the acerola wine production. *Food Sci. Technol.* **2014**, *34*, 292–297. [CrossRef]
33. Benvenga, M.A.C.; Librantz, A.F.H.; Santana, J.C.C.; Tambourgi, E.B. Genetic algorithm applied to study of the economic viability of alcohol production from Cassava root from 2002 to 2013. *J. Clean. Prod.* **2016**, *113*, 483–494. [CrossRef]
34. Curvelo Santana, J.C.; Araújo, S.A.; Librantz, A.F.H.; Tambourgi, E.B. Optimization of Corn Malt Drying by Use of a Genetic Algorithm. *Dry. Technol.* **2010**, *28*, 1236–1244. [CrossRef]

35. Severo Jr, J.B.; De Sá Oliveira, L.S.; Sardeiro, F.S.; De Souza, R.R.; Lopes, F.L.G.; Santana, J.C.C.; Tambourgi, E.B. Response surface methodology to evaluation the recovery of amylases by hollow fiber membrane. *Braz. Arch. Biol. Technol.* **2007**, *50*, 713–718. [CrossRef]
36. AOAC. *Official Methods of Analysis*, 17th ed.; Association of Official Analytical Chemists Inc: Arlington, TX, USA, 2000.
37. Santana, J.C.C.; Dias, C.G.; De Souza, R.R.; Tambourgi, E.B. Applying of neural network on the wine sensorial analysis from barbados cherry. *J. Food Process. Eng.* **2010**, *33*, 365–378. [CrossRef]
38. Severo, J.B., Jr.; Almeida, S.S.; Narain, N.; Souza, R.R.; Santana, J.C.C.; Tambourgi, E.B. Wine clarification from Spondias mombin L. pulp by hollow fiber membrane system. *Process. Biochem.* **2007**, *42*, 1516–1520. [CrossRef]
39. Samant, S.S.; Seo, H.-S. Quality perception and acceptability of chicken breast meat labeled with sustainability claims vary as a function of consumers' label-understanding level. *Food Qual. Prefer.* **2016**, *49*, 151–160. [CrossRef]
40. Berssaneti, F.T.; Assumpção, A.; Nakao, O.S. Engineering, procurement and construction (EPC): What are the variables that impact the success of the projects currently running in Brazil? *Gestao Producao* **2014**, *21*, 95–109. [CrossRef]
41. Farias, T.M.B.; Santana, J.C.C.; Calarge, F.A. Stability evaluation of amitraz (C19H23N3) in acaricides through analysis FT- IR Spectroscopy. *J. Anim. Vet. Adv.* **2012**, *11*, 3531–3535. [CrossRef]
42. Ferreira, J.F.; Santana, J.C.C.; Tambourgi, E.B. The effect of pH on bromelain partition from Ananas comosus by PEG4000/Phosphate ATPS. *Braz. Arch. Biol. Technol.* **2011**, *54*, 713–718. [CrossRef]
43. Cecchi, H.M. *Theoretical and Practical Foundations in Food Analysis*, 2nd ed.; Unicamp: Campinas, Brazil, 2003; 207p. (In Portuguese)
44. Lannes, S.C.S.; Medeiros, M.L. Cupuaçu chocolate drink powder processed by spray-dryer. *Braz. J. Pharm. Sci.* **2003**, *39*, 115–123.
45. Klepa, R.B.; Medeiros, M.F.; Franco, M.A.C.; Tamberg, E.T.; Farias, T.M.D.B.; Paschoalin Filho, J.A.; Berssaneti, F.T.; Santana, J.C.C. Reuse of construction waste to produce thermoluminescent sensor for use in highway traffic control. *J. Clean. Prod.* **2019**, *209*, 250–258. [CrossRef]
46. Watanabe, S.; Farias, T.M.B.; Gennari, R.F.; Ferraz, G.M.; Kunzli, R. Chemical process to separate iron oxides particles in pottery samples for EPR dating. *SSA* **2007**, *10*, 10–18. [CrossRef] [PubMed]
47. Habte, G.; Hwang, I.M.; Kim, J.S.; Hong, J.H.; Hong, Y.S.; Choi, J.Y.; Nho, E.Y.; Jamila, N.; Khan, N.; Kim, K.S. Elemental profiling and geographical differentiation of Ethiopian coffee samples through inductively coupled plasma-optical emission spectroscopy (ICP-OES), ICP-mass spectrometry (ICP-MS) and direct mercury analyzer (DMA). *Food Chem.* **2016**, *212*, 512–520. [CrossRef] [PubMed]
48. Oliveira, A.P. Determinação de Cu, Zn e Se em soro humano por ICP-MS. *Revista Analytica* **2006**, *25*, 12–18.
49. Bureau, S.; Ruiz, D.; Reich, M.; Gouble, B.; Bertrand, D.; Audergon, J.-M.; Renard, C.M.G.C. Rapid and non-destructive analysis of apricot fruit quality using FT-near-infrared spectroscopy. *Food Chem.* **2009**, *113*, 1323–1328. [CrossRef]
50. Doyle, B.B.; Blout, E.R.; Bendit, E.G. Infrared spectroscopy of collagen and collagen like polypeptides. *Biopolymers* **1975**, *14*, 937–957. [CrossRef]
51. Muyonga, J.H.; Cole, C.G.B.; Duodu, K.G. Fourier transform infrared (FTIR) spectroscopic study of acid soluble collagen and gelatin from skins and bones of young and adult Nile perch (*Lates niloticus*). *Food Chem.* **2004**, *86*, 325–333. [CrossRef]
52. Fernandez-Diaz, M.D.; Montero, P.; Gomez-Guillen, M.C. Gel properties of collagens from skins of cod (Gadus morhua) and hake (Merluccius merluccius) and their modification by the coenhancers magnesium sulphate, glycerol and transglutaminase. *Food Chem.* **2001**, *74*, 161–167. [CrossRef]
53. Norziah, M.H.; Al-hassan, A.; Khairulnizam, A.B.; Mordi, M.N.; Norita, M. Characterization of fish gelatin from surimi processing wastes: Thermal analysis and effect of transglutaminase on gel properties. *Food Hydrocoll.* **2009**, *23*, 1610–1616. [CrossRef]
54. Giménez, B.; Gómez-Estaca, J.; Alemán, A.; Gómez-Guillén, M.C.; Montero, M.P. Physico-chemical and film forming properties of giant squid (Dosidicus gigas) gelatin. *Food Hydrocoll.* **2009**, *23*, 585–592. [CrossRef]
55. Gabas, V.G.S.; Cavalcanti, O.A. Influência da adição de goma arábica em filmes isolados de polímero acrílico. Estudo das propriedades de intumescimento e de permeabilidade. *Rev. Bras. Ciên. Farm.* **2003**, *39*, 441–448. [CrossRef]

56. Cavalcanti, O.A.; Van Der Mooter, G.; Caramico-Soares, I.; Kinget, R. Polysaccharides as excipients for colon-specific coatings, permeability and swelling properties of casted films. *Drug Dev. Ind. Pharm.* **2002**, *28*, 157–164. [CrossRef] [PubMed]
57. Lim, J.; Oh, S.; Kim, K.-O. The effects of processing conditions on the properties of chicken feet gelatin. *Food Sci. Biotechnol.* **2001**, *10*, 638–645.
58. Hao, S.; Li, L.; Yang, X.; Cen, J.; Shi, H.; Bo, Q.; He, J. The characteristics of gelatin extracted from sturgeon (Acipenser baeri) skin using various pretreatments. *Food Chem.* **2009**, *115*, 124–128. [CrossRef]
59. Jamilah, B.; Harvinder, K.G. Properties of gelatins from skins of fish-black tilapia (Oreochromis mossambicus) and red tilapia (Oreochromis nilotica). *Food Chem.* **2002**, *77*, 81–84. [CrossRef]
60. Cheow, C.S.; Norizah, M.S.; Kyaw, Z.Y.; Howell, N.K. Preparation and characterisation of gelatins from the skins of sin croaker (*Johnius dussumieri*) and shortfin scad (*Decapterus macrosoma*). *Food Chem.* **2007**, *101*, 386–391. [CrossRef]
61. Cho, S.-H.; Jahncke, M.L.; Chin, K.-B.; Eun, J.-B. The effect of processing conditions on the properties of gelatin from skate (Raja Kenojei) skins. *Food Hydrocoll.* **2006**, *20*, 810–816. [CrossRef]
62. Cliché, S.; Amiot, J.; Avezard, C.; Gariépy, C. Extraction and characterization of collagen with or without telopeptides from chicken skin. *Poult. Sci.* **2003**, *82*, 503–509. [CrossRef]
63. Ahmad, M.; Benjakul, S. Characteristics of gelatin from the skin of unicorn leatherjacket (*Aluterus monoceros*) as influenced by acid pretreatment and extraction time. *Food Hydrocoll.* **2011**, *25*, 381–388. [CrossRef]
64. Nicolai, B.M.; Beullens, K.; Bobelyn, E.; Peirs, A.; Saeys, W.; Theron, K.I.; Lammertyn, J. Nondestructive measurement of fruit and vegetable quality by means of NIR spectroscopy: A review. *Postharvest Biol. Technol.* **2007**, *46*, 99–118. [CrossRef]
65. Clark, C.J.; McGlone, V.A.; Requejo, C.; White, A.; Woolf, A.B. Dry matter determination in 'Hass' avocado by NIR spectroscopy. *Postharvest Biol. Technol.* **2003**, *29*, 300–307. [CrossRef]
66. Wu, D.; He, Y.; Nie, P.; Cao, F.; Bao, Y. Hybrid variable selection in visible and near-infrared spectral analysis for non-invasive quality determination of grape juice. *Anal. Chim. Acta* **2010**, *659*, 229–237. [CrossRef] [PubMed]
67. Ziegler, F.F.; Sgarbieri, V.C. Caracterização químico-nutricional de um isolado proteico de soro de leite, um hidrolisado de colágeno bovino e misturas dos dois produtos. *Rev. Nutr.* **2009**, *22*, 61–70. [CrossRef]
68. Peng, Y.Y.; Glattauer, V.; Ramshaw, J.A.M.; Werkmeister, J.A. Evaluation of the immunogenicity and cell compatibility of avian collagen for biomedical application. *J. Biom. Mater. Res.* **2010**, *93*, 1235–1244. [CrossRef] [PubMed]

© 2020 by the authors. Licensee MDPI, Basel, Switzerland. This article is an open access article distributed under the terms and conditions of the Creative Commons Attribution (CC BY) license (http://creativecommons.org/licenses/by/4.0/).

polymers

MDPI

Article

Adsorptive Removal of Antibiotic Ciprofloxacin from Aqueous Solution Using Protein-Modified Nanosilica

Tien Duc Pham [1,*], Thi Ngan Vu [1], Hai Long Nguyen [2], Pham Hai Phong Le [2] and Thi Sim Hoang [1]

1 Faculty of Chemistry, VNU University of Science, Vietnam National University, Hanoi, 19 Le Thanh Tong, Hoan Kiem, Hanoi 100000, Vietnam; vuthingan_t62@hus.edu.vn (T.N.V.); simhoang.hus@gmail.com (T.S.H.)

2 HUS High School for Gifted Students, VNU University of Science, Vietnam National University, Hanoi, 182 Luong The Vinh, Thanh Xuan, Hanoi 100000, Vietnam; ceonguyenhailong@gmail.com (H.L.N.); lephamhaiphong@gmail.com (P.H.P.L.)

* Correspondence: tienduchphn@gmail.com or tienducpham@hus.edu.vn; Tel.: +84-243-825-3503; Fax: +84-243-824-1140

Received: 1 December 2019; Accepted: 20 December 2019; Published: 1 January 2020

Abstract: The present study aims to investigate adsorptive removal of molecular ciprofloxacin using protein-modified nanosilica (ProMNS). Protein was successfully extracted from Moringa seeds while nanosilica was synthesized from rice husk. Fourier-transform infrared (FTIR), ultraviolet visible (UV-Vis) and high-performance liquid chromatography (HPLC) were used to evaluate the characterization of protein. Adsorption of protein onto nanosilica at different pH and ionic strength was thoroughly studied to modify nanosilica surface. The removal efficiency of antibiotic ciprofloxacin (CFX) increased from 56.84% to 89.86% after surface modification with protein. Effective conditions for CFX removal using ProMNS were systematically optimized and found to be pH 7.0, adsorption time 90 min, adsorbent dosage 10 mg/mL, and ionic strength 1 mM KCl. A two-step model was successfully used to fit the adsorption isotherms of CFX onto ProMNS at different ionic strength while a pseudo-second-order model could fit adsorption kinetic of CFX onto ProMNS very well. Maximum adsorption capacity was very high that reached to 85 mg/g. Adsorption of CFX onto ProMNS decreased with increasing KCl concentration, suggesting that adsorption of CFX onto ProMNS is mainly controlled by electrostatic attraction between positively charged ProMNS surface and anionic species of CFX. Adsorption mechanisms of CFX onto ProMNS were discussed in detail based on adsorption isotherms, the change in surface charge by zeta potentail and the change in functional groups by FT-IR. The removal of CFX after three regenerations was greater than 73% while CFX removal from an actual hospital wastewater using ProMNS reached to 70%. Our results suggest that ProMNS is a new and eco-friendly adsorbent to remove antibiotics from aqueous solutions.

Keywords: ciprofloxacin; adsorption; protein; moringa seeds; nanosilica; two-step model

---

1. Introduction

Antibiotics, which are synthetic chemicals or natural products, are a powerful medicine to bacterial infections. Ciprofloxacin (CFX), which is one of the second generated group of synthetic Fluoroquinolones (FQs), is an important antibiotic for the treatment of both gram-negative and positive bacterial infections. The CFX makes a large broader antibacterial spectrum so that it can exhibit greater intrinsic antibacterial activity [1–3]. The CFX can also make a resonance with other antibiotic groups such as beta-lactam or aminoglycosid that have very high performance in disease treatment. Nevertheless, one unexpected thing is that CFX resistance is rapidly increasing to seriously high levels in developing countries. The antibiotic residual including CFX is also very toxic to organisms in an aquatic environment although the concentrations are low [4].

Removal of antibiotic is important for environmental concern, because antibiotic residual in aqueous solution can cause antibiotic resistance. Various techniques have been developed and used for antibiotic removal, such as membrane process [5–7], photocatalytic degradation [8], advanced oxidation [9], and adsorption [10–12]. Many studies indicate that adsorption is one of the most effective and suitable method for antibiotics removal toward green chemistry by using low eco-friendly adsorbents [13,14].

Protein is a mixture of various amino acids that is an origin of life [15,16]. Protein adsorption onto particle surface plays an important role in the biological environment so that numerous researches on protein adsorption onto solid surface have been investigated [17]. Conformation of stable protein on a negatively charged surface has been found, demonstrating that protein adsorption onto particles can be applied to modify a solid surface [18]. It is reported that Moringa (MO) seeds contain a high amount of protein, which have anti-macrobial potential [19]. Protein extracted from MO seeds as an effective coagulant reagent for water treatment was thoroughly studied [20,21]. Nevertheless, application of adsorption of protein MO seeds for antibiotic removal has not been reported.

Silica, which is one of the most popular oxide in nature, is basically negatively charge. Silica is also easily synthesized from agricultural sub-product rice husk [22]. Due to the low charge density and non-porous structure of nanosilica from rice husk, the low adsorption of antibiotic onto nanosilica is evident. Surface modification of nanosilica by protein adsorption may be useful to remove of antibiotic. Many studies investigated adsorption different kinds of proteins onto silica to find the characteristics of protein on silica surface [15,18]. However, protein adsorption onto nanosilica to enhance removal of antibiotic CFX has not been studied.

An understanding of adsorption mechanisms of antibiotic onto protein-modified nanosilica (ProMNS) is needed for further application to a real system in the presence of many interferences. For this purpose, adsorption isotherms by different modes is often evaluated. The two-step model proposed by Zhu et al. [23], with a general adsorption isotherm equation, was successfully applied to various types of antibiotic adsorbates for numerous systems [4,24,25]. Additionally, the two-step model is able to evaluate the growth of adsorbed layers [26,27]. Therefore, the two-step model is suitable to fit adsorption isotherms of CFX onto ProMNS.

In this paper, for the first time, we report adsorptive removal of CFX using nanosilica fabricated from rice husk with surface modification by adsorption of MO seeds protein. The effective conditions on CFX removal were systematically studied. Adsorption mechanisms are suggested based on surface modification by Fourier transform infrared spectroscopy (FTIR), the change in surface charge by ζ potential measurements, and adsorption isotherm. The application of the optimum condition for CFX removal for actual hospital wastewater is also investigated in the present study.

2. Experimental

2.1. Materials

The Moringa (MO) seeds used in the present study were purchased from Minh Tue Company (Hanoi, Vietnam) while rice husk was collected from Bacninh, Vietnam. Nanosilica was synthesized and characterized by the hydrothermal method according to our previously published papers [4,22,24]. Ciprofloxacin hydrochloride monohydrate (CFX) (CAS 86393-32-0) with purity higher than 98% (HPLC grade) was supplied by Tokyo Chemical Industry (Tokyo, Japan). The chemical structure of CFX is shown in Figure 1. Solid ammonium sulfate $(NH_4)_2SO_4$ and acetone to synthesize protein was purchased from Scharlau (Spain, EU). Amino acid standards were acquired from Sigma Aldrich (St. Louis, MO, USA). Methanol and Acetonitrile for HPLC were supplied from Merck. The KCl, HCl, and KOH (p.A, Merck) were used to study the effect of ionic strength and adjust pH solution. Solution pH was measured using an HI 2215 pH meter (Hanna, Woonsocket, RI, USA). Before use, the pH electrode was calibrated with three standard buffers of 4.01, 7.01, and 10.01 (Hanna). Other chemicals with analytical grade were also supplied from Merck. An ultrapure water produced from the ultrapure water system (Labconco, Kansas City, MO, USA) with a resistivity of 18.2 MΩcm was used to prepare all aqueous solutions.

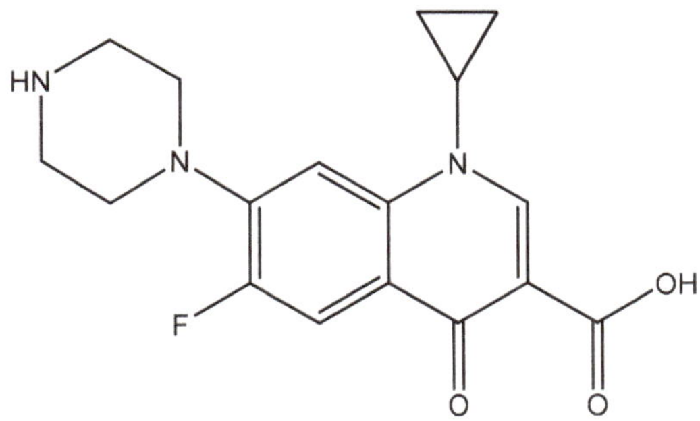

Figure 1. Chemical structure Cirofloxacin (CFX).

2.2. Fabrication and Purification of Protein from Moringa Seeds

Protein was extracted from Moringa (MO) seeds according to Kwaambwa et al. [16,20,28], with a modification. The MO seeds were shelled before drying at 40 °C for one week and milling to powder by mortar and pestle (Figure 2A). About 50.0 g of milled MO seed powder was mixed and extracted by with 100 mL petroleum ether to remove oil for 10 h using Orbital shaker. The extraction process was repeated 4 times with every 50 mL petroleum ether. The solids were dissolved in 200 mL pure water and filtered by filtration paper. To remove MO oil completely, liquid–liquid extraction was repeated 3 times with every 10 mL petroleum ether. Then, protein in aqueous filtration was precipitated by slowly adding solid $(NH_4)_2SO_4$. Protein was obtained by centrifuging. Finally, powder of protein was formed with acetone before freeze-drying (Figure 2B). Purification of protein was conducted with dialysis using cellulose membrane (Sigma Aldrich). The lyophilized protein powder was kept at a fridge in a dark glass bottle.

Figure 2. Photos of (**A**) milled Moringa seeds, and (**B**) synthesized protein.

2.3. Characterization and Analytical Methods

The protein synthesized from Moringa (MO) seeds was characterized by ultraviolet-visible spectroscopy (UV-Vis), Fourier transform infrared spectroscopy (FT-IR), and high-performance liquid chromatography (HPLC) with fluorescence detection (FLD) and photo diode array detection (PDA).

The UV-Vis measurements were carried out by a double-beam spectrophotometer using a couple of quartz cuvettes with a 1 cm optical path length using (UV-1650 PC, Shimadzu, Japan).

The FTIR spectra were conducted with an Affinity-1S spectrometer (Shimadzu, Japan). All spectra were obtained at a resolution of 4 cm^{-1} and at 25 °C and atmospheric pressure.

The HPLC-FLD and HPLC-PDA with two columns were used to quantify amino acids in MO seeds protein [29,30]. Amino acids were hydrolyzed by 6 M NaOH for 24 h at 125 °C. Then, the samples adjusted to neutral pH by HCl. After that, the solutions were filtered and diluted appropriated before injecting into HPLC systems. The reversal phase column RP18 AccQ Tag (150 mm × 4.6 mm × 3.9 µm) (Water, Milford, MA, USA) and mobile phase using acetate phosphate buffer (pH 5.05) and acetonitrile with gradient elution were used. The total time for separation with a flow rate of 1.0 mL/min was 23 min. Eluted peaks were monitored by PDA detector with a wavelength of 260 nm [30]. To quantify Tryptophan, the column Xbridge (150 mm × 4.6 mm × 5 µm) (Water, Milford, MA, USA) and mobile phase using H_2O:acetonitrile (*v/v*: 95:5) with isocratic elution were used. The excited wavelength and emission wavelength were λex = 295 nm and λem = 345 nm, respectively [31].

The surface charges of nanosilica, protein-modified nanosilica (ProMNS), and ProMNS after CFX adsorption were examined by zeta (ζ) potential measurements. The ζ potential was calculated from electrophoretic mobility with Smoluchowski's equation [32]:

$$\zeta = \frac{u_e\,\eta}{\varepsilon_{rs}\,\varepsilon_0} \tag{1}$$

where ζ is the zeta potential (mV), u_e the electrophoretic mobility (µms^{-1}/Vcm^{-1}), η is the dynamic viscosity of the liquid (mPa s), ε_{rs} is the relative permittivity constant of the electrolyte solution, and ε_0 is the electric permittivity of vacuum (8.854 × 10^{-12} F/m).

2.4. Adsorption Studies

All adsorption experiments were conducted by batch mode in 15 mL Falcon tubes at 25 ± 2 °C controlled by an air conditioner. Initial amount of protein was precisely weighed and then diluted with ultrapure water to a stock solution of 2500 mg/L. Then, the stock solution was appropriately diluted to prepare a series of protein concentration for protein adsorption isotherms.

A 10 mg/mL nanosilica was mixed in 10 mL of protein in the range of 50 to 2600 mg/L for 2 h. After that, the protein solutions were separated using ultra-centrifuging at 12,000 rpm (5 °C) for 10 min with a refrigerated centrifuge (MR23i, JOUAN, France). The effects of pH and ionic strength on adsorption of protein were systematically studied. The concentrations of protein were quantified by UV-Vis spectroscopy and HPLC-PDA. The adsorption capacity Γ (mg/g) of protein or CFX onto nanosilica, ProMNS was calculated by Equation (2):

$$\Gamma = \frac{C_i - C_f}{m} \tag{2}$$

where C_i (mg/L) and C_f (mg/L) are the initial and the final concentrations of protein or CFX, respectively, while m (mg/mL) is adsorbent dosage.

For adsorptive removal of CFX, a different adsorbent dosage was mixed well with 10 mL protein using an orbital shaker orbital OS-350D (Digisystem laboratory, Taiwan) under optimum conditions to modify the nanosilica surface. The adsorbent was then washed with ultrapure water before adding the concentrations of CFX. The concentrations of CFX were determined by UV–Vis spectroscopy. The relationship between the concentrations of CFX and measured absorbance and at a wavelength of 272 nm as standard calibration curves in different conditions with a correlation coefficient of at least 0.999 was confirmed.

The removal (% R) of CFX was calculated by Equation (3):

$$Removal\ (\%R) = \frac{C_i - C_e}{C_i} \times 100\%. \tag{3}$$

The adsorption isotherms of CFX onto ProMNS were fitted by two-step model using a general isotherm equation that was successfully described beta-lactam onto polyelectrolyte modified nanosilica [24]. The general isotherm equation [23] is

$$\Gamma = \frac{\Gamma_{\infty} k_1 C\left(\frac{1}{n} + k_2 C^{n-1}\right)}{1 + k_1 C\left(1 + k_2 C^{n-1}\right)} \quad (4)$$

where Γ_{∞} (mg/g) is the maximum CFX adsorption, Γ_{∞} can be determined from the data of adsorption isotherm at high CFX concentrations, k_1 (g/mg), and k_2 $(g/mg)^{n-1}$ are equilibrium constants for first layer adsorption and clusters of n molecules. C (mg/L) denotes the equilibrium concentrations of CFX in solution.

3. Results and Discussion

3.1. Characterizations of Protein from Moringa Seeds

The protein extracted from Moringa (MO) seeds was characterized by UV-Vis, FTIR, and HPLC methods.

The UV-Vis spectrum of MO seeds protein was indicated in Figure 3.

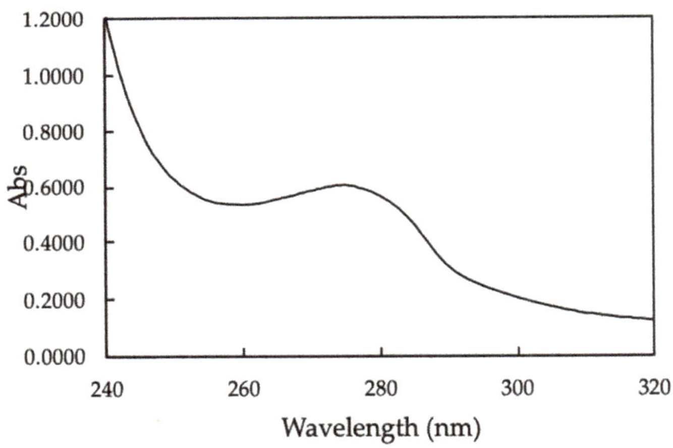

Figure 3. The UV-Vis spectrum of Moringa (MO) seeds protein.

The UV-Vis spectrum of MO seeds shows a broadened peak with maximum absorbance at 278 nm, which indicate the natural characteristic of protein [33].

The FTIR spectrum of the MO seeds protein indicated in Figure 4 indicates that the peaks of 1643.35, 1537.27, 1514.12, and 1409.96 cm^{-1} were assigned for C=O stretching amide I, NH amide II, NH amide I bending, and C-H stretching amide III, respectively. These results are in good agreement with previously published paper [34]. In addition, the peaks at 2983.88 and 2931.80 cm^{-1} associated with C-H alkyl in amino acids structures. The FT-IR results confirmed that the presence of amino acids with functional groups are evident.

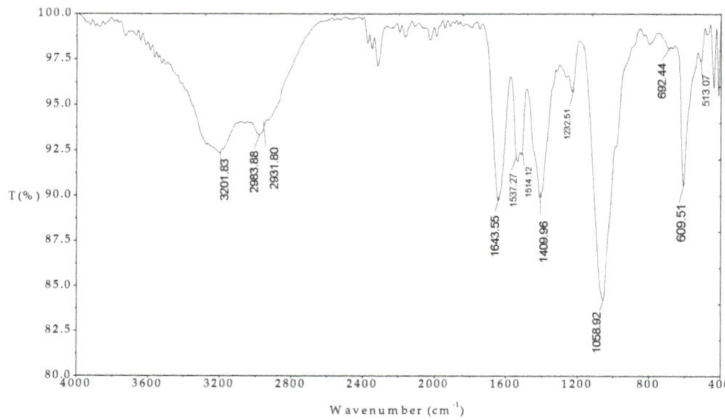

Figure 4. FTIR spectrum of Moringa (MO) seeds protein in the wave number range 400–4000 cm^{-1}.

The content of existed amino acids in MO seeds protein determined by HPLC-PDA and HPLC-FLD indicated Table 1 shows that the positive amino acids such as Arginine and polar side chain amino acid such as Glutamine have higher content of 4.41% and 5.00%, respectively. These results agree well with protein extracted from MO by Kwaambwa et al. [20]. It implies that high positive charge of amino acids in protein is dominant than negative ones. This results in high isoelectric point (IEP) of MO seeds protein, which is good for surface modification of nanosilica.

Table 1. The amino acids contents in protein extracted from Moringa (MO) seeds.

Amino Acid	Content (%)	Standard Deviation
Histidine	0.70	0.02
Serine	0.67	0.08
Glycine	1.14	0.19
Arginine	4.41	0.38
Aspartic	0.68	0.21
Glutamine	5.00	0.62
Threonine	0.56	0.07
Alanine	0.88	0.12
Proline	1.55	0.12
Cystine	0.38	0.04
Lysine	0.35	0.10
Tyrosine	0.57	0.01
Methionine	0.52	0.01
Valine	0.92	0.07
Isoleucine	0.91	0.11
Leucine	1.49	0.13
Phenylalanine	1.40	0.03
Tryptophan	0.36	0.03
Total	22.10	0.20

3.2. *Adsorption of Protein onto Nanosilica*

3.2.1. Effect of pH on Protein Adsorption

Figure 5 indicates that MO seed protein adsorption capacity calculated form Equation (2) increases with increasing in the pH range of 3–10 because of an increase of absolute negative charge of nanosilica with increasing pH. From pH 10 to 11, adsorption decreased dramatically due to the dissolution of silica at high pH [35]. Another reason is due to the IEP of protein is about 10 [28], so that protein becomes negative charge at pH > 10. The electrical repulsion force causes the less adsorption between protein and nanosilica. Therefore, pH 10 is optimum to modify nanosilica surface that is kept for further studies.

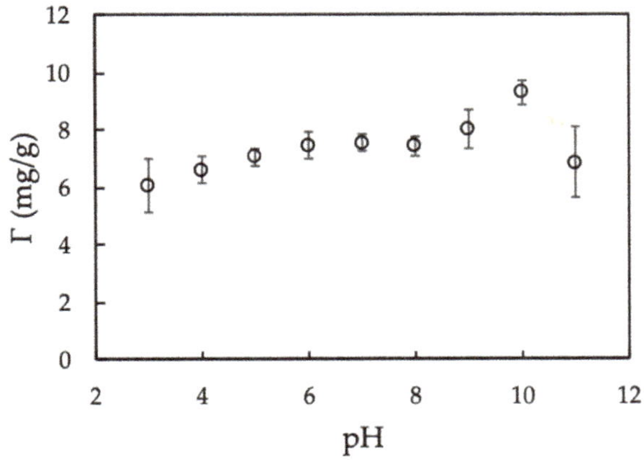

Figure 5. Effect of pH on Moringa seeds protein adsorption onto nanosilica. (C_i (protein) = 100 mg/L, adsorbent dosage 10 mg/mL, 10 mM KCl). Error bars show standard deviations of three replicates.

3.2.2. Effect of Ionic Strength on Protein Adsorption

Adsorption isotherms at different salt concentrations can demonstrate the effect of ionic strength because the influence of salt strongly induce the electrostatic interactions [36].

Figure 6 shows that at pH 10, adsorption capacities of MO seeds protein onto nanosilica using Equation (2) is independent on ionic strength. Although KCl concentration increased 10 times, the adsorption capacity did not change for different initial concentrations of protein. This trend is similar to adsorption of β-lactoglobulin onto nanosilica at pH 6 [18]. Also, all experimental data are highly repeatable with very small standard deviations of replications. When increasing ionic strength, the number of counter cations increase on the negatively charged nanosilica surface. As a result, a decrease of electrostatic attraction between of protein nanosilica occurred. Nevertheless, protein adsorption onto nanosilica still remains so that other interactions, such as hydrogen bonding, hydrophobic, and lateral interaction can control adsorption [37], demonstrating that protein adsorption on nanosilica is induced by both electrostatic and nonelectrostatic interactions.

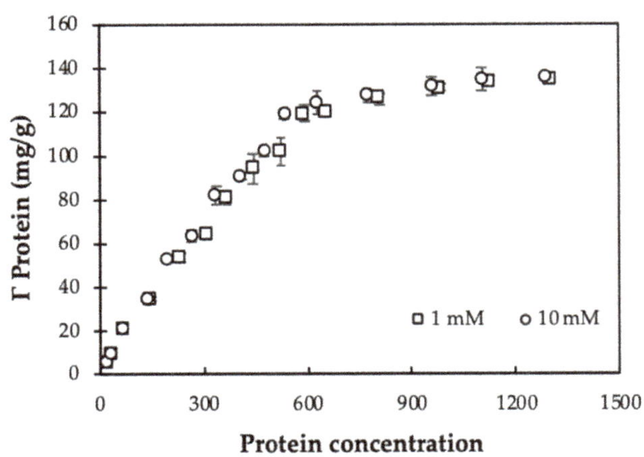

Figure 6. Adsorption isotherms of Moringa (MO) seeds protein onto nanosilica at different KCl concentrations (pH 10). Error bars show standard deviations of three replicates.

Adsorption of MO seeds protein onto nanosilica at different ionic strengths reaches equilibrium when protein concentration is 2000 mg/L. Thus, the initial protein concentration of 2000 mg/L is suitable for modification of nanosilica surface to enhance removal of CFX.

3.3. *Adsorptive Removal of Ciprofloxacin (CFX) Using Protein-Modified Nanosilica (ProMNS)*

3.3.1. Effect of pH

The solution pH plays the most important role for CFX adsorption onto ProMNS. The removal of CFX calculated by Equation (3) is strongly influenced by pH because of CFX charging behavior and surface charge of ProMNS. Figure 7 shows the effect of initial pH on CFX removal from pH 3 to 11 in 1 mM KCl.

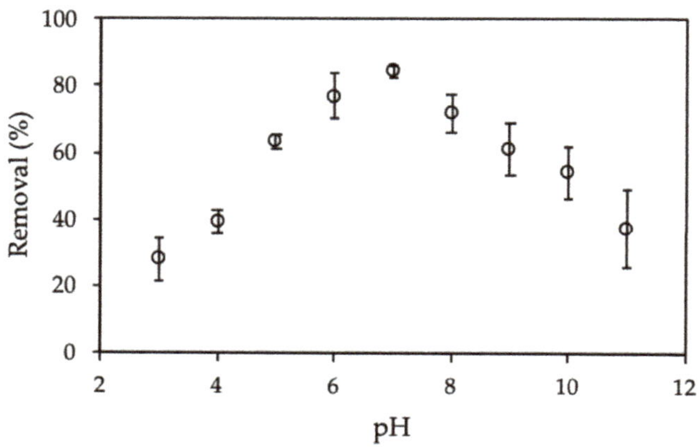

Figure 7. Effect of pH on removal of CFX using protein-modified nanosilica (ProMNS). (C_i (CFX) = 20 mg/L, contact time 90 min, adsorbent dosage 10 mg/mL, 1 mM KCl). Error bars show standard deviations of three replicates.

As seen in Figure 7, CFX removal increased dramatically with an increase of pH in the range of 3–7, then the removal decreased significantly from pH 7 to 11. At pH < 6.09 ($pK_{a,1}$), CFX has positive charge while at pH higher than 8.62 ($pK_{a,1}$), CFX has a negative charge [38]. It implies that at pH 7, CFX is zwitterionic form. However, the maximum removal of CFX is achieved at pH 7, suggesting that CFX adsorption onto ProMNS due to non-electrostatic interactions. On the one hand, at pH < 6.0, CFX removal is low due to the repulsive force between positive CFX species and positively charged ProMNS surface. On the other hand, at pH > 9.0, CFX removal is also low due to the desorption of protein and the dissolution of silica at high pH [39,40]. It should be noted that isoelectric point (IEP) of MO seeds protein is about 9.0 so that the negative form of protein can be taken place at pH > 9.0 [28]. Therefore, solution pH 7.0 is optimum for CFX removal using ProMNS and pH 7.0 was fixed for further investigation.

3.3.2. Effect of Adsorption Time

Adsorption time influences the equilibrium process of CFX onto ProMNS. The effect of adsorption time on CFX removal using ProMNS is indicated in Figure 8. As can be seen in Figure 8, CFX removal using Equation (3) increased sharply with increasing adsorption time from 5 min to 30 min. Then, CFX removal still increased when adsorption time increased to 90 min. After 90 min, removal of CFX changes insignificantly, indicating that adsorption equilibrium reaches 90 min. The adsorption time for CFX using ProMNS in this case is faster than CFX adsorption onto tea leaves biochar in which the equilibrium adsorption time is 540 min. Thus, adsorption time 90 min is kept for further study on CFX removal using ProMNS.

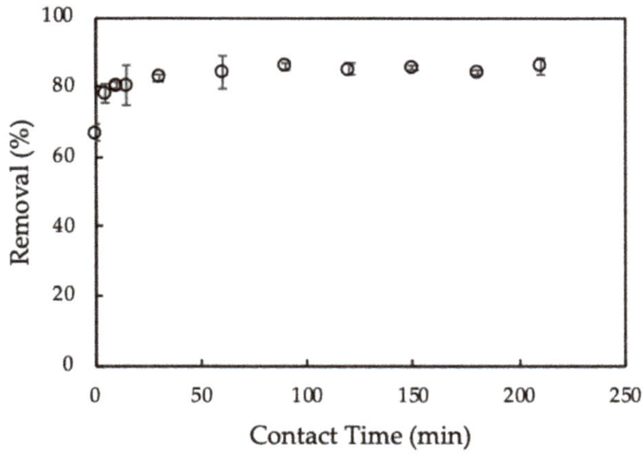

Figure 8. Effect of adsorption time on removal of ciprofloxacin (CFX) using protein-modified nanosilica (ProMNS). C_i (CFX) = 20 mg/L, pH 7.0, adsorbent dosage 10 mg/mL, 1 mM KCl). Error bars show standard deviations of three replicates.

3.3.3. Effect of Adsorbent Dosage

The adsorbent dosage is effective effect on the adsorption process, because it influences the total surface area and charge density of the adsorbent [4,41–43]. The dosage of ProMNS varied from 1.0 to 50.0 mg/mL (Figure 9).

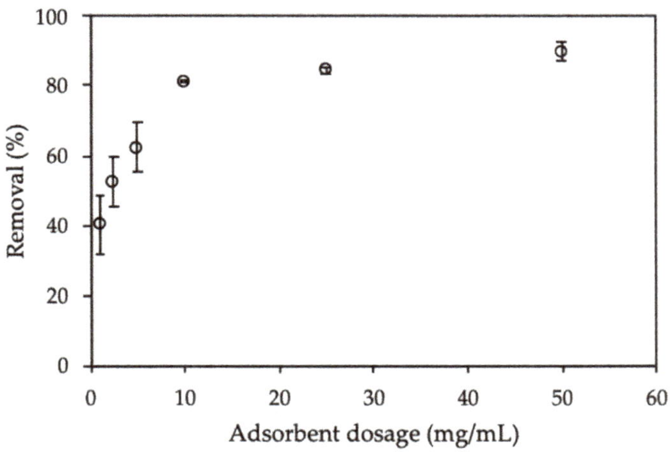

Figure 9. Effect of adsorbent dosage on removal of CFX using protein-modified nanosilica (ProMNS). C_i (CFX) = 20 mg/L, pH 7.0, adsorption time 90 min, 1 mM KCl). Error bars show standard deviations of three replicates.

Figure 9 shows that CFX removal using ProMNS calculated by Equation (3) increased dramatically with an increase of adsorbent dosage from 1.5 to 10 mg/mL. This phenomenon is due to a large number of binding sites for adsorption or the enhancement of specific surface area with increasing adsorbent amount [44]. Nevertheless, when increasing adsorbent dosage higher than 10 mg/mL, CFX removal changed insignificantly. The error bar showing the deviations of three replicates with 10 mg/mL is also smallest comparing with other dosages. It implies that an adsorbent dosage of 10 mg/mL is suitable for CFX removal through adsorption technique using ProMNS.

3.3.4. Effect of Ionic Strength

Ionic strength influences the electrostatic interaction between the positively charged ProMNS surface and CFX molecular. Figure 10 shows the results of CFX removal using ProMNS calculated by Equation (3) at KCl concentration from 0 to 100 mM. As can be seen, at $C_{KCl} < 1$ mM, CFX removal change slightly. However, CFX removal decreases dramatically when increasing salt from 1 to 100 mM KCl. It implies that the electrostatic interaction between CFX molecular and ProMNS is screened with an increase of KCl concentration. As a result, CFX removal using ProMNS is highly influenced by the screening of electrostatic force. The effect of ionic strength is investigated and discussed in detail by adsorption isotherms described below.

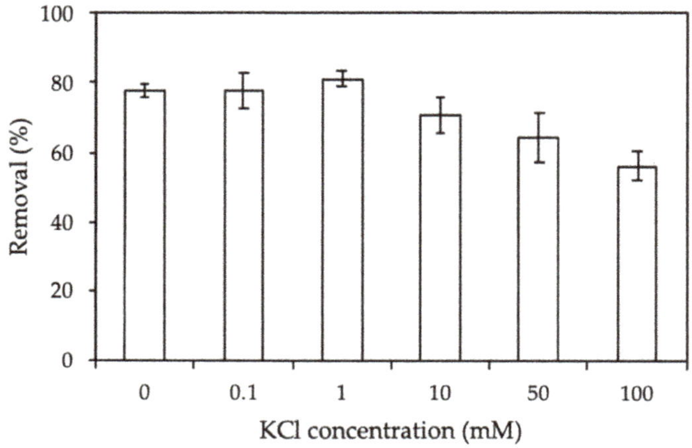

Figure 10. Effect of ionic strength on removal of CFX using protein-modified nanosilica (ProMNS). (C_i (CFX) = 20 mg/L, pH 7.0, adsorption time 90 min, adsorbent dosage 10 mg/mL). Error bars show standard deviations of three replicates.

3.4. *Adsorption Isotherms of CFX onto Protein-Modified Nanosilica (ProMNS)*

Adsorption isotherms of CFX onto ProMNS were achieved with initial concentrations of antibiotic in the range of 20–1600 mg/L. The two-step adsorption model was used to fit adsorption isotherms of CFX onto ProMNS (Figure 11).

Figure 11 shows that the higher KCl concentration is, the lower adsorption capacity is. At different initial CFX concentrations, adsorption at 100 mM KCl is always lower than that at 10 mM and 10 mM is lower than 1 mM. The CFX adsorption increased with decreasing KCl concentration due to a decrease of various cations K^+ on the ProMNS surface with positive charge. As a result, an increase of the electrostatic attraction was obtained. Figure 11 also indicates that adsorption isotherms of CFX onto ProMNS at three KCl concentrations could be reasonably represented by two-step adsorption model with Equation (4) using the fitting parameters in Table 2. The fitting values for k_1 and Γ_{CFX} were calculated from the Langmuir isotherms while other fitting parameters (k_2 and n) at different salt concentration are calculated by trials and error method using OriginPro 8. Although some experimental points have quite high deviations compared with modeling, almost experimental values matched the calculated one from the model. In this case, the deviations are higher than beta lactam cefixime adsorption onto strong polycation, polydiallyldimethylammonium chloride (PDADMAC) modified nanosilica rice husk because the charging properties of protein is less than that of PDADMAC. However, the two-step model used in the present study to represent adsorption isotherms of CFX onto ProMNS is much better than the fit of cellulose-based polymer adsorption isotherms onto cotton fibers [45]. It can be seen that all error bars show standard deviations of three replicates are suitable and close to the solid lines fitted by two-step adsorption model. The Table 2 shows that the value of k_1 decreased with increasing KCl concentrations from 1 to 100 mM while the k_2 values are constant for 1 and 10 mM and slightly decrease for 100 mM. Furthermore, the values of n increased slightly with increasing ionic strength for CFX adsorption. It implies that k_1 and n are useful parameters to evaluate the influence of ionic strength on CFX adsorption onto ProMNS.

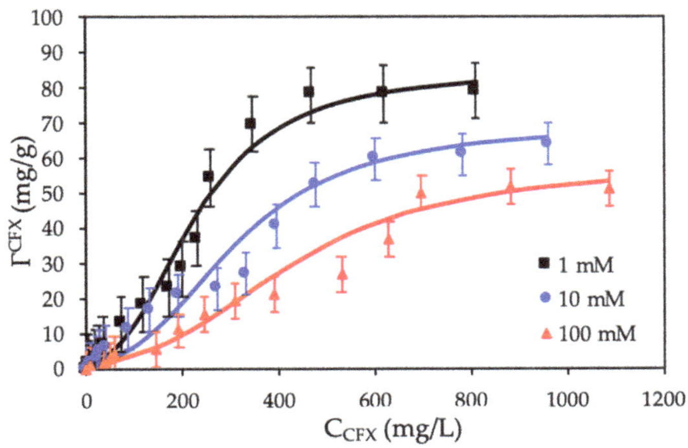

Figure 11. Adsorption isotherms of CFX onto protein-modified nanosilica (ProMNS) at different KCl concentrations. Points are experimental results and solid lines are fitted by two-step adsorption model. Error bars show standard deviations of three replicates.

Table 2 also shows that the maximum adsorption capacity of CFX reaches to 85 mg/g at 1 mM KCl. Adsorption capacity in our case is highest compared with different adsorbents for CFX removal [46], demonstrating that ProMNS is a novel, eco-friendly material for CFX removal from aqueous solution due to the low cost of nanosilica synthesized from rice husk and natural protein extracted from MO seeds.

Table 2. The fitting parameters for CFX adsorption onto protein-modified nanosilia (ProMNS) at different KCl concentrations.

C_{KCl} (mM)	Γ_{CFX} (mg/g)	k_1 (g/mg)	k_2 (g/mg) $^{n-1}$	n
1	85	10.0×10^2	4.0×10^6	3.0
10	70	9.0×10^2	4.0×10^6	3.1
100	58	7.0×10^2	3.5×10^6	3.2

3.5. Adsorption Mechanisms of CFX onto Protein-Modified Nanosilica (ProMNS)

In this section, adsorption mechanism of CFX onto ProMNS are discussed in detail on the basis of the surface charge change by ζ potential, the change in surface functional group by FT-IR, and adsorption isotherm of CFX onto ProMNS.

Surface charge change by monitoring ζ potential calculated by Equation (1) before and after adsorption of protein was thoroughly studied [18,47]. In the present study, the ζ potential was used to evaluate charging behavior of nanosilica rice husk without adsorption, after protein adsorption, and CFX adsorption to suggest the adsorption mechanism of CFX onto ProMNS. Figure 12 shows that charge reversal occurred after modification of nanosilica by protein adsorption. Nanosilica with negative ζ potential −35.9 mV changed to positive (ζ = 31.3 mV) after modification with protein (ProMNS). However, after CFX adsorption at pH 7.0 ($pKa_1 < pH < pKa_2$), with the zwitterionic species of CFX, a decrease in positive charge of ProMNS is observed. These results are in good agreement with the effect of pH on the CFX removal (Section 3.3.1).

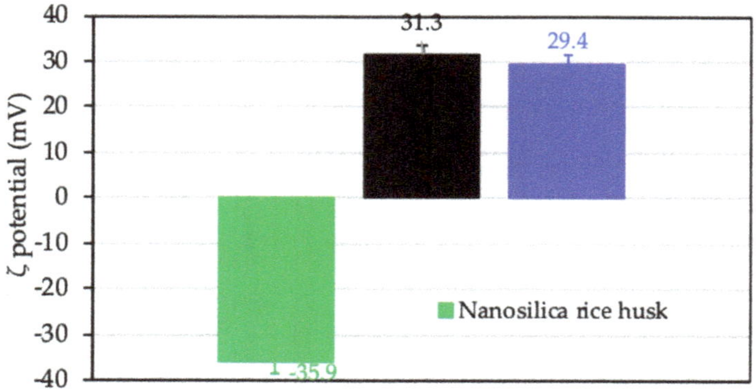

Figure 12. The ζ potential of nanosilica rice husk, protein-modified nanosilica (ProMNS), and ProMNS after CFX adsorption. Error bars show the standard deviations of three replicates.

The FT-IR is useful tool to evaluate the surface functional groups in adsorption technique [48]. Figure 13 shows the spectrum of protein-modified nanosilica (ProMNS) after CFX adsorption in the range of wavenumber 400–4000 cm^{-1}.

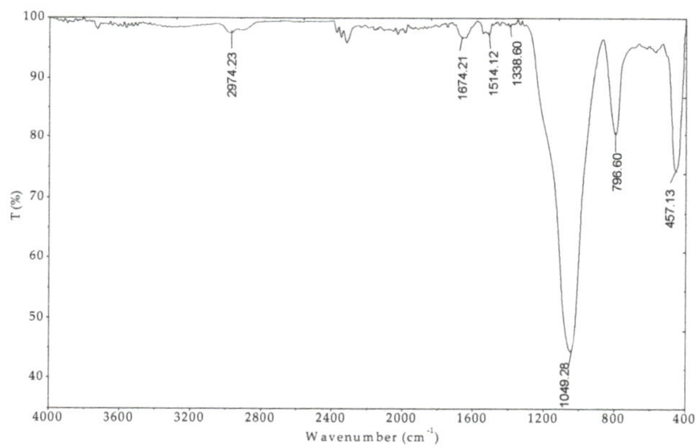

Figure 13. FTIR spectrum of nanosilica with protein modification after Ciprofloxacin (CFX) adsorption in the wave number range 400–4000 cm^{-1}.

The FT-IR spectrum of nanosilica synthesized from rice husk [24]. The surface modification of nanosilica by protein adsorption (ProMNS) was achieved in the presence of the peaks with the wavenumbers of 1645.28, 1539.20, and 1514.12 cm^{-1} (not shown here). Nevertheless, after CFX adsorption, only the peak of 1514.12 cm^{-1} assigned for NH amide I bending occurred. In addition, a small peak appeared at 1338.60 cm^{-1} indicating aromatic nitro compound of CFX in the FT-IR spectra of ProMNS after CFX adsorption [49]. Also, the characteristics of stretching vibration C=O at 1045.42 cm^{-1} and phenolic C-OH stretch at 1267.23 cm^{-1} of molecular CFX disappear after adsorption indicate that CFX adsorption occurs onto ProMNS surface by carboxyl groups. Therefore, the less positive charge of ProMNS was obtained. These results agree with the changes in surface charge of ProMNS after CFX adsorption are further in accordance to adsorption isotherms presented above. The driving force inducing the CFX adsorption is mainly by electrostatic attraction between negative CFX species with positively charged ProMNS surface.

3.6. Adsorption Kinetics of CFX onto Protein-Modified Nanosilica (ProMNS)

The adsorption kinetics of CFX onto ProMNS were carried out at the three initial CFX concentrations of 20, 200, and 1200 mg/L from 0 to 210 min.

The pseudo-second-order was used to predict the adsorption kinetic.

$$\frac{t}{q_t} = \frac{1}{k_k \cdot q_e^2} + \frac{1}{q_e} t \tag{5}$$

where q_e and q_t (mg/g) are adsorption capacity of RhB onto SML at equilibrium and time t, respectively; k_k (g/mg·min) is reaction rate constant of pseudo-second-order adsorption kinetic.

Figure 14 shows that the pseudo-second-order fitted experimental kinetic data for three concentrations of CFX concentrations very well. All excellent of R^2 (greater than 0.997) indicates that the adsorption kinetics of CFX onto ProMNS are in good agreement with the pseudo-second-order model. These results are similar to CFX adsorption onto biocomposite fibers of graphene oxide/calcium alginate [50] in which pseudo-second-order achieved the best fit comparing with other kinetic models.

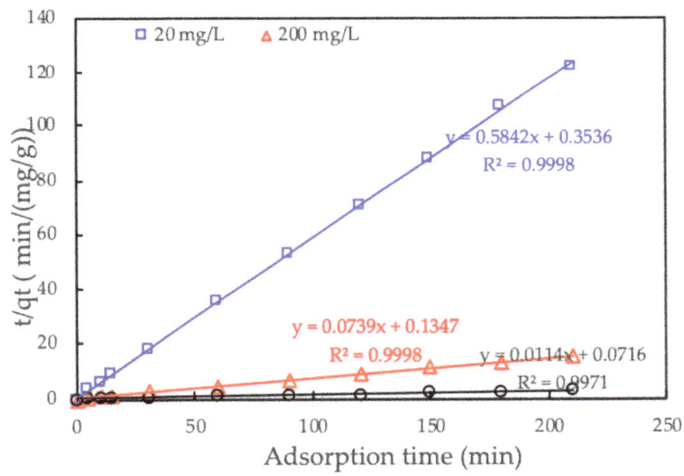

Figure 14. Adsorption kinetics of CFX onto protein-modified nanosilica (ProMNS) for three initial CFX concentrations. Points are experimental results and solid lines are fitted by pseudo-second-order model.

3.7. Adsorptive Removal of CFX Using Nanosilica without and with Surface Modification by Protein

To emphasize the role of surface modification with protein adsorption, we compared the removal of CFX using nanosilica without and with protein. Figure 15 indicates that CFX removal with an initial concentration of 20 mg/L in 1 mM KCl using the same adsorbent dosage of 10 mg/mL increases about 1.6 times from 56.84% to 89.85% with protein modification. The enhancement of CFX removal due to the increase of electrostatic attraction between the molecular CFX and the positively charged ProMNS surface. This implies that ProMNS is better than nanosilica synthesized from rice husk without any modification in term of CFX removal.

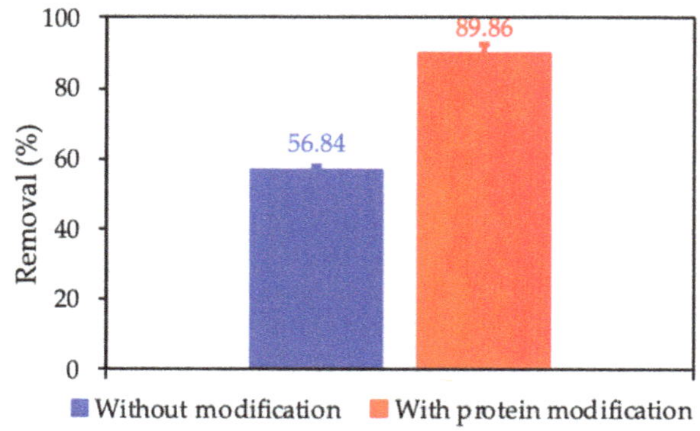

Figure 15. Removal of CFX using nanosilica without and with protein modification. Error bars show standard deviations of three replicates.

3.8. Comparison of the Effectiveness of ProMNS and Other Adsorbents for Removal of CFX

Recently, many studies have reported various adsorbents for CFX removal. However, adsorptive removal of CFX using ProMNS has not been investigated. In addition, the ProMNS in this study has highest removal efficiency and capacity compared to other adsorbents (Table 3). Another feature is that ProMNS is a green and low-cost adsorbent because MO seeds protein is natural product while rice husk is agricultural sub-product. This implies that ProMNS is not only a new material, but also an eco-friendly material for antibiotics removal from aqueous solutions.

Table 3. Adsorption capacity and removal efficiency of protein-modified nanosilica (ProMNS) and other absorbents for ciprofloxacin (CFX) removal.

Adsorbent	Adsorption Capacity (mg/g)	Removal Efficiency (%)	References
Graphene oxide/calcium alginate biocomposite	39.06	NI	[50]
Activated sludge	10.87	59	[51]
Kaolinte	7.95	50	[52]
Nanoscale zerovalent iron (nZVI) -Cu	NI	81.6	[53]
Silica nanoparticle	30	78	[54]
Hydrous oxides of Al (HAO)	13.6	NI	[55]
ProMNS	85	89.86	This study

NI: no information.

Although nanosilica synthesized from rice husk with surface modification with protein extracted from MO seeds is low-cost adsorbent, the adsorption cycles needed to evaluate the reuse potential and stability of ProMNS. The regeneration of ProMNS by using 0.2M HCl was repeated three times. Figure 16 shows the CFX removal using ProMNS after three cycles. Although the removal was decreased, it was still higher than 73% after three regenerations. The error bars show the standard deviations for all cycles are very small, demonstrating that ProMNS is novel and reusable adsorbent.

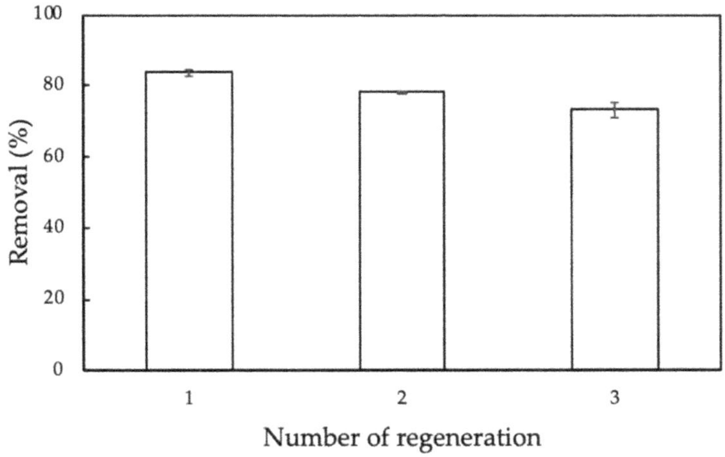

Figure 16. Removal of CFX using protein-modified nanosilica (ProMNS) after three regenerations. Error bars show standard deviations of two replicates.

3.9. Removal of CFX from Hospital Wastewater Using Protein-Modified Nanosilica (ProMNS)

The removal of antibiotics from actual wastewater sample is important to understand the implication for real system. We used ProMNS to removal CFX from hospital. An actual wastewater sample, which was selected from a big hospital in Hanoi, Vietnam was used to attempt to remove CFX using ProMNS under optimum conditions. An actual hospital wastewater sample remained in a dark bottle in a refrigerator before carrying out removal experiment in our laboratory.

For CFX removal with an actual sample, the experiments were conducted within 48 h. The optimum conditions for CFX removal using ProMNS were pH 7, adsorption time 90 min, and adsorbent dosage 10 mg/mL. Due to the very low CFX residual concentration to evaluate the performance, a 10 mg/L CFX was added into an actual sample to evaluate removal efficiency. Figure 17 indicates the UV-Vis spectra of CFX in the actual hospital wastewater samples before and after treatment using ProMNS. For the actual wastewater sample without addition, the spectrum is dramatically decreased, indicating that not only CFX, but also other organic pollutants were treated. In addition, the specific wavelength of 272 nm for actual sample with 10 mg/L CFX decreases sharply. By basic calculation at the maximum absorbance, the removal efficiency of CFX using ProMNS was achieved about 70%. This results again demonstrate that ProMNS is a novel, eco-friendly adsorbent for antibiotics removal from hospital wastewater.

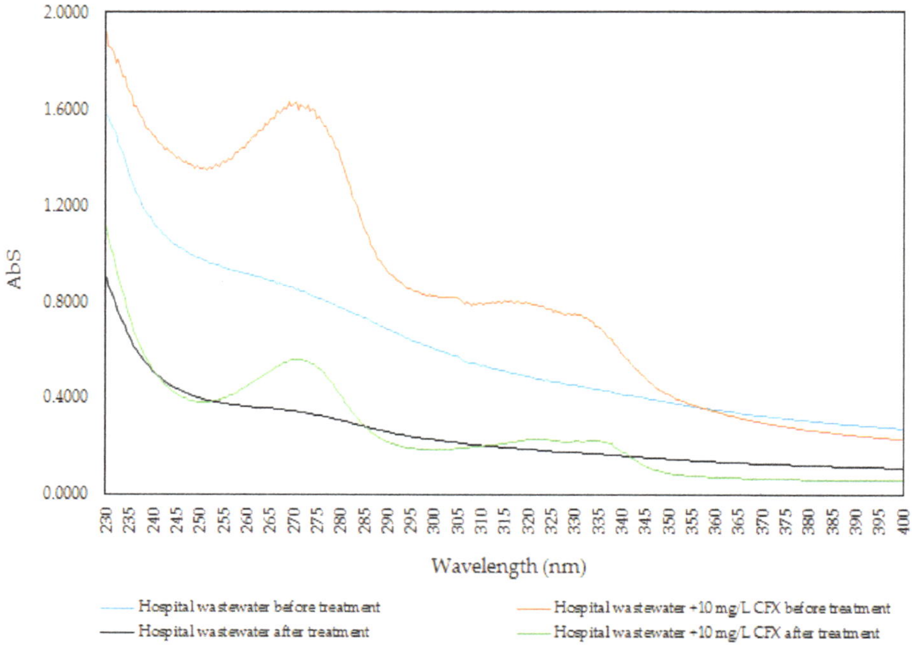

Figure 17. UV-Vis spectrum of CFX of the actual hospital wastewater samples before and after treatment using protein-modified nanosilica (ProMNS).

4. Conclusions

The removal of emerging pollutant, antibiotic ciprofloxacin (CFX) using the Moringa (MO) seeds protein-modified nanosilica rice husk (ProMNS) was systematically investigated in the present study. Protein extracted from MO seeds had 18 amino acids with high purity, which was characterized by FTIR, UV-Vis, and HPLC. Adsorption of protein onto nanosilica was induced by both electrostatic and non-electrostatic interaction at pH 10. The CFX removal increased about 1.6 times using nanosilica after surface modification with protein. Optimum conditions for CFX removal from aqueous solution using ProMNS were thoroughly studied and found to be pH 7.0, adsorption time 90 min, adsorbent dosage 10 mg/mL, and ionic strength 1 mM KCl. Under optimum conditions, the maximum adsorption capacity reached 85 mg/g, which is much higher than many common adsorbents. Adsorption isotherms of CFX onto ProMNS at three KCl concentrations were fitted well by a two-step adsorption model while adsorption kinetics are in good agreement with the pseudo-second-order model. Adsorption of CFX onto ProMNS was mainly controlled by electrostatic attraction between anionic species of CFX and positively charged ProMNS surface. After three regenerations, the CFX removal was still higher than 73% and a high CFX removal of about 70% was achieved with an actual hospital wastewater when using ProMNS. We indicate that ProMNS is an excellent and new eco-friendly material to remove antibiotics from hospital wastewater.

Author Contributions: Conceptualization, T.D.P.; methodology, T.D.P.; software, T.D.P.; validation, T.D.P.; formal analysis, T.N.V., H.L.N. P.H.P.L., and T.S.H.; investigation, T.D.P.; resources, T.N.V., T.S.H.; data curation, T.N.V., H.L.N. P.H.P.L., and T.S.H.; writing—original draft preparation, T.D.P.; writing—review and editing, T.D.P.; visualization, T.D.P.; supervision, T.D.P.; project administration, T.D.P.; All authors have read and agreed to the published version of the manuscript.

Funding: This research received no external funding.

Conflicts of Interest: The authors declare no conflict of interest.

References

1. Kapetanovic, V.; Milovanovic, L.; Erceg, M. Spectrophotometric and polarographic investigation of the Ofloxacin-Cu(II) complexes. *Talanta* **1996**, *43*, 2123–2130. [CrossRef]
2. Khaliq, Y.; Zhanel, G.G. Fluoroquinolone-Associated Tendinopathy: A Critical Review of the Literature. *Clin. Infect. Dis.* **2003**, *36*, 1404–1410. [CrossRef]
3. Zhanel, G.G.; Ennis, K.; Vercaigne, L.; Walkty, A.; Gin, A.S.; Embil, J.; Smith, H.; Hoban, D.J. A Critical Review of the Fluoroquinolones. *Drugs* **2002**, *62*, 13–59. [CrossRef] [PubMed]
4. Pham, T.D.; Bui, T.T.; Trang Truong, T.T.; Hoang, T.H.; Le, T.S.; Duong, V.D.; Yamaguchi, A.; Kobayashi, M.; Adachi, Y. Adsorption characteristics of beta-lactam cefixime onto nanosilica fabricated from rice HUSK with surface modification by polyelectrolyte. *J. Mol. Liq.* **2019**. [CrossRef]
5. Radjenović, J.; Petrović, M.; Ventura, F.; Barceló, D. Rejection of pharmaceuticals in nanofiltration and reverse osmosis membrane drinking water treatment. *Water Res.* **2008**, *42*, 3601–3610. [CrossRef] [PubMed]
6. Tran, N.H.; Chen, H.; Reinhard, M.; Mao, F.; Gin, K.Y.H. Occurrence and removal of multiple classes of antibiotics and antimicrobial agents in biological wastewater treatment processes. *Water Res.* **2016**, *104*, 461–472. [CrossRef] [PubMed]
7. Tran, N.H.; Gin, K.Y.H. Occurrence and removal of pharmaceuticals, hormones, personal care products, and endocrine disrupters in a full-scale water reclamation plant. *Sci. Total Environ.* **2017**, *599*, 1503–1516. [CrossRef] [PubMed]
8. Dao, T.H.; Tran, T.T.; Nguyen, V.R.; Pham, T.N.M.; Vu, C.M.; Pham, T.D. Removal of antibiotic from aqueous solution using synthesized TiO2 nanoparticles: Characteristics and mechanisms. *Environ. Earth Sci.* **2018**, *77*, 359. [CrossRef]
9. Karthikeyan, S.; Gupta, V.K.; Boopathy, R.; Titus, A.; Sekaran, G. A new approach for the degradation of high concentration of aromatic amine by heterocatalytic Fenton oxidation: Kinetic and spectroscopic studies. *J. Mol. Liq.* **2012**, *173*, 153–163. [CrossRef]
10. Grimes, K.L.; Dunphy, L.J.; Loudermilk, E.M.; Melara, A.J.; Kolling, G.L.; Papin, J.A.; Colosi, L.M. Evaluating the efficacy of an algae-based treatment to mitigate elicitation of antibiotic resistance. *Chemosphere* **2019**, *237*, 124421. [CrossRef]
11. Le, T.H.; Ng, C.; Tran, N.H.; Chen, H.; Gin, K.Y.H. Removal of antibiotic residues, antibiotic resistant bacteria and antibiotic resistance genes in municipal wastewater by membrane bioreactor systems. *Water Res.* **2018**, *145*, 498–508. [CrossRef] [PubMed]
12. Tran, N.H.; Hoang, L.; Nghiem, L.D.; Nguyen, N.M.H.; Ngo, H.H.; Guo, W.; Trinh, Q.T.; Mai, N.H.; Chen, H.; Nguyen, D.D.; et al. Occurrence and risk assessment of multiple classes of antibiotics in urban canals and lakes in Hanoi, Vietnam. *Sci. Total Environ.* **2019**, *692*, 157–174. [CrossRef] [PubMed]
13. Ahmed, M.B.; Zhou, J.L.; Ngo, H.H.; Guo, W. Adsorptive removal of antibiotics from water and wastewater: Progress and challenges. *Sci. Total Environ.* **2015**, *532*, 112–126. [CrossRef] [PubMed]
14. Homem, V.; Santos, L. Degradation and removal methods of antibiotics from aqueous matrices—A review. *J. Environ. Manag.* **2011**, *92*, 2304–2347. [CrossRef]
15. Coglitore, D.; Janot, J.M.; Balme, S. Protein at liquid solid interfaces: Toward a new paradigm to change the approach to design hybrid protein/solid-state materials. *Adv. Colloid Interface Sci.* **2019**, *270*, 278–292. [CrossRef]
16. Kwaambwa, H.M.; Hellsing, M.S.; Rennie, A.R.; Barker, R. Interaction of Moringa oleifera seed protein with a mineral surface and the influence of surfactants. *J. Colloid Interface Sci.* **2015**, *448*, 339–346. [CrossRef]
17. Shemetov, A.A.; Nabiev, I.; Sukhanova, A. Molecular Interaction of Proteins and Peptides with Nanoparticles. *ACS Nano* **2012**, *6*, 4585–4602. [CrossRef]
18. Meissner, J.; Prause, A.; Bharti, B.; Findenegg, G.H. Characterization of protein adsorption onto silica nanoparticles: Influence of pH and ionic strength. *Colloid Polym. Sci.* **2015**, *293*, 3381–3391. [CrossRef]
19. Chen, R.; Wang, X.J.; Zhang, Y.Y.; Xing, Y.; Yang, L.; Ni, H.; Li, H.H. Simultaneous extraction and separation of oil, proteins, and glucosinolates from Moringa oleifera seeds. *Food Chem.* **2019**, *300*, 125162. [CrossRef]
20. Kwaambwa, H.M.; Hellsing, M.; Rennie, A.R. Adsorption of a Water Treatment Protein from Moringa oleifera Seeds to a Silicon Oxide Surface Studied by Neutron Reflection. *Langmuir* **2010**, *26*, 3902–3910. [CrossRef]

21. Moulin, M.; Mossou, E.; Signor, L.; Kieffer-Jaquinod, S.; Kwaambwa, H.M.; Nermark, F.; Gutfreund, P.; Mitchell, E.P.; Haertlein, M.; Forsyth, V.T.; et al. Towards a molecular understanding of the water purification properties of Moringa seed proteins. *J. Colloid Interface Sci.* **2019**, *554*, 296–304. [CrossRef] [PubMed]
22. Pham, T.D.; Vu, C.M.; Choi, H.J. Enhanced fracture toughness and mechanical properties of epoxy resin with rice husk-based nano-silica. *Polym. Sci. Ser. A* **2017**, *59*, 437–444. [CrossRef]
23. Zhu, B.Y.; Gu, T. Surfactant adsorption at solid-liquid interfaces. *Adv. Colloid Interface Sci.* **1991**, *37*, 1–32. [CrossRef]
24. Pham, T.D.; Bui, T.T.; Nguyen, V.T.; Bui, T.K.V.; Tran, T.T.; Phan, Q.C.; Pham, T.D.; Hoang, T.H. Adsorption of Polyelectrolyte onto Nanosilica Synthesized from Rice Husk: Characteristics, Mechanisms, and Application for Antibiotic Removal. *Polymers* **2018**, *10*, 220. [CrossRef] [PubMed]
25. Pham, T.D.; Tran, T.T.; Le, V.A.; Pham, T.T.; Dao, T.H.; Le, T.S. Adsorption characteristics of molecular oxytetracycline onto alumina particles: The role of surface modification with an anionic surfactant. *J. Mol. Liq.* **2019**, *287*, 110900. [CrossRef]
26. Guzmán, E.; Ritacco, H.A.; Ortega, F.; Rubio, R.G. Growth of Polyelectrolyte Layers Formed by Poly(4-styrenesulfonate sodium salt) and Two Different Polycations: New Insights from Study of Adsorption Kinetics. *J. Phys. Chem. C* **2012**, *116*, 15474–15483. [CrossRef]
27. Pham, T.D.; Do, T.U.; Pham, T.T.; Nguyen, T.A.H.; Nguyen, T.K.T.; Vu, N.D.; Le, T.S.; Vu, C.M.; Kobayashi, M. Adsorption of poly(styrenesulfonate) onto different-sized alumina particles: Characteristics and mechanisms. *Colloid Polym Sci* **2019**, *297*, 13–22. [CrossRef]
28. Kwaambwa, H.M.; Maikokera, R. A fluorescence spectroscopic study of a coagulating protein extracted from Moringa oleifera seeds. *Colloids Surf. B Biointerfaces* **2007**, *60*, 213–220. [CrossRef]
29. Castellanos, M.; Van Eendenburg, C.V.; Gubern, C.; Sanchez, J.M. Ethyl-bridged hybrid column as an efficient alternative for HPLC analysis of plasma amino acids by pre-column derivatization with 6-aminoquinolyl-N-hydroxysuccinimidyl carbamate. *J. Chromatogr. B* **2016**, *1029*, 137–144. [CrossRef]
30. Szkudzińska, K.; Smutniak, I.; Rubaj, J.; Korol, W.; Bielecka, G. Method validation for determination of amino acids in feed by UPLC. *Accredit. Qual. Assur.* **2017**, *22*, 247–252. [CrossRef]
31. Naffa, R.; Holmes, G.; Zhang, W.; Maidment, C.; Shehadi, I.; Norris, G. Comparison of liquid chromatography with fluorescence detection to liquid chromatography-mass spectrometry for amino acid analysis with derivatisation by 6-aminoquinolyl-N-hydroxysuccinimidyl-carbamate: Applications for analysis of amino acids in skin. *Arab. J. Chem.* **2019**. [CrossRef]
32. Delgado, A.V.; González-Caballero, F.; Hunter, R.J.; Koopal, L.K.; Lyklema, J. Measurement and interpretation of electrokinetic phenomena. *J. Colloid Interface Sci.* **2007**, *309*, 194–224. [CrossRef] [PubMed]
33. Vance, S.J.; Desai, V.; Smith, B.O.; Kennedy, M.W.; Cooper, A. Aqueous solubilization of C60 fullerene by natural protein surfactants, latherin and ranaspumin-2. *Biophys. Chem.* **2016**, *214–215*, 27–32. [CrossRef] [PubMed]
34. Kebede, T.G.; Dube, S.; Nindi, M.M. Removal of Multi-Class Antibiotic Drugs from Wastewater Using Water-Soluble Protein of Moringa stenopetala Seeds. *Water* **2019**, *11*, 595. [CrossRef]
35. Pham, A.L.T.; Sedlak, D.L.; Doyle, F.M. Dissolution of mesoporous silica supports in aqueous solutions: Implications for mesoporous silica-based water treatment processes. *Appl. Catal. B* **2012**, *126*, 258. [CrossRef] [PubMed]
36. Guzman, E.; Ritacco, H.; Rubio, J.E.F.; Rubio, R.G.; Ortega, F. Salt-induced changes in the growth of polyelectrolyte layers of poly(diallyl-dimethylammonium chloride) and poly(4-styrene sulfonate of sodium). *Soft Matter* **2009**, *5*, 2130–2142. [CrossRef]
37. Mészáros, R.; Thompson, L.; Bos, M.; de Groot, P. Adsorption and Electrokinetic Properties of Polyethylenimine on Silica Surfaces. *Langmuir* **2002**, *18*, 6164–6169. [CrossRef]
38. Sharma, P.C.; Jain, A.; Jain, S.; Pahwa, R.; Yar, M.S. Ciprofloxacin: Review on developments in synthetic, analytical, and medicinal aspects. *J. Enzym. Inhib. Med. Chem.* **2010**, *25*, 577–589. [CrossRef]
39. Buergisser, C.S.; Scheidegger, A.M.; Borkovec, M.; Sticher, H. Chromatographic Charge Density Determination of Materials with Low Surface Area. *Langmuir* **1994**, *10*, 855–860. [CrossRef]
40. Kobayashi, M.; Juillerat, F.; Galletto, P.; Bowen, P.; Borkovec, M. Aggregation and Charging of Colloidal Silica Particles: Effect of Particle Size. *Langmuir* **2005**, *21*, 5761–5769. [CrossRef]

41. Chu, T.P.M.; Nguyen, N.T.; Vu, T.L.; Dao, T.H.; Dinh, L.C.; Nguyen, H.L.; Hoang, T.H.; Le, T.S.; Pham, T.D. Synthesis, Characterization, and Modification of Alumina Nanoparticles for Cationic Dye Removal. *Materials* **2019**, *12*, 450. [CrossRef] [PubMed]
42. Pham, T.D.; Do, T.T.; Ha, V.L.; Doan, T.H.Y.; Nguyen, T.A.H.; Mai, T.D.; Kobayashi, M.; Adachi, Y. Adsorptive removal of ammonium ion from aqueous solution using surfactant-modified alumina. *Environ. Chem.* **2017**, *14*, 327–337. [CrossRef]
43. Pham, T.D.; Nguyen, H.H.; Nguyen, N.V.; Vu, T.T.; Pham, T.N.M.; Doan, T.H.Y.; Nguyen, M.H.; Ngo, T.M.V. Adsorptive Removal of Copper by Using Surfactant Modified Laterite Soil. *J. Chem.* **2017**, *2017*, 1986071. [CrossRef]
44. Mazloomi, F.; Jalali, M. Ammonium removal from aqueous solutions by natural Iranian zeolite in the presence of organic acids, cations and anions. *J. Environ. Chem. Eng.* **2016**, *4*, 1664–1673. [CrossRef]
45. Hoffmann, I.; Oppel, C.; Gernert, U.; Barreleiro, P.; von Rybinski, W.; Gradzielski, M. Adsorption Isotherms of Cellulose-Based Polymers onto Cotton Fibers Determined by Means of a Direct Method of Fluorescence Spectroscopy. *Langmuir* **2012**, *28*, 7695–7703. [CrossRef]
46. Fei, Y.; Li, Y.; Han, S.; Ma, J. Adsorptive removal of ciprofloxacin by sodium alginate/graphene oxide composite beads from aqueous solution. *J. Colloid Interface Sci.* **2016**, *484*, 196–204. [CrossRef]
47. Huang, Y.; Yamaguchi, A.; Pham, T.D.; Kobayashi, M. Charging and aggregation behavior of silica particles in the presence of lysozymes. *Colloid Polym. Sci.* **2018**, *296*, 145–155. [CrossRef]
48. Hind, A.R.; Bhargava, S.K.; McKinnon, A. At the solid/liquid interface: FTIR/ATR—the tool of choice. *Adv. Colloid Interface Sci.* **2001**, *93*, 91–114. [CrossRef]
49. Li, J.; Yu, G.; Pan, L.; Li, C.; You, F.; Xie, S.; Wang, Y.; Ma, J.; Shang, X. Study of ciprofloxacin removal by biochar obtained from used tea leaves. *J. Environ. Sci.* **2018**, *73*, 20–30. [CrossRef]
50. Wu, S.; Zhao, X.; Li, Y.; Zhao, C.; Du, Q.; Sun, J.; Wang, Y.; Peng, X.; Xia, Y.; Wang, Z.; et al. Adsorption of ciprofloxacin onto biocomposite fibers of graphene oxide/calcium alginate. *Chem. Eng. J.* **2013**, *230*, 389–395. [CrossRef]
51. Ferreira, V.R.A.; Amorim, C.L.; Cravo, S.M.; Tiritan, M.E.; Castro, P.M.L.; Afonso, C.M.M. Fluoroquinolones biosorption onto microbial biomass: Activated sludge and aerobic granular sludge. *Int. Biodeterior. Biodegrad.* **2016**, *110*, 53–60. [CrossRef]
52. MacKay, A.A.; Seremet, D.E. Probe Compounds to Quantify Cation Exchange and Complexation Interactions of Ciprofloxacin with Soils. *Environ. Sci. Technol.* **2008**, *42*, 8270–8276. [CrossRef] [PubMed]
53. Chen, L.; Ni, R.; Yuan, T.; Gao, Y.; Kong, W.; Zhang, P.; Yue, Q.; Gao, B. Effects of green synthesis, magnetization, and regeneration on ciprofloxacin removal by bimetallic nZVI/Cu composites and insights of degradation mechanism. *J. Hazard. Mater.* **2020**, *382*, 121008. [CrossRef]
54. Nassar, M.Y.; Ahmed, I.S.; Raya, M.A. A facile and tunable approach for synthesis of pure silica nanostructures from rice husk for the removal of ciprofloxacin drug from polluted aqueous solutions. *J. Mol. Liq.* **2019**, *282*, 251–263. [CrossRef]
55. Gu, C.; Karthikeyan, K.G. Sorption of the Antimicrobial Ciprofloxacin to Aluminum and Iron Hydrous Oxides. *Environ. Sci. Technol.* **2005**, *39*, 9166–9173. [CrossRef] [PubMed]

© 2020 by the authors. Licensee MDPI, Basel, Switzerland. This article is an open access article distributed under the terms and conditions of the Creative Commons Attribution (CC BY) license (http://creativecommons.org/licenses/by/4.0/).

polymers

MDPI

Article

The Use of a Biopolymer Conjugate for an Eco-Friendly One-Pot Synthesis of Palladium-Platinum Alloys

Daniele Silvestri [1], Stanisław Wacławek [1,*], Rohith K. Ramakrishnan [1], Abhilash Venkateshaiah [1], Kamil Krawczyk [1], Vinod V. T. Padil [1], Bartłomiej Sobel [2] and Miroslav Černík [1,*]

1 Institute for Nanomaterials, Advanced Technologies and Innovation, Technical University of Liberec, 46117 Liberec, Czech Republic; daniele.silvestri@tul.cz (D.S.); rohith.kunjiparambil.ramakrishnan@tul.cz (R.K.R.); abhilash.venkateshaiah@tul.cz (A.V.); kamil.krawczyk@tul.cz (K.K.); vinod.padil@tul.cz (V.V.T.P.)

2 Institute of Engineering Materials and Biomaterials, Faculty of Mechanical Engineering, Silesian University of Technology, 44–100 Gliwice, Poland; bartlomiej.sobel@gmail.com

* Correspondence: stanislaw.waclawek@tul.cz (S.W.); miroslav.cernik@tul.cz (M.Č.)

Received: 6 November 2019; Accepted: 25 November 2019; Published: 27 November 2019

Abstract: Raising health and environmental concerns over the nanoparticles synthesized from hazardous chemicals have urged researchers to focus on safer, environmentally friendlier and cheaper alternatives as well as prompted the development of green synthesis. Apart from many advantages, green synthesis is often not selective enough (among other issues) to create shape-specific nanoparticle structures. Herein, we have used a biopolymer conjugate and Pd and Pt precursors to prepare sustainable bimetallic nanoparticles with various morphology types. The nanoparticles were synthesized by a novel green approach using a bio-conjugate of chitosan and polyhydroxybutyrate (Cs-PHB). The bio-conjugate plays the simultaneous roles of a reducing and a capping agent, which was confirmed by attenuated total reflection Fourier transform infrared spectroscopy (ATR-FTIR) and energy dispersive X-ray spectrometry (EDS) analysis, proving the presence of a Cs-PHB layer on the surface of the prepared nanoparticles. The EDS profile also revealed the elemental structure of these nanoparticles and confirmed the formation of a Pd/Pt alloy. TEM morphological analysis showed the formation of star-like, octahedron or decahedron Pd/Pt nanoparticles, depending on the synthesis conditions. The bimetallic Pd/Pt nanoparticles synthesized with various Pd/Pt molar ratios were successfully applied for the catalytic reduction of 4-nitrophenol to 4-aminophenol by borohydride. The calculated κ_C values (ratio of k_{app} to the concentration of the catalyst) revealed that the decahedron nanoparticles (size of 15 ± 4 nm), synthesized at the molar ratio of 2:1 (Pd/Pt), temperature of 130 °C, 10 g/L of Cs-PHB conjugate and time of 30 min, exhibited excellent catalytic activity compared to other bimetallic nanoparticles reported in the literature.

Keywords: green synthesis; biopolymers; bimetallic nanoparticles; catalytic reduction; 4-nitrophenol

1. Introduction

The raising health and environmental concerns over nanoparticles synthesized from hazardous chemicals, which are also often economically unfeasible, have urged researchers to focus on safer, environmentally-friendlier and cheaper alternatives. These reasons have prompted the development of green nanoparticle syntheses, which are safe and adhere to the green chemistry approach [1]. Biopolymers, which are abundantly available and easily biodegradable, are promising materials for providing an environmentally-benign synthesis of nanomaterials. These natural polymers have been successfully used as reducing, stabilizing and capping agents in the synthesis of nanoparticles [2–4], allowing alterations in the nanoparticle size [5] and shape [6]. Furthermore, the different functional groups present in these biopolymers can actively contribute to the improvement of metallic nanoparticle catalytic reactions [7,8]. A typical example of a simple monometallic and bimetallic nanoparticle synthesis is the one-step reduction and stabilization of Au and Ag nanoparticles [9,10].

Chitosan is considered one of the most studied biopolymers in the literature, and has been successfully used in different applications such as the synthesis and stabilization of various different nanoparticles [11–13], bio-medical applications [14], drug delivery [15], water treatment [16] and many other uses [17–19]. Chitosan is obtained from chitin, which is mainly extracted from crustacean shell wastes [20]. While its non-toxicity for mammals and biodegradability make it popular, its insolubility in water is one of its drawbacks.

Another biopolymer that is starting to gain interest in different scientific fields is poly(3-hydroxybutyrate) (PHB), which can be produced by different bacteria [21] and also from waste materials [22,23]. It is usually used as a carbon source in *in-situ* bioremediation [24], a drug delivery carrier [25], a biodegradable bioplastic [26], and a stabilizing agent for nanoparticles [27]. However, difficulties such as the solubility of PHB in only organic solvents, which are toxic to both humans and the environment [28,29], need to be addressed to achieve good dispersions for synthesis.

Motivated by the above situation, our group developed a water-soluble conjugate of chitosan and PHB, which was successfully applied to control not only the growth and aggregation of the Au nanoparticles but also their surface properties [30].

Due to interactions between two metals and their unique and more flexible surface structures in comparison to monometallic nanoparticles, bimetallic nanoparticles have gained precedence over traditional heterogeneous catalysts due to their excellent catalytic activity [31]. The nanoparticle surface area plays a key role in heterogeneous catalysis because it is directly correlated to the catalyst active sites on which the catalytic reactions are taking place. Moreover, the nanoparticles are often easily recovered from the reaction medium, and they possess steric environments within their active sites, both features that can positively influence the catalytic activity [32]. Among the noble metals, both palladium (Pd) and platinum (Pt) are well known for their unique characteristics, and both are used successfully in different scientific fields, including catalysis [33–38]. Due to the fact they have similar face-centred cubic (fcc) crystal structures and a high lattice match (lattice mismatch of 0.77%), palladium and platinum are highly miscible [39,40].

We propose a one-pot, quick and green synthesis of decahedral Pd/Pt using solely Pd and Pt precursors and a Cs-PHB bio-conjugate as a reducing reagent. Based on our previous studies, we hypothesize that Cs-PHB cannot only help to control the growth and aggregation of the nanoparticles but also to tailor their catalytic activity. To the best of our knowledge, this is the first report to use Cs-PHB for the green synthesis of bimetallic nanoparticles. In addition, we believe in the simplicity of this procedure for obtaining decahedron Pd/Pt bimetallic nanoparticles. The synthesized nanoparticles were characterized by ATR-FTIR, TEM, and EDS, and successfully tested on the standard reduction reaction of the 4-nitrophenol (4-NP) to 4-aminophenol (4-AP).

2. Materials and Methods

2.1. Reagents and Solutions

Chitosan (low M_w of 50–190 kDa, 75%–85% deacetylated), sodium borohydride (98%), 4-nitrophenol (ReagentPlus, >99%), K_2PdCl_4 (98%), $PtCl_4$ (96%) were purchased from Sigma–Aldrich (Saint Louis, MO, USA); polyhydroxybutyrate (PHB, Biomer P209) from Biomer (Krailing, Germany); nitric acid (65%) from Lach-ner (Neratovice, Czech Republic). Deionized water (DI; 18.2 $M\Omega \cdot cm^{-1}$, ELGA, Veolia Water, Marlow, UK) was used in all of the experiments.

2.2. Analytical Methods

ATR-FTIR spectra were recorded at a resolution of 4 cm^{-1} over the 4000–700 cm^{-1} range using a NICOLET IZ10 spectrometer (Thermo Scientific, Waltham, MA, USA) equipped with a germanium ATR crystal and a single reflection angle 45° horizontal ATR accessory. The UV-Vis spectroscopic analysis was performed using a DR 3900 UV-Vis spectrophotometer (Hach Lange, Loveland, CO, USA) equipped with 1 cm quartz cuvettes. High-resolution transmission electron microscopy (HR-TEM) analysis was carried out using transmission electron microscopy/scanning transmission electron microscopy (TEM/STEM) system (Titan 80-300, FEI, city, state abbrev if USA, country) with a super twin-lens operated at 300 kV and equipped with an annular dark field detector. The presence of various elements in the obtained nanoparticles was analysed using energy-dispersive X-ray spectroscopy (EDX, Aztec, Oxford Instruments, Abingdon, UK). ICP-MS (Elan 6000, Perkin Elmer, Waltham, MA, USA) was used to determine the Pd/Pt concentration.

2.3. Preparation of Cs-PHB Conjugate

The conjugate was prepared following the procedure reported previously by our group [30]. Briefly, a chitosan solution was made by adding chitosan (0.5 g) to acidified deionized water (100 mL) and stirring to obtain a homogeneous solution. Subsequently, PHB (0.12 g) was added to the mixture and stirred overnight at 80 °C. The resulting solution was sonicated for 30 min at 80 °C, purified by a dialysis tube, and finally freeze-dried.

2.4. Synthesis of Bimetallic Nanoparticles

Pd/Pt bimetallic nanoparticles were synthesized following a modified co-reduction method of Lim et al. [41]. Briefly, K_2PdCl_4 and $PtCl_4$ were dissolved in DI to get two (Pd and Pt) stock solutions with a concentration of 10 mM each. Both solutions were stirred for 5 minutes in order to dissolve the salts. Cs-PHB was dissolved in DI to get a stock solution of 20 g/L. Subsequently, a certain amount of palladium and platinum precursor stock solutions (0.5 mL of Pd and 0.5 mL of Pt precursor stock solution for the Pd:Pt ratio of 1:1; 0.25 mL of Pd and 0.5 mL of Pt precursor stock solution for the Pd:Pt ratio of 1:2; 0.5 mL of Pd and 0.25 mL of Pt precursor stock solution for the Pd:Pt ratio 2:1) were added to the Cs-PHB solution (2.5 mL), the volume was raised to 5 mL by adding DI. The reactor was heated (130–150 °C) for 30 min following the procedure reported by Venkateshaiah et al. [42]. The reaction was stopped by cooling down the samples in cold water. The obtained nanoparticles were washed three times with deionized water and stored in a refrigerator (4 °C) for future use.

2.5. Catalytic Test

The catalytic test of 4-NP reduction to 4-AP by $NaBH_4$ was carried out in a standard 1 cm path length quartz cuvette. The procedure was reported previously by Baruah et al. [43]. A typical test involves the mixing of 24 μL of 4-NP (5 mM), and an excess of $NaBH_4$ (120 μL of 0.1 M) in an Eppendorf tube (1.5 mL). A certain amount of nanoparticles was added, and the volume was adjusted to 1 mL using DI water. Then the solution was immediately transferred into a quartz cuvette and the absorbance was recorded by UV-Vis at regular intervals. All of the tests were carried out at room temperature (25 °C) in triplicate. An excess of $NaBH_4$ (12 mM of $NaBH_4$ and 0.12 mM of 4-NP) was used in the reduction process.

3. Results and Discussion

Pd/Pt nanoparticles were synthesized under different conditions (temperature from 130 to 150 °C) and using different ratios of Pd and Pt precursors (from 1:2 to 2:1). The resulting particles were characterised by ATR-FTIR, HR-TEM and EDS. Three types of nanoparticles synthetized at a constant temperature but at different metallic ratio were also compared for their catalytic activity.

3.1. Characterization of the Nanoparticles

3.1.1. ATR-FTIR

An ATR-FTIR analysis was performed to examine the functional groups located on the Pd/Pt bimetallic nanoparticles (Figure 1). The peak observable at ~1724 cm^{-1} in the PHB spectrum (Figure 1a) may be attributed to the ester group present in the PHB. The chitosan spectrum shows a peak at ~3300 cm^{-1} due to the O–H and N–H bonds, whereas the peak at ~2900 cm^{-1} may be ascribed to the symmetric or asymmetric CH_2 stretching vibrations. The peak at ~1600 cm^{-1} may be assigned to the NH_2 groups, while at ~1380 cm^{-1} the peak may be ascribed to CH_3 symmetrical deformations [44]. The last representative peak at ~1100 cm^{-1} may be attributed to C–O–C glycosidic linkage. The conjugate spectrum (Figure 1c) shows differences when comparing the PHB (Figure 1a) to the chitosan (Figure 1b) spectra. A decrease in the intensity of the NH_2 group at ~1600 cm^{-1} [30] was observed, while an increase in the intensity at ~1555 cm^{-1} was observed, which may correspond to the amide type II bond formation [30]. This suggests that the amino group of chitosan reacts with the C–O–C group of PHB to form the amide bond.

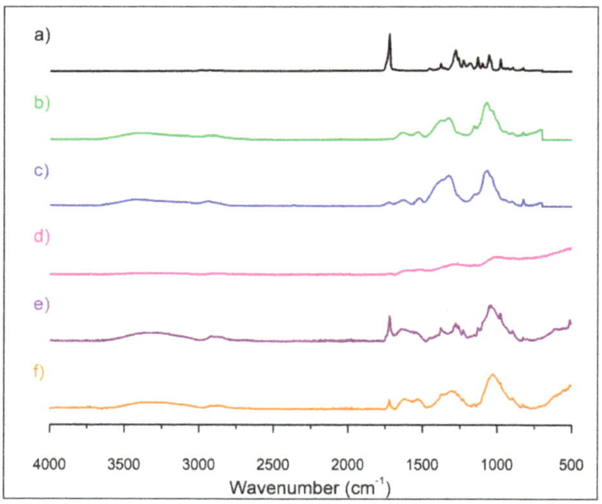

Figure 1. ATR-FTIR analysis of (**a**) PHB, (**b**) chitosan, (**c**) Cs-PHB, (**d**) Pd/Pt ratio 1:1 (zoom on the region of the 4000–1500 cm^{-1} spectrum part is available in Figure S1 in Supplementary Materials), (**e**) Pd/Pt ratio 1:2 and (**f**) Pd/Pt ratio 2:1 (synthesis temperature of Pd/Pt: 130 °C).

The ATR-FTIR spectra of Pd/Pt nanoparticles (Figure 1d–f) showed several bands. The first one (~3330 cm^{-1}) may be attributed to the NH/OH bond, whereas the one at ~2926 cm^{-1} is compatible with asymmetric or symmetric CH_2 stretching vibration. The peak at 1724 cm^{-1} may be related to the ester group, while the one at ~1555 cm^{-1} to the amide type II bond. The hydroxyl groups present in the conjugate may assist in the reduction of the precursor as reported by Dang et al. [45] and by Dorjnamjin et al. [46], while the rest of the polymer may coat the nanoparticles. The differences in intensity between the variously synthesized Pd/Pt nanoparticles may indicate different amounts of organic and inorganic material that could be found in samples. For example, Pd/Pt = 1:1 nanoparticles were synthesized with the highest ratio of metal precursors concentration (2 mM overall) to the polymer conjugate concentration, and in the low frequencies (<1500 cm^{-1}) their spectrum exhibits a (high absorbance) baseline sloping down to the left (typical for some metal nanoparticles; similar phenomena could be observed e.g. in the work of Hu et al. [47]).

3.1.2. HR-TEM

In order to obtain more information about the morphology of the synthesized Pd/Pt nanoparticles, a HR-TEM analysis was performed. Figure 2 shows the different shapes obtained by altering the synthesis temperature (from 130 to 150 °C) and Pd/Pt molar ratio (1:1, 1:2 and 2:1).

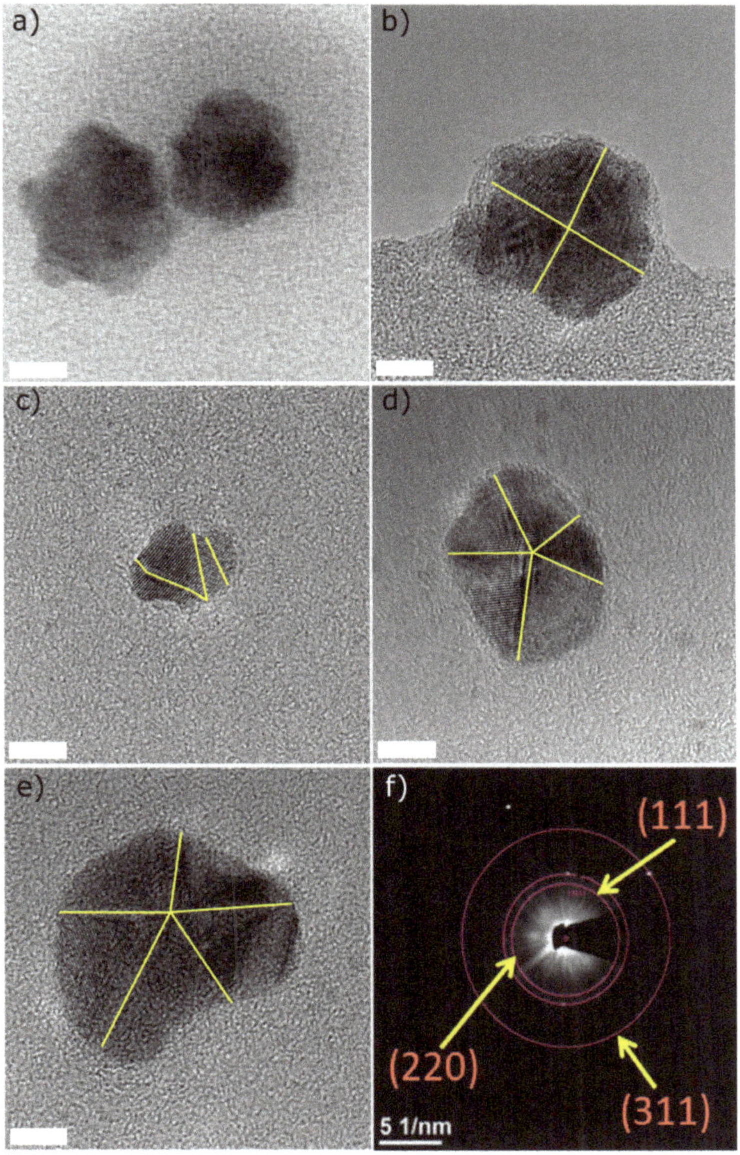

Figure 2. HR-TEM images of characteristic Pd/Pt nanoparticles synthesized with different molar ratios and temperatures (**a**) Pd/Pt 1:1 at 150 °C, (**b**) 1:1 at 140 °C, (**c**) 1:1 at 130 °C, and (**d**) 2:1 at 130 °C; (**e**) 1:2 at 130 °C and (**f**) SAED pattern of Pd/Pt (2:1). For all of the samples, the scale bar stands for 5 nm.

The sample synthesized at 150 °C and with a molar ratio 1:1 (Pd/Pt) (Figure 2a) shows a star-like structure, while lowering the temperature of synthesis to 140 °C (Figure 2b) changed the morphology of the nanoparticles to an octahedron. An additional decrease of temperature to 130 °C (Figure 2c) caused a synthesis of smaller nanoparticles with a surface characterized by different, randomly-orientated faces. Moreover, when the molar ratio changed to 2:1 and 1:2 (Pd/Pt) at the remaining temperature (130 °C), nanoparticles with decahedral morphology were observed (Figure 2d,e). The decahedral shape of the nanoparticles occurs only under strict conditions [48]. When certain conditions are applied, the ions specifically interact to form a Cs-PHB/precursor complex, which determines the formation of decahedron shapes upon reduction. However, when the conditions and the ratios vary, other shapes are formed. Zhang et al. [49] reported that hydroxyl groups may affect the shape of the nanoparticles. Ghosh et al. [50] showed the possibility of obtaining flower-shaped zero-valent iron by controlling the amount of hydroxyl groups during the synthesis process, which suggests that the presence of hydroxyl groups may influence the formation of decahedral morphology. The SAED pattern for a Pd/Pt ratio of 2:1 indicates the polycrystalline nature of an as-synthesized Pd/Pt bimetallic alloy (Figure 2f). The SAED analysis identified (111), (220) and (311) planes of fcc. For this sample, the nanoparticle size distribution was calculated from TEM micrographs, and the mean size of these nanoparticles was found to be 15 ± 4 nm (Figure S2 in Supplementary Materials).

The various synthesis strategies used to obtain bimetallic Pd/Pt nanoparticles with varying morphologies are shown in Table 1. As stated earlier, changes to the synthesis procedure may result in the formation of structurally different nanoparticles, e.g. nanocubes are obtained by reduction with poly(vinylpyrrolidone) (PVP) while nanotetrahedra are formed when $Na_2C_2O_4$ and formaldehyde are used [39]. Conventionally for these kind of reactions, high temperatures [51] and prolonged synthesis times [52,53] are required. Very often the reducing agents used are hazardous, e.g. sodium borohydride [44,45].

Table 1. Synthesis procedures reported in the literature for obtaining Pd/Pt nanoparticles with different shapes.

Shape	Solvent	Precursors	Molar Pd/Pt Ratio	Reducing Agent	Temperature (°C)	Synthesis Time (min)	Reference
Cube	DMF	Na_2PdCl_4 K_2PtCl_6	1:1	-	130	300	[54]
Nanosponges	Water	H_2PdCl_4 K_2PtCl_6	1:1	$NaBH_4$	Room temperature	~5	[55]
Tetrahedron	Water	Na_2PdCl_4 K_2PtCl_6	1:1	HCHO	180	120	[56]
Octahedron	Water	Na_2PdCl_4 H_2PtCl_6	1:1	Glycerol	100	180	[57]
Corallite-like structure	Water	K_2PdCl_4 $K_2Pt(CN)_4$	2.05:1	$NaBH_4$	Room temperature	120	[58]
Branched Dandelion-like	Water	Na_2PdCl_4 K_2PtCl_6	1:7	Ascorbic acid	Room temperature	30	[55]
Nanocages	Water	K_2PdBr_4 Na_2PtBr_6	1:2	Ascorbic acid	Room temperature	480	[40]
Irregular polyhedron	Water	K_2PdCl_4 $PtCl_4$	1:1	Cs-PHB	130	30	This work
Decahedron	Water	K_2PdCl_4 $PtCl_4$	1:2	Cs-PHB	130	30	This work
Decahedron	Water	K_2PdCl_4 $PtCl_4$	2:1	Cs-PHB	130	30	This work

Wang et al. [59] reported the possibility to synthesize decahedral Pd/Pt, wherein the synthesis procedure can be divided into two steps: the first is to obtain Pd decahedraln structures, followed by the platinum deposition. Nano star-shaped Pd/Pt particles were reported by Lim et al. [41] using a co-reduction method involving Na_2PdCl_4 and K_2PtCl_4 in a PVP aqueous medium at 80 °C for 18 h. Another example of shape-specific synthesis of Pd/Pt was a seeded growth method using palladium truncated octahedral seeds for the synthesis of Pd/Pt nanodendrites [60].

3.1.3. EDS, Mapping and Profile

The EDS analysis of the bimetallic nanoparticles shows that all of the samples contain both palladium and platinum metals (Figure 3). Moreover, carbon and oxygen were also present in all of the analysed samples. The presence of both C and O may be attributed to the existence of a conjugate on the surface of these nanoparticles, which may act as a stabilizing agent.

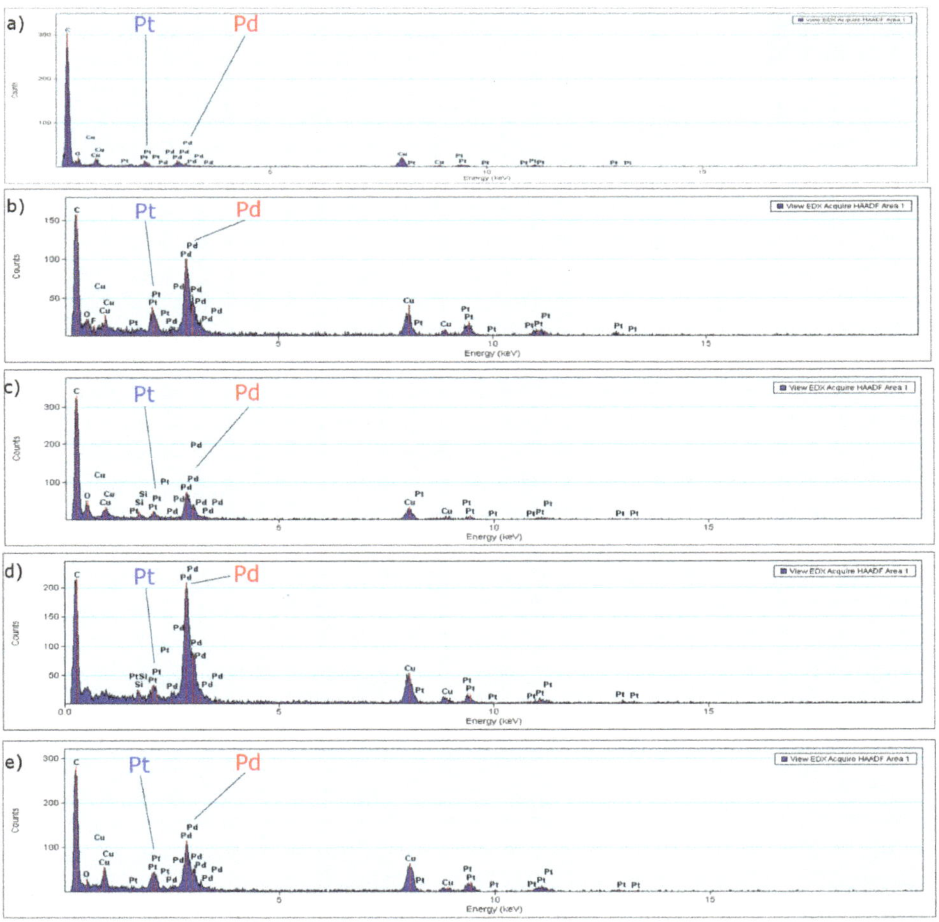

Figure 3. EDS analysis of Pd/Pt nanoparticles synthesized at different temperatures and molar ratios (**a**) 150 °C and 1:1, (**b**) 140 °C and 1:1, (**c**) 130 °C and 1:1, (**d**) 130 °C and 2:1, and (**e**) 130 °C and 1:2.

The EDS mapping analysis of the bimetallic nanoparticles (molar ratio of Pd/Pt of 2:1 and temperature of 130 °C) clearly shows the presence of both metals ubiquitously on the surface of the nanoparticle (Figure 4a,b). The EDS mapping also determined the presence of Pd/Pt alloy. The EDS profile analysis (Figure 4c) shows the presence of carbon (due to the presence of the conjugate), and it is in accordance with the previous ATR-FTIR analysis (see above). The profile also confirmed the predominant presence of Pd in almost all of the particle regions due to the initial Pd/Pt ratio of 2:1.

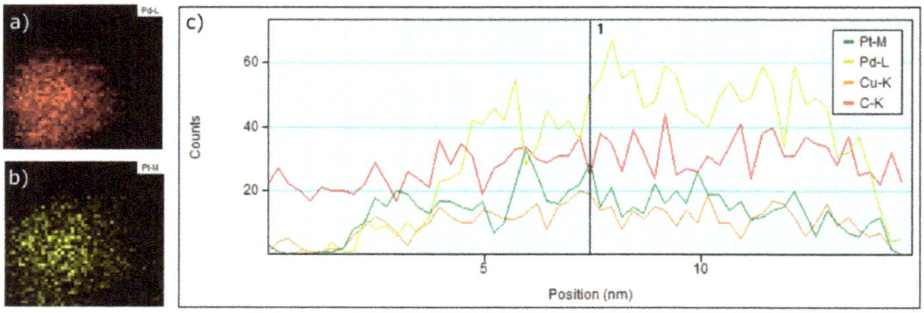

Figure 4. EDS mapping of (**a**) Pd and (**b**) Pt, and (**c**) EDS profile analysis of Pd/Pt nanoparticles (synthesis condition: Pd/Pt ratio 2:1, 10 g/L of Cs-PHB and temperature of 130 °C).

3.2. *Catalysis*

The catalytic performance of the Pd/Pt nanoparticles was proven by employing the reduction of 4-NP to 4-AP by $NaBH_4$ as a model [61]. The aqueous 4-NP solution shows a maximum absorption at ~317 nm, which upon addition of sodium borohydride shifts to 401 nm, indicating the formation of 4-nitrophenolate, and the solution turns from pale yellow to bright yellow. The reduction does not take place in the absence of a catalyst (kinetic barrier), which was verified by the unchanged intensity of the maximum absorption at 401 nm in the absence of the catalysts for 40 min (data not shown).

When the Pd/Pt nanoparticles were added to the solution, the intensity at 401 nm gradually decreased until it disappeared. Because an excess of $NaBH_4$ was used, the pseudo first-order kinetics model was applied to evaluate the catalytic performance of the Pd/Pt nanoparticles [62]. Due to the fact that the absorbance at 401 nm was linearly dependent on the 4-NP concentration (through 4-nitrophenolate), the rate constant k of the reaction can be calculated from the linear plot of $\ln(A_t/A_0)$ versus the reaction time t(min) [63–67]:

$$\ln(\mathrm{A_t/A_0}) = -k_{\mathrm{app}}t \tag{1}$$

where A_t and A_0 is absorbance at time t and 0, respectively. The pseudo-first-order kinetic rate constants (k_{app}) of the 4-NP reduction calculated based on Equation (1) for the various concentrations of nanoparticles synthesized with different molar (Pd/Pt) ratios of 1:1, 1:2 and 2:1, respectively are summarized in Table 2.

The reduction of 4-NP to 4-AP with borohydride catalysed by Pd/Pt nanoparticles may be explained by an electrochemical reaction, where the nanoparticles facilitate the electron transfer from BH_4^- to 4-NP. The mechanism is divided into the following steps: first both borohydride and 4-NP are adsorbed on the surface of the Pd/Pt nanoparticles, then electrons are transferred from BH_4^- to the nanoparticles with the formation of a negatively charged layer on their surface, later the electrons are transferred to 4-NP with a consequent formation of reduced products (Figure 5).

Table 2. The pseudo-first-order kinetic rate constants (k_{app}) of Pd/Pt synthesized in different ratios and κ_c value obtained by linear approximation of k_{app} (s^{-1}) vs concentration of catalysts (g/L).

Catalysts	Synthesis Temperature (°C)	Concentration (mg/L)	k_{app} (min^{-1})	κ_c ($L\ s^{-1}\ g^{-1}$)
Pd/Pt (1:1)	130	0.379	0.038	12 ± 4
		0.757	0.546	
		1.515	0.897	
Pd/Pt (1:2)	130	0.147	0.066	9 ± 1
		0.293	0.152	
		0.586	0.305	
Pd/Pt (2:1)	130	0.202	0.198	51 ± 11
		0.404	0.424	
		0.809	1.967	

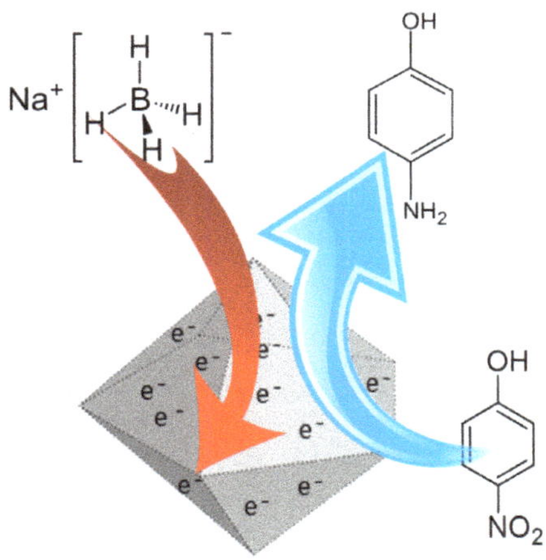

Figure 5. Electron transfer mechanism for reduction of 4-NP to 4-AP.

4-NP is often used for testing the catalytic activities of nanoparticles; nevertheless, catalytic performance comparisons are not easy. Most studies report only the k_{app}, but the k_{app} is strongly dependent on the concentration of reactants and the catalyst used for the reaction. Increasing the amount of catalyst in the reactor increases the total surface area available for the reaction, which means that a higher reaction rate is facilitated, and the time needed for reduction is shortened. Also, it was not possible in our study to add the same amounts of catalysts at different Pd/Pt ratios during the experiments. To overcome this, the activity parameter (κ_c) was employed to compare the efficiencies. This was determined by calculation of the slope of k_{app} (s^{-1}) as a function of the catalyst concentration (g/L) [65].

To the best of our knowledge, κ_c is the most appropriate parameter for comparing the catalytic activity of catalysts reported in the literature [65,67]. The κ_c values were calculated for these three different Pd/Pt ratios and, therefore, different morphologies. While for the same Pd/Pt ratio (1:1) and excess of Pt (1:2), the κ_c value is approximately 10, for the excess of Pd (2:1), the κ_c value of 51 is significantly higher. This is probably caused by a higher Pd ratio and not by morphology, since both excesses of one of the metals (1:2 and 2:1) have the same morphology of a decahedron, but significantly different κ_c values (Figure 6).

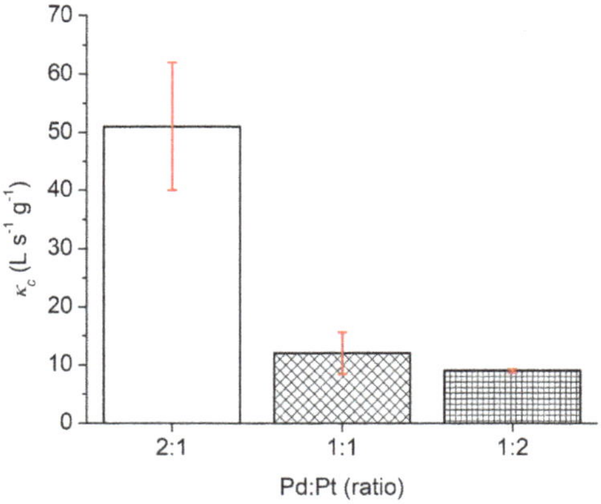

Figure 6. Comparison between the κ_c values of nanoparticles obtained by different ratios of Pd and Pt precursors (the red error bars represent the slope error).

Table 3 shows a comparison of the κ_c values obtained by different bimetallic catalysts for the reduction of 4-NP. The value determined by this study for a Pd/Pt ratio of 2:1 is one of the highest known to-date.

Table 3. Comparison of different bimetallic catalysts on the reduction of 4-NP reported in the literature.

Catalysts	Catalyst Concentration (mg/L)	4-NP Concentration (mM)	$NaBH_4$ Concentration (mM)	k_{app} (s^{-1})	κ_c (L s^{-1} g^{-1})	Ref.
Pd/Au	8	0.07	21	0.258	32	[68]
$Au_{53}Pd_{47}$/graphene nanosheets	0.06	0.05	5	0.014	240	[69]
Cu/Ag	0.48	0.096	11.2	0.0003	7.18	[70]
PdCuY	20	0.72	1.5	0.002	0.12	[71]
Pd/Pt nanotubes	3.4	0.09	100	0.008	25	[72]
Pd/Pt (2:1)	0.809	0.12	12	0.033	51 ± 11	This work
Pd/Pt (1:1)	0.757	0.12	12	0.009	12 ± 4	This work
Pd/Pt (1:2)	0.586	0.12	12	0.005	9 ± 1	This work

4. Conclusions

The present research describes a facile mediated green synthesis of bimetallic Pd/Pt nanoparticles of various morphologies. The nanoparticles were synthesized from K_2PdCl_4 and $PtCl_4$ precursor salts by a co-reduction with a Cs-PHB conjugate. Depending on the temperature and metal ratio, nanoparticles with a star-like structure, and octahedral or decahedral morphology were formed. The optimal conditions to obtain a decahedral shape were found to be: (Pd/Pt) molar ratio of 2:1, synthesis temperature of 130 °C, 10 g/L of Cs-PHB conjugate and time of 30 min. While ATR-FTIR and EDS confirmed the presence of a Cs-PHB layer on the surface of the nanoparticles, EDS verified the formation of a Pd/Pt alloy. TEM analysis confirmed the different shapes and sizes of the nanoparticles by changing the temperature and molar metal ratio. The decahedral bimetallic nanoparticles prepared at a molar ratio of 2:1 show an excellent catalytic performance for the catalytic reduction of 4-NP by borohydride. The pseudo-first-order kinetic constant for the nanoparticles synthesized with molar ratios (Pd/Pt) of 1:1, 1:2 and 2:1 were found to be 0.898 min^{-1} (catalyst concentration: 1.515 mg/L), 0.305 min^{-1} (0.586 mg/L) and 1.968 min^{-1} (0.809 mg/L), respectively. Compared with the other bimetallic nanoparticles, the (2:1) decahedral Pd/Pt nanoparticles exhibit excellent catalytic performance, which is demonstrated by the high κ_c value (51 ± 11 L s^{-1} g^{-1}).

Supplementary Materials: The following are available online at http://www.mdpi.com/2073-4360/11/12/1948/s1, Figure S1 FTIR analysis of Pd/Pt ratio 1:1, Figure S2: Size distribution of Pd/Pt synthesized at 130 °C and molar ratio 2:1 (Pd:Pt), Figure S3: pseudo-first-order kinetics of sample synthesized with molar ratio of 1:1 with differet concentration of nanoparticles (a) 0.379 mg/L (R^2 0.993), (b) 0.757 mg/L (R^2 0.992), (c) 1.515 mg/L (R^2 0.982) and (d) HRTEM image of Pd/Pt decahedron nanoparticle (molar ratio 1:1 (Pd/Pt) 130 °C), Figure S4: pseudo-first-order kinetics of sample synthesized with molar ratio of 1:2 with differet concentration of nanoparticles (a) 0.147 mg/L (R^2 0.992), (b) 0.293 mg/L (R^2 0.998), (c) 0.586 mg/L (R^2 0.984) and (d) HRTEM image of Pd/Pt decahedron nanoparticle (molar ratio 1:2 (Pd/Pt) 130 °C), Figure S5: pseudo-first-order kinetics of sample synthesized with molar ratio of 2:1 with different concentration of nanoparticles (a) 0.202 mg/L (R^2 0.997) (b) 0.404 mg/L (R^2 0.986) (c) 0.809 g/L (R^2 0.979) and d) HRTEM image of Pd/Pt decahedron nanoparticle (molar ratio 2:1 (Pd/Pt) 130 °C).

Author Contributions: Conceptualization, D.S. and S.W.; Funding acquisition, M.Č.; Investigation, D.S. and R.K.R.; Methodology, R.K.R., A.V. and B.S.; Supervision, S.W.; Writing—original draft, D.S.; Writing—review & editing, S.W., K.K., V.V.T.P., B.S. and M.Č.

Funding: This work was supported by the Student Grant Scheme 2019 project of the Technical University of Liberec. The authors would also like to acknowledge the assistance provided by the Research Infrastructure NanoEnviCz, supported by the Ministry of Education, Youth and Sports of the Czech Republic in the framework of Project No. LM2015073, and the project Pro-NanoEnviCz (Reg. No. CZ.02.1.01/0.0/0.0/16_013/0001821) supported by the Ministry of Education, Youth and Sports of the Czech Republic and the European Union - European Structural and Investments Funds in the frame of Operational Programme Research Development and Education.

Conflicts of Interest: The authors declare no conflict of interest.

References

1. Virkutyte, J.; Varma, R.S. Green synthesis of metal nanoparticles: Biodegradable polymers and enzymes in stabilization and surface functionalization. *Chem. Sci.* **2011**, *2*, 837–846. [CrossRef]
2. Moulton, M.C.; Braydich-Stolle, L.K.; Nadagouda, M.N.; Kunzelman, S.; Hussain, S.M.; Varma, R.S. Synthesis, characterization and biocompatibility of "green" synthesized silver nanoparticles using tea polyphenols. *Nanoscale* **2010**, *2*, 763–770. [CrossRef] [PubMed]
3. Vinod, V.T.P.; Saravanan, P.; Sreedhar, B.; Devi, D.K.; Sashidhar, R.B. A facile synthesis and characterization of Ag, Au and Pt nanoparticles using a natural hydrocolloid gum kondagogu (Cochlospermum gossypium). *Colloids Surf. B Biointerfaces* **2011**, *83*, 291–298. [CrossRef] [PubMed]
4. Padil, V.V.T.; Černík, M.; Thekkae Padil, V.V.; Černík, M. Green synthesis of copper oxide nanoparticles using gum karaya as a biotemplate and their antibacterial application. *Int. J. Nanomed.* **2013**, *8*, 889–898. [CrossRef]
5. Nadagouda, M.N.; Speth, T.F.; Varma, R.S. Microwave-assisted green synthesis of silver nanostructures. *Acc. Chem. Res.* **2011**, *44*, 469–478. [CrossRef]
6. Hebbalalu, D.; Lalley, J.; Nadagouda, M.N.; Varma, R.S. Greener techniques for the synthesis of silver nanoparticles using plant extracts, enzymes, bacteria, biodegradable polymers, and microwaves. *ACS Sustain. Chem. Eng.* **2013**, *1*, 703–712. [CrossRef]
7. Seo, E.; Kim, J.; Hong, Y.; Kim, Y.S.; Lee, D.; Kim, B.S. Double hydrophilic block copolymer templated au nanoparticles with enhanced catalytic activity toward nitroarene reduction. *J. Phys. Chem. C* **2013**, *117*, 11686–11693. [CrossRef]
8. Machmudah, S.; Sato, T.; Wahyudiono; Sasaki, M.; Goto, M. Silver nanoparticles generated by pulsed laser ablation in supercritical CO_2 medium. *High Press. Res.* **2012**, *32*, 60–66. [CrossRef]
9. Zhao, L.; Song, J.; Xue, Y.; Zhao, X.; Deng, Y.; Li, Q.; Xia, Y. Green synthesis of Ag–Au bimetallic nanoparticles with alginate for sensitive detection of H_2O_2. *Catal. Lett.* **2018**, *148*, 3248–3256. [CrossRef]
10. Padil, V.V.T.; Wacławek, S.; Černík, M. Green Synthesis: Nanoparticles and Nanofibres Based on Tree Gums for Environmental Applications. *Ecol. Chem. Eng. S* **2016**, *23*, 533–557. [CrossRef]
11. Sun, L.; Li, J.; Cai, J.; Zhong, L.; Ren, G.; Ma, Q. One pot synthesis of gold nanoparticles using chitosan with varying degree of deacetylation and molecular weight. *Carbohydr. Polym.* **2017**, *178*, 105–114. [CrossRef] [PubMed]
12. Chen, X.; Xu, X.J.; Zheng, X.C.; Guan, X.X.; Liu, P. Chitosan supported palladium nanoparticles: The novel catalysts for hydrogen generation from hydrolysis of ammonia borane. *Mater. Res. Bull.* **2018**, *103*, 89–95. [CrossRef]

13. Oliveira, Â.A.S.; Medeiros, R.L.B.A.; Figueredo, G.P.; Macedo, H.P.; Braga, R.M.; Maziviero, F.V.; Melo, M.A.F.; Melo, D.M.A.; Vieira, M.M. One-step synthesis of $LaNiO_3$ with chitosan for dry reforming of methane. *Int. J. Hydrogen Energy* **2018**, *43*, 9696–9704. [CrossRef]
14. Anitha, A.; Sowmya, S.; Kumar, P.T.S.; Deepthi, S.; Chennazhi, K.P.; Ehrlich, H.; Tsurkan, M.; Jayakumar, R. Chitin and chitosan in selected biomedical applications. *Prog. Polym. Sci.* **2014**, *39*, 1644–1667. [CrossRef]
15. Ali, A.; Ahmed, S. A review on chitosan and its nanocomposites in drug delivery. *Int. J. Biol. Macromol.* **2018**, *109*, 273–286. [CrossRef]
16. Hahn, T.; Zibek, S. Sewage Polluted Water Treatment via Chitosan: A Review. In *Chitin-Chitosan—Myriad Functionalities in Science and Technology*; InTechOpen: London, UK, 2018.
17. Logithkumar, R.; Keshavnarayan, A.; Dhivya, S.; Chawla, A.; Saravanan, S.; Selvamurugan, N. A review of chitosan and its derivatives in bone tissue engineering. *Carbohydr. Polym.* **2016**, *151*, 172–188. [CrossRef]
18. Baig, R.B.N.; Varma, R.S. Copper on chitosan: A recyclable heterogeneous catalyst for azide-alkyne cycloaddition reactions in water. *Green Chem.* **2013**, *15*, 1839. [CrossRef]
19. Sedghi, R.; Heidari, B.; Shahmohamadi, H.; Zarshenas, P.; Varma, R.S. Pd Nanocatalyst Adorned on Magnetic Chitosan@N-Heterocyclic Carbene: Eco-Compatible Suzuki Cross-Coupling Reaction. *Molecules* **2019**, *24*, 3048. [CrossRef]
20. Devi, R.; Dhamodharan, R. Pretreatment in Hot Glycerol for Facile and Green Separation of Chitin from Prawn Shell Waste. *ACS Sustain. Chem. Eng.* **2018**, *6*, 846–853. [CrossRef]
21. Al Rowaihi, I.S.; Paillier, A.; Rasul, S.; Karan, R.; Grötzinger, S.W.; Takanabe, K.; Eppinger, J. Poly(3-hydroxybutyrate) production in an integrated electromicrobial setup: Investigation under stress-inducing conditions. *PLoS ONE* **2018**, *13*, e0196079. [CrossRef]
22. Silva, F.; Campanari, S.; Matteo, S.; Valentino, F.; Majone, M.; Villano, M. Impact of nitrogen feeding regulation on polyhydroxyalkanoates production by mixed microbial cultures. *New Biotechnol.* **2017**, *37*, 90–98. [CrossRef] [PubMed]
23. Valentino, F.; Morgan-Sagastume, F.; Campanari, S.; Villano, M.; Werker, A.; Majone, M. Carbon recovery from wastewater through bioconversion into biodegradable polymers. *New Biotechnol.* **2017**, *37*, 9–23. [CrossRef] [PubMed]
24. Baric, M.; Pierro, L.; Pietrangeli, B.; Papini, M.P. Polyhydroxyalkanoate (PHB) as a slow-release electron donor for advanced in situ bioremediation of chlorinated solvent-contaminated aquifers. *New Biotechnol.* **2014**, *31*, 377–382. [CrossRef] [PubMed]
25. Michalak, M.; Marek, A.A.; Zawadiak, J.; Kawalec, M.; Kurcok, P. Synthesis of PHB-based carrier for drug delivery systems with pH-controlled release. *Eur. Polym. J.* **2013**, *49*, 4149–4156. [CrossRef]
26. Getachew, A.; Woldesenbet, F. Production of biodegradable plastic by polyhydroxybutyrate (PHB) accumulating bacteria using low cost agricultural waste material. *BMC Res. Notes* **2016**, *9*, 1–9. [CrossRef] [PubMed]
27. Wacławek, S.; Chronopoulou, L.; Petrangeli Papini, M.; Vtp, V.; Palocci, C.; Kupčík, J.; Černík, M. Enhancement of stability and reactivity of nanosized zero-valent iron with polyhydroxybutyrate. *Desalin. Water Treat.* **2017**, *69*. [CrossRef]
28. Torkelson, T.R.; Oyen, F.; Rowe, V.K. The toxicity of chloroform as determined by single and repeated exposure of laboratory animals. *Am. Ind. Hyg. Assoc. J.* **1976**, *37*, 697–705. [CrossRef]
29. Rannug, U. Genotoxic effects of 1,2-dibromoethane and 1,2-dichloroethane. *Mutat. Res. Rev. Genet. Toxicol.* **1980**, *76*, 269–295. [CrossRef]
30. Silvestri, D.; Wacławek, S.; Sobel, B.; Torres-Mendieta, R.; Novotný, V.; Nguyen, N.H.A.; Ševců, A.; Padil, V.V.T.; Müllerová, J.; Stuchlík, M.; et al. A poly(3-hydroxybutyrate)-chitosan polymer conjugate for the synthesis of safer gold nanoparticles and their applications. *Green Chem.* **2018**, *20*, 4975–4982. [CrossRef]
31. Sharma, G.; Kumar, A.; Sharma, S.; Naushad, M.; Prakash Dwivedi, R.; ALOthman, Z.A.; Mola, G.T. Novel development of nanoparticles to bimetallic nanoparticles and their composites: A review. *J. King Saud Univ. Sci.* **2019**, *31*, 257–269. [CrossRef]
32. Shifrina, Z.B.; Matveeva, V.G.; Bronstein, L.M. Role of polymer structures in catalysis by transition metal and metal oxide Nanoparticle Composites. *Chem. Rev.* **2019**. [CrossRef] [PubMed]
33. Mei, Y.; Lu, Y.; Polzer, F.; Ballauff, M.; Drechsler, M. Catalytic activity of palladium nanoparticles encapsulated in spherical poly electrolyte brushes and core-shell microgels. *Chem. Mater.* **2007**, *19*, 1062–1069. [CrossRef]

34. Mei, Y.; Sharma, G.; Lu, Y.; Ballauff, M.; Drechsler, M.; Irrgang, T.; Kempe, R. High catalytic activity of platinum nanoparticles immobilized on spherical polyelectrolyte brushes. *Langmuir* **2005**, *21*, 12229–12234. [CrossRef] [PubMed]
35. Li, Z.; Yao, C.; Wang, Y.C.; Mikael, S.; Gunasekaran, S.; Ma, Z.; Cai, Z.; Wang, X. High-density platinum nanoparticle-decorated titanium dioxide nanofiber networks for efficient capillary photocatalytic hydrogen generation. *J. Mater. Chem. A* **2016**, *4*, 11672–11679. [CrossRef]
36. Wu, M.C.; Hsiao, K.C.; Chang, Y.H.; Chan, S.H. Photocatalytic hydrogen evolution of palladium nanoparticles decorated black TiO_2 calcined in argon atmosphere. *Appl. Surf. Sci.* **2018**, *430*, 407–414. [CrossRef]
37. Elsey, J.; Bubley, J.A.; Zhu, L.; Rao, S.; Sasaki, M.; Pollack, B.P.; Yang, L.; Arbiser, J.L. Palladium based nanoparticles for the treatment of advanced melanoma. *Sci. Rep.* **2019**, *9*, 3255. [CrossRef] [PubMed]
38. Samadi, A.; Klingberg, H.; Jauffred, L.; Kjær, A.; Bendix, P.M.; Oddershede, L.B. Platinum nanoparticles: A non-toxic, effective and thermally stable alternative plasmonic material for cancer therapy and bioengineering. *Nanoscale* **2018**, *10*, 9097–9107. [CrossRef]
39. Lai, J.; Luque, R.; Xu, G. Recent Advances in the Synthesis and Electrocatalytic Applications of Platinum-Based Bimetallic Alloy Nanostructures. *ChemCatChem* **2015**, *7*, 3206–3228. [CrossRef]
40. Wang, L.; Yamauchi, Y. Metallic nanocages: Synthesis of bimetallic Pt-Pd hollow nanoparticles with dendritic shells by selective chemical etching. *J. Am. Chem. Soc.* **2013**, *135*, 16762–16765. [CrossRef]
41. Lim, B.; Wang, J.; Camargo, P.H.C.; Cobley, C.M.; Kim, M.J.; Xia, Y. Twin-induced growth of palladium-platinum alloy nanocrystals. *Angew. Chem. Int. Ed.* **2009**, *48*, 6304–6308. [CrossRef]
42. Venkateshaiah, A.; Silvestri, D.; Ramakrishnan, R.K.; Wacławek, S.; Padil, V.V.T.; Černík, M.; Varma, R.S. Gum Kondagogu/Reduced Graphene Oxide Framed Platinum Nanoparticles and Their Catalytic Role. *Molecules* **2019**, *24*, 3643. [CrossRef] [PubMed]
43. Baruah, B.; Gabriel, G.J.; Akbashev, M.J.; Booher, M.E. Facile synthesis of silver nanoparticles stabilized by cationic polynorbornenes and their catalytic activity in 4-nitrophenol reduction. *Langmuir* **2013**, *29*, 4225–4234. [CrossRef] [PubMed]
44. Queiroz, M.F.; Melo, K.R.T.; Sabry, D.A.; Sassaki, G.L.; Rocha, H.A.O. Does the use of chitosan contribute to oxalate kidney stone formation? *Mar. Drugs* **2015**, *13*, 141–158. [CrossRef] [PubMed]
45. Dang Nguyen Vô, K.; Kowandy, C.; Dupont, L.; Coqueret, X. Evidence of chitosan-mediated reduction of Au(III) to Au(0) nanoparticles under electron beam by using OH and e-aq scavengers. *Chem. Commun.* **2015**, *51*, 4017–4020. [CrossRef]
46. Dorjnamjin, D.; Ariunaa, M.; Shim, Y.K. Synthesis of silver nanoparticles using hydroxyl functionalized ionic liquids and their antimicrobial activity. *Int. J. Mol. Sci.* **2008**, *9*, 807–820. [CrossRef]
47. Hu, P.; Song, Y.; Rojas-Andrade, M.D.; Chen, S. Platinum Nanoparticles Functionalized with Ethynylphenylboronic Acid Derivatives: Selective Manipulation of Nanoparticle Photoluminescence by Fluoride Ions. *Langmuir* **2014**, *30*, 5224–5229. [CrossRef]
48. Mendoza-Pérez, R.; Guisbiers, G. Bimetallic Pt-Pd nano-catalyst: Size, shape and composition matter. *Nanotechnology* **2019**, *30*, 305702. [CrossRef]
49. Zhang, Q.; Li, N.; Goebl, J.; Lu, Z.; Yin, Y. A systematic study of the synthesis of silver nanoplates: Is citrate a "magic" reagent? *J. Am. Chem. Soc.* **2011**, *133*, 18931–18939. [CrossRef]
50. Ghosh, A.; Dutta, S.; Mukherjee, I.; Biswas, S.; Chatterjee, S.; Saha, R. Template-free synthesis of flower-shaped zero-valent iron nanoparticle: Role of hydroxyl group in controlling morphology and nitrate reduction. *Adv. Powder Technol.* **2017**, *28*, 2256–2264. [CrossRef]
51. Long, N.V.; Hien, T.D.; Asaka, T.; Ohtaki, M.; Nogami, M. Synthesis and characterization of Pt–Pd nanoparticles with core-shell morphology: Nucleation and overgrowth of the Pd shells on the as-prepared and defined Pt seeds. *J. Alloys Compd.* **2011**, *509*, 7702–7709. [CrossRef]
52. Tuo, Y.; Liu, G.; Dong, B.; Yu, H.; Zhou, J.; Wang, J.; Jin, R. Microbial synthesis of bimetallic PdPt nanoparticles for catalytic reduction of 4-nitrophenol. *Environ. Sci. Pollut. Res.* **2017**, *24*, 5249–5258. [CrossRef]
53. Li, H.; Han, L.; Cooper-White, J.; Kim, I. Palladium nanoparticles decorated carbon nanotubes: Facile synthesis and their applications as highly efficient catalysts for the reduction of 4-nitrophenol. *Green Chem.* **2012**, *14*, 586. [CrossRef]
54. Huang, X.; Li, Y.; Li, Y.; Zhou, H.; Duan, X.; Huang, Y. Synthesis of PtPd bimetal nanocrystals with controllable shape, composition, and their tunable catalytic properties. *Nano Lett.* **2012**, *12*, 4265–4270. [CrossRef] [PubMed]

55. Zhu, C.; Guo, S.; Dong, S. Rapid, general synthesis of pdpt bimetallic alloy nanosponges and their enhanced catalytic performance for ethanol/methanol electrooxidation in an alkaline medium. *Chem. A Eur. J.* **2013**, *19*, 1104–1111. [CrossRef] [PubMed]
56. Yin, A.X.; Min, X.Q.; Zhang, Y.W.; Yan, C.H. Shape-selective synthesis and facet-dependent enhanced electrocatalytic activity and durability of monodisperse Sub-10 nm Pt-Pd tetrahedrons and cubes. *J. Am. Chem. Soc.* **2011**, *133*, 3816–3819. [CrossRef] [PubMed]
57. Lee, Y.W.; Ko, A.R.; Han, S.B.; Kim, H.S.; Park, K.W. Synthesis of octahedral Pt–Pd alloy nanoparticles for improved catalytic activity and stability in methanol electrooxidation. *Phys. Chem. Chem. Phys.* **2011**, *13*, 5569. [CrossRef]
58. Liu, X.Y.; Zhang, Y.; Gong, M.X.; Tang, Y.W.; Lu, T.H.; Chen, Y.; Lee, J.M. Facile synthesis of corallite-like Pt-Pd alloy nanostructures and their enhanced catalytic activity and stability for ethanol oxidation. *J. Mater. Chem. A* **2014**, *2*, 13840–13844. [CrossRef]
59. Scott, R.W.J.; Datye, A.K.; Crooks, R.M. Bimetallic Palladium-Platinum Dendrimer-Encapsulated Catalysts. *J. Am. Chem. Soc.* **2003**, *125*, 3708–3709. [CrossRef]
60. Datta, K.J.; Datta, K.K.R.; Gawande, M.B.; Ranc, V.; Čépe, K.; Malgras, V.; Yamauchi, Y.; Varma, R.S.; Zboril, R. Pd@Pt Core-Shell Nanoparticles with Branched Dandelion-like Morphology as Highly Efficient Catalysts for Olefin Reduction. *Chem. A Eur. J.* **2016**, *22*, 1577–1581. [CrossRef]
61. Wang, X.; Vara, M.; Luo, M.; Huang, H.; Ruditskiy, A.; Park, J.; Bao, S.; Liu, J.; Howe, J.; Chi, M.; et al. Pd@Pt Core—Shell Concave Decahedra: A Class of Catalysts for the Oxygen Reduction Reaction with Enhanced Activity and Durability. *J. Am. Chem. Soc.* **2015**, *137*, 15036–15042. [CrossRef]
62. Lim, B.; Jiang, M.; Camargo, P.H.C.; Cho, E.C.; Tao, J.; Lu, X.; Zhu, Y.; Xia, Y. Pd-Pt bimetallic nanodendrites with high activity for oxygen reduction. *Science* **2009**, *324*, 1302–1305. [CrossRef]
63. Wacławek, S.; Gončuková, Z.; Adach, K.; Fijałkowski, M.; Černík, M. Green synthesis of gold nanoparticles using Artemisia dracunculus extract: Control of the shape and size by varying synthesis conditions. *Environ. Sci. Pollut. Res.* **2018**, *25*, 24210–24219. [CrossRef] [PubMed]
64. Stumm, W.; Morgan, J.J. *Aquatic Chemistry: Chemical Equilibria and Rates in Natural Waters*; Wiley: Hoboken, NJ, USA, 2012; ISBN 1118591488.
65. Kästner, C.; Thünemann, A.F. Catalytic Reduction of 4-Nitrophenol Using Silver Nanoparticles with Adjustable Activity. *Langmuir* **2016**, *32*, 7383–7391. [CrossRef] [PubMed]
66. Gangula, A.; Podila, R.; Karanam, L.; Janardhana, C.; Rao, A.M. Catalytic reduction of 4-nitrophenol using biogenic gold and silver nanoparticles derived from breynia rhamnoides. *Langmuir* **2011**, *27*, 15268–15274. [CrossRef] [PubMed]
67. Lara, L.R.S.; Zottis, A.D.; Elias, W.C.; Faggion, D.; Maduro De Campos, C.E.; Acuña, J.J.S.; Domingos, J.B. The catalytic evaluation of in situ grown Pd nanoparticles on the surface of Fe3O4@dextran particles in the p-nitrophenol reduction reaction. *RSC Adv.* **2015**, *5*, 8289–8296. [CrossRef]
68. Ma, T.; Liang, F.; Chen, R.; Liu, S.; Zhang, H. Synthesis of Au-Pd bimetallic nanoflowers for catalytic reduction of 4-nitrophenol. *Nanomaterials* **2017**, *7*, 239. [CrossRef]
69. Chen, X.; Cai, Z.; Chen, X.; Oyama, M. AuPd bimetallic nanoparticles decorated on graphene nanosheets: Their green synthesis, growth mechanism and high catalytic ability in 4-nitrophenol reduction. *J. Mater. Chem. A* **2014**, *2*, 5668–5674. [CrossRef]
70. Wu, W.; Lei, M.; Yang, S.; Zhou, L.; Liu, L.; Xiao, X.; Jiang, C.; Roy, V.A.L. A one-pot route to the synthesis of alloyed Cu/Ag bimetallic nanoparticles with different mass ratios for catalytic reduction of 4-nitrophenol. *J. Mater. Chem. A* **2015**, *3*, 3450–3455. [CrossRef]
71. El-Bahy, Z.M.; Hanafy, A.I.; El-Bahy, S.M. Preparation of Pt, Pd and Cu nano single and bimetallic systems-supported NaY zeolite and test their activity in p-nitrophenol reduction and as anticancer agents. *J. Environ. Chem. Eng.* **2019**, *7*, 103117. [CrossRef]
72. Wang, Y.; Li, Q.; Zhang, P.; O'Connor, D.; Varma, R.S.; Yu, M.; Hou, D. One-pot green synthesis of bimetallic hollow palladium-platinum nanotubes for enhanced catalytic reduction of p-nitrophenol. *J. Colloid Interf. Sci.* **2019**, *539*, 161–167. [CrossRef]

© 2019 by the authors. Licensee MDPI, Basel, Switzerland. This article is an open access article distributed under the terms and conditions of the Creative Commons Attribution (CC BY) license (http://creativecommons.org/licenses/by/4.0/).

polymers

Article

High Efficiency Gas Permeability Membranes from Ethyl Cellulose Grafted with Ionic Liquids

Jingyu Xu [1], Hongge Jia [1,*], Nan Yang [1,*], Qingji Wang [2], Guoxing Yang [3], Mingyu Zhang [1], Shuangping Xu [1], Yu Zang [1], Liqun Ma [1], Pengfei Jiang [1], Hailiang Zhou [1] and Honghan Wang [1]

1 College of Materials Science and Engineering, Heilongjiang Province Key Laboratory of Polymeric Composition, College of Architecture and Civil Engineering, Qiqihar University, Wenhua Street 42, Qiqihar 161006, China; xjy951011@163.com (J.X.); zhangmingyuno1@163.com (M.Z.); xshp_1979_1999@163.com (S.X.); zangyu.25@163.com (Y.Z.); maliqun6166@163.com (L.M.); jpf848185@163.com (P.J.); zhouhailiang95@163.com (H.Z.); wanghonghan0628@163.com (H.W.)

2 Daqing Oilfield Construction Design and Research Institute, XiLing Road 32, Daqing 1637241, China; wangqingji@petrochina.com.cn

3 Daqing Petrochemical Research Center, Petrochemical Research Institute, China National Petroleum Corporation, Chengxiang Road 2, Daqing 163714, China; ygx459@petrochina.com.cn

* Correspondence: jiahonggell@hotmail.com (H.J.); nanyoung1980@nefu.edu.cn (N.Y.); Tel.: +86-452-2738752 (H.J.); +86-452-2725797 (N.Y.)

Received: 5 October 2019; Accepted: 13 November 2019; Published: 18 November 2019

Abstract: Ethyl cellulose was grafted with ionic liquids in optimal yields (62.5–64.1%) and grafting degrees (5.93–7.90%) by the esterification of the hydroxyl groups in ethyl cellulose with the carboxyl groups in ionic liquids. In IR spectra of the ethyl cellulose derivatives exhibited C=O bond stretching vibration peaks at 1760 or 1740 cm^{-1}, confirming the formation of the ester groups and furnishing the evidence of the successful grafting of ethyl cellulose with ionic liquids. The ethyl cellulose grafted with ionic liquids could be formed into membranes by using the casting solution method. The resulting membranes exhibited good membrane forming ability and mechanical properties. The EC grafted with ionic liquids-based membranes demonstrated PCO_2/PCH_4 separation factors of up to 18.8, whereas the PCO_2/PCH_4 separation factor of 9.0 was obtained for pure EC membrane (both for CO_2/CH_4 mixture gas). The membranes also demonstrated an excellent gas permeability coefficient PCO_2, up to 199 Barrer, which was higher than pure EC (PCO_2 = 46.8 Barrer). Therefore, it can be concluded that the ionic liquids with imidazole groups are immensely useful for improving the gas separation performances of EC membranes.

Keywords: ethyl cellulose; liquid; imidazole; membrane; gas permeation properties

1. Introduction

The separation membrane represents the core of the separation technology, and the performance of the separation membrane depends largely on the membrane material and formation process. Therefore, a good gas separation membrane material must have optimal gas permeability coefficient, separation factor, chemical stability, mechanical strength and film forming ability. The membrane materials currently used in the field of gas separation are mainly organic polymer materials, including poly(dimethyl siloxane) (PDMS) [1], polysulfone (PSF) [2], poly(vinylidene fluoride) (PVDF) [3], polyimide (PI) [4], polyetherimide (PEI) [5], cellulose acetate (CA) [6], ethyl cellulose (EC) [7], etc.

EC is a functional, non-ionic cellulose ether obtained by reacting cellulose with CH_3CH_2Cl [8]. EC is renewable and abundantly available. In addition, EC also has unique properties otherwise not possessed by other cellulose ethers, such as chemical stability, film forming ability and good mechanical properties. The hydroxyl groups in EC are replaced by a large number of ethyl groups, forming a typical hydrophobic cellulose. The unsubstituted hydroxyl groups are connected by hydrogen bonds to form a tight coral-like network structure, which gives EC excellent mechanical properties [9–11]. The non-toxic, hydrophobic, mechanical, thermoplastic and film-forming properties of EC have applications in many fields, such as food, microencapsulation, filtration and medicine [9,12]. EC has also been used as a separation membrane material [13].

As early as the 1990s, literature studies reported the preparation of ethyl cellulose membranes for separating oxygen and nitrogen [14]. S.M. Chen et al. studied the gas permeability of homogeneous [15] and porous EC membranes [16] by structural modification and solvent optimization. The O_2/N_2 separation factors of the homogeneous membrane reached 6.2, indicating that the EC membranes have high selectivity for O_2 [15]. Although the pores supporting EC membranes are currently for oxygen/nitrogen separation, having higher gas permeability, a lower selectivity for CO_2/N_2 and CO_2/CH_4 is generally observed [17]. In order to expand the further application of ethyl cellulose, modified EC films have also been prepared [18–23].

In another study, silyl ethers of ethyl cellulose were synthesized by the reaction of various chlorosilanes with the residual hydroxy groups of ethyl cellulose [19]. The PCO_2/PN_2 permselectivity values of the polymers were observed to be in the range of 15–19. The increased diffusion coefficients resulting from the introduction of silyl moiety in ethyl cellulose were observed for gas permeability; thus, their good separation performance for CO_2 /N_2 and CO_2 /CH_4 was discerned [19]. X.G. Li et al. prepared binary blend membranes of EC and poly (4-vinylpyridine) using a solution casting technique with chloroform as the solvent [20]. Remarkable and continuously enhanced selectivity was achieved for important gas pairs, including oxygen/nitrogen, carbon dioxide/methane and hydrogen/nitrogen with increasing poly (4-vinylpyridine) content [20]. The blends had higher gas separation factors and comparable gas permeabilities. Such behavior can be envisaged for all microscopically immiscible polymer blends and has important practical implications [20]. The membrane blend of EC and maleic anhydride end-capped poly(propylene carbonate) was reported to be difficult for actual gas separation due to unstable gas separation performance [21]. Q. Hu [22] and M. Moaddeb [23] also introduced nano-scale TiO_2 or SiO_2 particles into the EC membrane materials, which improved the gas selectivity of the membranes, but the improvement effect on gas permeability was not obvious.

Ionic liquids (IL) refer to a class of ionic compounds consisting of organic cations and organic (inorganic) anions at room temperature or near room temperature [24]. The current applications of ionic liquids in gas membrane technology are mainly reflected in the separation of CO_2, which is mainly due to the excellent solubility and selectivity of ionic liquids towards CO_2, especially in the case of ionic liquids containing functional imidazole groups [25]. Nikolaeva et al. [26]. have synthesized a new cellulose-derived poly(ionic liquid) (PIL) and characterized it for CO_2 separation. The ideal CO_2/N_2 adsorption selectivity of poly(diallyldimethyl ammonium)-bis-(triflfluoromethylsulfonyl) imide (P (CA) (Tf_2N)) was constantly below 10 bar. The mixed gas permeation test showed that the P[CA][Tf_2N] based film with a 5 μm thick selection layer had twice the CO_2 flux of the conventional cellulose acetate (CA). These results indicate that the modified CA with IL is a successful method to increase permeating flow and improve process stability over a higher CO_2 /N_2 and CO_2/CH_4 gas mixture concentration and pressure range.

In this study, membranes from ethyl cellulose grafted with ionic liquids with imidazole groups with high gas permeabilities and selectivities were prepared. An in-depth study of the relationship between the membrane structure and CO_2 separation performance was studied subsequently.

2. Experimental Section

2.1. Materials

Ethyl cellulose and ionic liquids were purchased from Chembee (Shanghai, China) and Greenchem ILs (Lanzhou, China), and were used as reactants. The ethyl cellulose we used was 45-55 MPa.s, 95% pure, and contained 5% toluene/isopropanol = 80:20. This is different from the purity and type of ethyl cellulose used in the literature, resulting in a certain deviation from the literature. Toluene and isopropanol are not plasticizers. Tetrahydrofuran was purchased from Kaitong Chemical Reagent Co. Ltd. (Tianjin, China) and was employed after distillation. EC grafted with ionic liquids (EC-g1–EC-g4 in Scheme 1) and EC blended with ionic liquids (EC-b1–EC-b4 in Scheme 2) were prepared according to Schemes 1 and 2. The details of the synthesis procedure and analytical analysis are presented in the following sections.

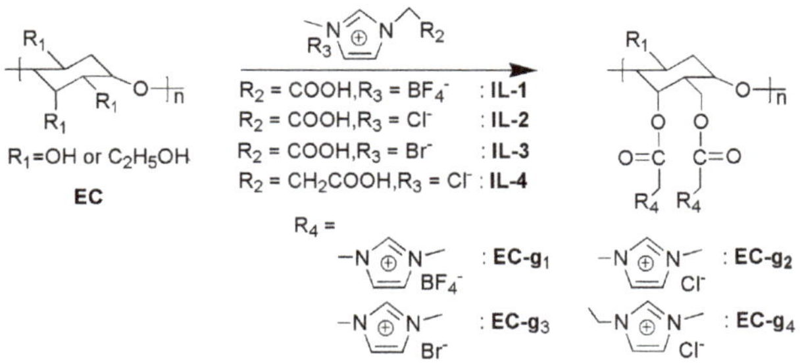

Scheme 1. Synthetic route for obtaining ethyl cellulose (EC) grafted with ionic liquids (EC-g1–EC-g4).

2.2. Preparation of Ethyl Cellulose Grafted with 1-Carboxymethyl-3-Methylimidazolium Tetrafluoroborate (EC-g1)

Ethyl cellulose (1.45 g, 6.10 mmol), and pyridine (0.80 g, 10.12 mmol) were placed in a flask, evacuated for a 30 min, flushed with nitrogen and dissolved in THF (50 mL) at room temperature. Subsequently, 1-carboxymethyl-3-methylimidazolium tetrafluoroborate (1.20 g, 4.50 mmol) was injected dropwise, and stirred for 48 h at 65 °C. After the reaction mixture was cooled to room temperature, it was added to excess methanol (1000 mL) and precipitated for a week. The precipitate was concentrated by centrifugal separation and dried under vacuum to afford the desired product (1.02 g), obtained as a white solid, with a yield of 64.1%. IR (KBr, cm^{-1}) bands were observed at 3080, 3020, 2960, 2930, 2860, 1760, 1640, 1620, 1500, 1470, 1400, 1380, 1348, 1250, 1100, 920, 890, and 620 (see Figure 1). The yield was calculated by this formula:

$$\text{Yield} = \frac{m_s}{\frac{m_o}{W_u} \times [d \times (W_u + W_{IL}) + (1 - d) \times W_u]} \times 100\%, \quad (1)$$

where m_s is the mass of the grafted product obtained, m_o is the mass of EC added, W_u is the molecular weight of repetitive unit in EC, W_{IL} is the molecular weight of ionic liquid, d is the grafting degree in EC and d can be calculated from Equation (7).

2.3. Preparation of Ethyl Cellulose Blended Ionic Liquids (EC-b1– EC-b4)

According to Scheme 2, ethyl cellulose (1.45 g, 6.10 mmol) was added to anhydrous ethanol (50 mL) and stirred for 12 h at room temperature. Subsequently, ionic liquids (1.20 g, 4.50 mmol) were added to the solution and stirred for 12 h at room temperature.

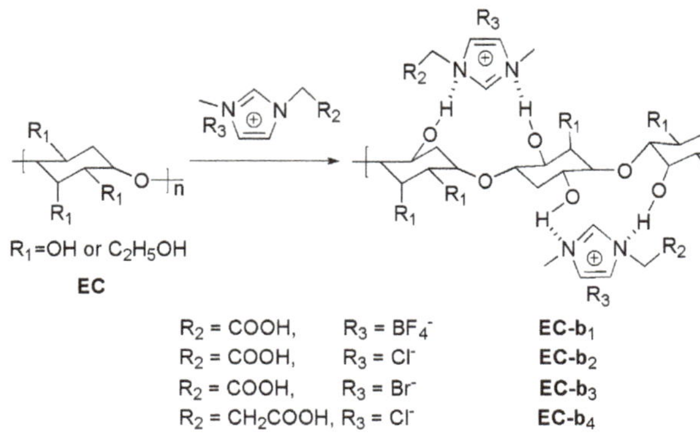

Scheme 2. Preparation of EC blended with ionic liquids (EC-b1 - EC-b4).

2.4. Preparation of Gas Separation Membranes and Mechanical Testing

The dried products were dissolved in tetrahydrofuran at room temperature for 12 h. The uniformly mixed casting solution was carefully poured on a clean glass plate and was placed at room temperature for 24 h. The thicknesses of the films were measured by a thickness gauge. Thereafter, the solution was poured evenly on a clean glass plate to genedegree the membrane. The Spin Coater was used to prepare the membrane (the Spin Coater was purchased from Shanghai Sanyan Technology Co., Ltd., Shanghai, China; model SYSC-50). The casting membrane temperature was about 25 °C, the evaporation membrane temperature was about 25 °C, the evaporation time was 12 h, the casting thickness was about 0.5 mm and the relative humidity was 52%. The thicknesses of the membranes were measured by a thickness gauge (thickness gauge was purchased from Shanghai Liuling Instrument Factory, model was CH-1-B hand-type millimeter thickness gauge, the graduation value was 0.001mm, the measurement range was 0–1 mm and the error was about ≤0.001 mm). Mechanical properties were analyzed with a film tensile testing machine (XLW(PC)-500N, Sumspring, Jinan, China) at 25 °C.

2.5. Measurements and Calculations of the Gas Permeability Coefficient and the Separation Factor

The permeability coefficients of the mixed gases were measured by gas chromatographic method using the differential pressure gas transmission instrument (GTR-11MH type). The gas permeability coefficient P was calculated by the following, Equation (2) [27]:

$$P = \frac{\mathrm{q} \times K \times \mathrm{L}}{\mathrm{a} \times \mathrm{p} \times \mathrm{t}} \left(\mathrm{ml} \bullet \mathrm{cm} \bullet \mathrm{cm}^{-2} \bullet \mathrm{s}^{-1} \bullet \mathrm{cmHg}^{-1}\right), \tag{2}$$

where, q is transmission volume (mL); K is auxiliary positive coefficient (the fixed value is 2), meaning that it is the setting point instrument by factory; L is film thickness (cm); p is permeability pressure (cmHg); t is measurement time (s); and a is the area of the gas permeation film (the fixed value is 0.785 cm^2).

In this experiment, the gas separation factor was calculated by the Equation (3) [27]:

$$\alpha_{\mathrm{PCO_2/PN_2}} = \frac{P_{\mathrm{CO_2}}}{P_{\mathrm{N_2}}}, \tag{3}$$

$$\alpha_{\mathrm{PCO_2/PCH_4}} = \frac{P_{\mathrm{CO_2}}}{P_{\mathrm{CH_4}}}, \tag{4}$$

where $P_{\mathrm{CO_2}}$, $P_{\mathrm{N_2}}$ and $P_{\mathrm{CH_4}}$ can be calculated from Equation (2).

The diffusion coefficients (D) and the solubility coefficients (S) were calculated by the following equations [28,29]:

$$D = \frac{L^2}{6T}, \tag{5}$$

$$S = \frac{P}{D}, \tag{6}$$

where L and T are the amount of the thickness of the membrane and the time lag.

2.6. Analysis of the Grafting Degree in EC

The grafting degree is defined as the mass fraction of ionic liquid in EC. Using organic elemental analyzer, it was calculated by the following Equation (7) [30]:

$$d = \frac{1}{2}N\% \times \frac{M_{IL}}{M_N} \times 100\%, \tag{7}$$

where, d is the grafting degree in EC; N% is the mass fraction of N which was determined by using elemental analyzer. N content is expressed in wt%. M_{IL} is the molecular weight of ionic liquid and M_N is the molecular weight of nitrogen element.

3. Instruments

The gas permeability was measured at 25 °C using a differential pressure gas transmission instrument (GTR-11MH type, GTR TEC Corporation, Kyoto, Japan; the test area was 0.785 cm^2. The instrument test temperature was 34 °C. And the test pressure was maintained at 49 KPa. The gas was a mixed gas, the content of the two components was the same, the test pressure was 0.1 MPa. The carrier gas was H_2, and the pressure was 0.5 MPa). Infrared spectra were recorded on a Fourier transform infrared spectrometer (Spectrum Two, PE company, Waltham, MA, USA). The content of elemental N in grafted product was measured using an organic elemental analyzer (PE2400 SERIES II CHNS/O, PerkinElmer, Waltham, MA, USA).

4. Results and discussion

4.1. The Esterification of Ethyl Cellulose

Under appropriate reaction conditions, carboxyl groups in ionic liquid and hydroxyl groups in EC were reacted by esterification reaction (Scheme S1 in Supporting Information), which led to stable EC grafted with ionic liquids (Scheme 1, EC-g1–EC-g4).

Fourier transform infrared spectrometry (FTIR) was used to determine the molecular structure of the grafted ethyl cellulose. As shown in Figure 1, there is a strong peak at 1760 cm^{-1}, which was attributed to C=O bond stretching vibration of ester group generate by grafting reaction between hydroxyl groups in EC and carboxyl groups in ionic liquid. The absorption peaks at 1620 cm^{-1} and 1640 cm^{-1} can be attributed to C=N and C=C bond stretching vibration of imidazole groups in ionic liquid, correspondingly. There are two weak spectral bands at 2960 cm^{-1} and 2860 cm^{-1}, which were generated by the C–H bond stretching vibration of methyl and methylene groups in EC respectively. The C–H stretching vibrations in the imidazole-based double bond in the ionic liquid appear at 3080 cm^{-1} and 3020 cm^{-1}, and its C–H bending vibration also appeared at 1400 cm^{-1}. The C–N stretching vibrations of the imidazole group are observed at 1470 cm^{-1}, 1250 cm^{-1} and 1348 cm^{-1}. The other EC derivatives (EC-g2–EC-g4) also showed the similar IR spectra (Figure S1–S3 in Supporting Information). On the basis of the observed IR spectra, it was concluded that the EC grafted with ionic liquids were synthesized successfully.

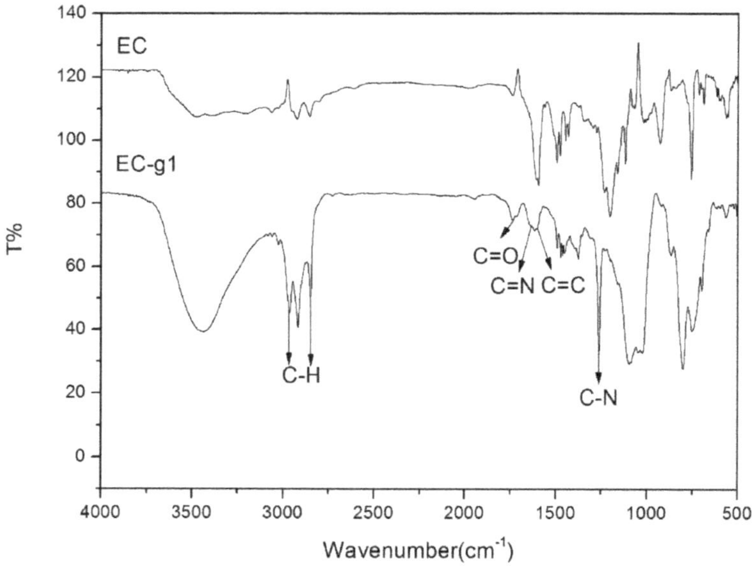

Figure 1. FT-IR spectrum of EC-g1 and EC.

The yields of the grafted products were calculated by weighing the products after drying (Table 1, EC-g1–EC-g4). The grafting degree of the final products was determined by N element content in the grafted products. The yields and grafting degrees of the products obtained in the experiment are shown in Table 1. The yields of EC grafted with ionic liquids were about 62.5–64.1% (Table 1, 2–5). As the EC derivatives were washed by an excess methanol, the yields were moderate, and pure products were obtained. The EC grafted with ionic liquids exhibited moderate grafting degrees (Table 1, 5.93–7.90%), which showed that the ionic liquids were successfully grafted into ethyl cellulose.

Table 1. Molecular weight, yield and grafting degree of pure ethyl cellulose and grafted with ethyl cellulose.

No.	EC Derivatives	Yield (%)	N-element Content (%)	Grafting Degree (%)	$\overline{M}_w{}^a (\times 10^5)$	$\overline{Mw}/\overline{Mn}^a$
1	EC	-	-	-	1.71	2.68
2	EC-g1	64.1	0.97	7.90	1.21	2.32
3	EC-g2	62.5	0.94	5.93	1.39	2.20
4	EC-g3	63.7	0.98	7.74	1.70	2.62
5	EC-g4	62.9	0.96	6.54	1.69	2.64

[a] Determined by GPC correlating polystyrene standard with THF eluent.

In this experiment, 1000 mL methanol solution was used to wash the grafted products for a week, which lost some unreacted ethyl cellulose to the methanol solution. That resulted in a decrease in the yield of the reaction.

EC is dissolved in solvent, but it is still an agglomerated macromolecule, and some hydroxyl groups are enclosed. The ionic liquid itself has a strong polarity, which affects the progress of grafting reaction. The above reasons can lead to low grafting degrees.

Through the GPC test, we found that the pure ethyl cellulose had an average molecular weight of 1.7×10^5 and a molecular weight distribution of 2.68. The molecular weight of the grafted ethyl cellulose was lower than that of pure ethyl cellulose, and the molecular weight distribution was narrowed.

4.2. Gas Permeation Properties of Membranes

The EC and the modified EC were stirred and fully dissolved in tetrahydrofuran to achieve a uniform and transparent casting solution. The casting solutions were subsequently used to generate gas separation membranes with similar thicknesses. Overall, the materials exhibited good membrane forming abilities. In order to compare the effects of ionic liquids in the EC membranes on their carbon dioxide permselectivities, the gas separation performances for taking CO_2 from CO_2/N_2 and CO_2/CH_4 mixtures were tested. It can be seen from Table 2 that the CO_2/CH_4 separation factors (12.0–18.8) in the membranes with EC grafted or blended with ionic liquids (Table 2, 2–9) were almost two times those of the pure EC membranes. However, the PCO_2/PCH_4 permselectivity values of silyl ethers of ethyl cellulose were observed to be in the range of 6.0–8.7 [19]. For a CO_2/N_2 gas mixture, the EC membranes modified by ionic liquids improved the separation factor to 31.9 from 20.6 for the pure EC membrane (Table 2). The PCO_2/PN_2 permselectivity values of silyl ethers of ethyl cellulose were observed to be in the range of 15–19 [19]. Thus, it can be confirmed that the imidazole ionic liquids are useful to improve the gas separation performance of the EC based membranes. This is because the imidazolium group in the ionic liquid interacts with CO_2, which promotes the penetration of CO_2 gas molecules and continues to undergo adsorption regulation within the membrane, ultimately resulting in high CO_2 permeability. On the other hand, CH_4 belongs to a regular tetrahedral structure and a non-polar molecule, but the ionic liquids added belong to a polar substance. According to the principle of similar compatibility, the solubility of methane in the membrane is poor, which causes a low permeation of CH_4. Therefore there was a high CO_2/CH_4 selectivity [31].

Table 2. Gas permeation properties of EC membranes modified with ionic liquids.

NO.	Membrane	PCO_2(Bar) [a] (PCO_2/PCH_4)	DCO_2 [b] (DCO_2/DCH_4) [c]	SCO_2 [d] (SCO_2/SCH_4) [e]	PCO_2(Bar) [a] (PCO_2/PN_2)	DCO_2 [b] (DCO_2/DN_2) [c]	SCO_2 [d] (SCO_2/SN_2) [e]
1	EC	44.1 (9.01)	9.70 (0.21)	4.61 (42.9)	43.2 (20.6)	14.6 (1.82)	3.01 (11.8)
2	EC-g1	102 (17.8)	10.8 (0.07)	9.42 (262)	96.0 (25.9)	41.2 (1.31)	2.31 (20.5)
3	EC-g2	61.5 (15.8)	35.6 (0.07)	1.71 (228)	67.7 (22.6)	31.8 (1.41)	2.11 (15.9)
4	EC-g3	165 (18.8)	5.31 (0.08)	31.0 (246)	157 (26.2)	18.4 (0.73)	8.6 (38.7)
5	EC-g4	199 (12.0)	3.32 (0.08)	60.8 (147)	194 (22.5)	13.6 (0.71)	14.2 (30.7)
6	EC-b1	110 (18.6)	9.41 (0.91)	11.7 (20.0)	119 (25.8)	19.3 (2.21)	6.21 (11.7)
7	EC-b2	62.2 (18.8)	9.22 (0.71)	6.81 (26.7)	70.8 (26.2)	48.1 (2.21)	1.51 (11.7)
8	EC-b3	76.0 (19.5)	23.5 (0.92)	3.22 (21.9)	79.7 (31.9)	13.2 (3.71)	6.02 (8.71)
9	EC-b4	46.0 (19.2)	10.3 (0.81)	4.51 (23.8)	47.8 (23.9)	37.2 (2.22)	1.31 (11.1)

[a] 1 barrer = 10^{-10} cm^3(STP)*cm $cm^{-2}s^{-1}cmHg^{-1}$, [b] In 10^{-8} cm^2 s^{-1}, [c] In 10^{-2}, [d] In 10^{-2} cm^3 (STP)*cm^{-3} $cmHg^{-1}$, [e] In 10^2.

The most important feature of the grafted EC membranes (Table 2, Nos. 2–5) is not only their high separation factor (12.0–18.8), but their excellent gas permeability coefficients (61.5–199 Barrer) for CO_2/CH_4 mixtures gas, which is better than pure EC. For instance, EC-g4 containing a flexible ethyl spacer in the ionic liquid exhibited the highest CO_2 permeability coefficient of 199 Barrer with a separation factor of 12 for CO_2/CH_4 mixture gas. For an CO_2/N_2 mixture gas, EC-g4 also demonstrated the highest PCO_2 value (Table 2, 194 Barrer). In our previous study [32], the flexible silyloxyl spacers in copoly (substituted acetylene) membranes were reported to be effective in enhancing their permeability coefficients. Therefore, it can be opined that the flexible ethyl spacer between imidazole and ester groups (EC-g4) largely governs the motion of imidazole, which strongly affects the CO_2 permeability coefficient.

The EC-g3 derivative containing bromide exhibited the highest separation factor (18.8 Barrer) with a permeability coefficient of 165 Barrer (Table 2, No. 4), which was located above 1991 Robeson's upper bound [17,33–35] and near to 2008 Robeson's upper bound (Figure 2a). The separation performance of the gas separation membrane EC-g2 was lower than that of 1991 Prior Upper Bound; however, it was relatively close (Figure 2a). Overall, the gas permeation properties of the grafted EC membranes approximately displayed the following order: EC-g3 > EC-g1 > EC-g4 > EC-g2 > EC. It can be mentioned that halogen atoms are highly electronegative and have electron-absorbing induction effect, F > Cl > Br > I [36]. On the other hand, BF_4^- has four fluorine boron bonds which are linked by SP_3 hybrid orbital to form a stable tetrahedral molecule, so it has weak electron donating capacity. In this study, the electron donating capacities of anions were $Br^- > Cl^- > BF_4^-$. The free electrons on bromide can interact with Lewis acid CO_2, which can make the imidazole group more stable. Therefore, the gas separation performance of the EC-g3 membrane containing bromide was observed to be the best.

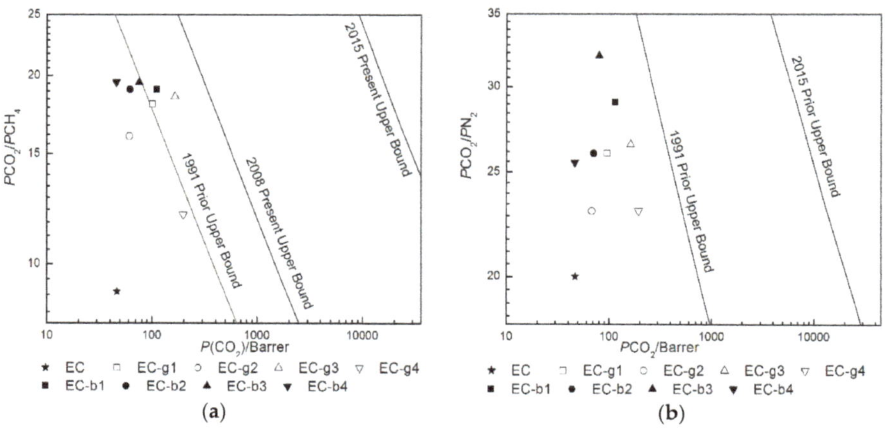

Figure 2. Plot of permselectivity vs permeability for the gas pairs (**a**) CO_2/CH_4 and (**b**) CO_2/N_2.

On the other hand, EC and ionic liquids were also blended to develop composite films. There are pairs of lone electrons in the imidazole ring of the ionic liquid, which can form hydrogen bonds with hydrogen atoms on the hydroxyl group of ethyl cellulose. Due to the action of the hydrogen bonding, the ionic liquid and ethyl cellulose can interact firmly, forming relatively stable composite films. The EC and ionic liquid-blended membranes exhibited high separation factors (18.6–19.5, Table 2, Nos. 6–9) and low permeability coefficients (46.0–110 Barrer). As a large amount of ionic liquid is added to the EC membranes, fractional free volume may become small, resulting in low permeability coefficients [37]. However, it also causes high permselectivity for CO_2. Because blending is a physical mixture without chemical reaction, the phase structure of blending membrane is two phasic. It can be observed from scanning electron microscopy (Figure S4 in Supporting Information) that some ionic liquids are agglomerated, so it is difficult to have a good interaction with CO_2, resulting in low permeability coefficients. Grafting is a chemical reaction that can have a homogeneous phase structure. IL dispersed well in the EC, effectively promoting the permeability of CO_2, thus, improves CO_2 permeation efficiency. Based on the findings observed, it can be concluded that the ionic liquid improved the separation factor of CO_2/CH_4 significantly.

For the mixture of CO_2 and CH_4, the CO_2 molecule as a whole has no polarity, but the O in CO_2 has a unique pair of electrons, which can form a hydrogen bond with H in CH_4 (see Scheme S2 in Supporting Information). When the mixed gas contacts the surface of the membrane, it may be adsorbed and dissolved on the surface of the membrane in the form of one molecule pair. In EC-g1, EC-g3 and EC-g4, SCO_2 is larger in CO_2/CH_4 than those in CO_2/N_2 (See Table 2). In the diffusion process, the presence of ionic liquids broke the hydrogen bonds. The interaction between CO_2 and IL makes CO_2 permeate preferentially, which can achieve the purpose of separation.

CO_2/N_2 permeability of the modified EC membranes was also determined, and the results are shown in Table 2 and Figure 2b. CO_2 permeability coefficients were observed to improve from 43.2 for pure EC membrane upto 194 Barrer for the modified EC membranes, indicating an improvement of 4.6 times over EC membrane. Among the grafted EC membranes (Table 2, numbers 2–5), EC-g3 containing bromide ion exhibited the highest CO_2 permselectivity (PCO_2/PN_2 26.2). EC-b3 also demonstrated the best PCO_2/PN_2 value of 31.9 among the blended EC membranes (Table 2, numbers 6–9). The separation performance of all gas separation membranes was observed to be under the 1991 prior upper bound (Figure 2b). Overall, the conclusion is similar to the case of CO_2/CH_4 mixture gas; i.e., bromide ion can enhance CO_2 permselectivity (PCO_2/PN_2 or PCO_2/PCH_4). In other words, 1-carboxymethyl-3-methylimidazolium bromide was confirmed to be effective in enhancing the CO_2 permselectivity of EC membranes.

4.3. Mechanical Properties of Polymer Membranes

The mechanical properties of the EC and modified EC membranes are shown in Table 3. Compared to EC, the elasticity moduli of the grafted membranes increased from 207 MPa for EC to 560–648 MPa; however, the blended membranes exhibited a decrease in modulus to 117–177 MPa. The results indicate that the grafted membranes (Table 3, numbers 2–5) are homogeneous, whose molecules are connected by covalent bonds. On the other hand, the blended membranes (Table 3, numbers 6–9) represent binary composites, which have weak physical connection, leading to reduced mechanical strength.

Table 3. Mechanical properties of the membranes of EC modified with ionic liquids [a].

No.	Membrane	Thickness (mm)	Elongation (%)	Elasticity Modulus (MPa)
1	EC	0.1391	21.6	207
2	EC-g1	0.1448	4.22	648
3	EC-g2	0.1138	8.61	560
4	EC-g3	0.1591	3.92	644
5	EC-g4	0.1282	5.53	593
6	EC-b1	0.1338	16.2	117
7	EC-b2	0.0986	10.2	177
8	EC-b3	0.1158	8.71	119
9	EC-b4	0.0892	15.4	143

[a] Tested at 5.00 mm/min speed. The standard spline had a length of 50 mm and a width of 10 mm.

In the phenomenon shown in Table 3, we believe that in the blending system, the interior exists is two phases. EC is the continuous phase, the ionic liquid is the dispersed phase. The ionic liquid does have not much influence on the EC molecular structure. Although the content of ionic liquid is high, its caking property is weak, which has little effect on its mechanical properties. However, when EC is grafted with ionic liquid, the crystal structure of EC may change. The molecules are covalently bonded and intermolecular interactions are strong. With the different grafting degrees, the intermolecular forces will be different, showing different mechanical properties. The chains of ethyl cellulose were broken by grafting to decrease in molecular weight (See Table 1), which caused an increase in brittleness.

5. Conclusions

ECs grafted with ionic liquids were synthesized. Based on the infrared spectra, ionic liquids containing imidazole groups were successfully grafted into ethyl cellulose. The yields of ECs grafted with ionic liquids were about 62.5–64.1%, and the grafting degrees were 5.93–7.90%. The grafted products exhibited good membrane forming abilities. The CO_2/CH_4 separation factor in the membranes with EC grafted or blended with ionic liquids was almost two times compared to the pure EC membranes.

Among these modified EC membranes, EC-g3 membrane was the best: the PCO_2/PCH_4 separation coefficient was 18.8 and the permeability coefficient (PCO_2) was 199 Barrer. The bromide ion in EC-g3 can interact well with CO_2, which promotes CO_2 permeability. Thus, it can be confirmed that the imidazole ionic liquids are useful for improving the gas separation performances of the EC-based membranes.

Supplementary Materials: The following are available online at http://www.mdpi.com/2073-4360/11/11/1900/s1, Scheme S1. Mechanism of pyridine-catalyzed esterification, Scheme S2. The coupling effect occurs of CO_2 in the presence of CH_4, Figure S1. FT-IR spectrum of EC-g2, Figure S2. FT-IR spectrum of EC-g3, Figure S3. FT-IR spectrum of EC-g4, Figure S4. SEM images of ethyl cellulose blended 1-carboxymethyl-3-methylimidazolium gas separation membrane.

Author Contributions: H.J. contributed to the conception of the study; N.Y. contributed significantly to the analysis and manuscript preparation; J.X. performed the data analyses and wrote the manuscript; Q.W., G.Y., M.Z., S.X., Y.Z., L.M., P.J., H.Z. and H.W. helped perform the analysis with constructive discussions.

Funding: Financial and facility support for this research came from the Fundamental Research Funds in Heilongjiang provincial universities (YSTSXK201862, 135309110, and 135309503), the Natural Science Foundation of Heilongjiang province, China (LH2019B032) and the Scientific Research Project of Qiqihar University (number 135309357).

Conflicts of Interest: The authors declare no conflict of interest.

References

1. Klemm, D.; Heublein, B.; Fink, H.; Bohn, A. Cellulose: Fascinating biopolymer and sustainable raw materia. *Angew. Chem. Int. Edit.* **2005**, *36*, 3358–3393. [CrossRef] [PubMed]
2. Houde, A.Y.; Stern, S.A. Solubility and diffusivity of light gases in ethyl cellulose at elevated pressures Effects of ethoxy content. *J. Membr. Sci.* **1997**, *127*, 171–183. [CrossRef]
3. Choi, S.H.; Tasselli, F.; Jansen, J.C. Effect of the preparation conditions on the formation of asymmetric poly (vinylidene fluoride) hollow fibre membranes with a dense skin. *Eur. Polym. J.* **2010**, *46*, 1713–1725. [CrossRef]
4. Shen, Y.; Wang, H.; Liu, J.D. Enhanced performance of a novel polyvinyl amine/chitosan/graphene oxide mixed matrix membrane for CO_2 capture. *ACS Sustain. Chem. Eng.* **2015**, *3*, 1819–1829. [CrossRef]
5. Yampolskii, Y.; Alentiev, A.; Bondarenko, G. Intermolecular interactions: New way to govern transport properties of membrane materials. *Ind. Eng. Chem. Res.* **2010**, *49*, 12031–12037. [CrossRef]
6. Li, H.; Huang, Q.; Li, D. The generation of a molecular imprinted membrane by coating cellulose acetate onto a ZrO_2-modified alumina membrane for the chiral separation of mandelic acid enantiomers. *Org. Process Res. Dev.* **2018**, *22*, 278–285. [CrossRef]
7. Veronika, V.; Ciahotny, K. Upgrading biogas to biomethane using membrane separation. *Energy Fuel.* **2017**, *31*, 9393–9401.
8. Coombs Obrien, J.; Torrente-Murciano, L.; Mattia, D. Continuous production of cellulose microbeads via membrane emulsification. *ACS Sustain. Chem. Eng.* **2017**, *5*, 5931–5939. [CrossRef]
9. Sirkar, K.K.; Shanbhag, P.V.; Kovvali, A.S. Membrane in a reactor: a functional perspective. *Ind. Eng. Chem. Res.* **1999**, *38*, 3715–3737. [CrossRef]
10. Josephine, O.M.; Kenneth, B.; John, F. Molecular sieving realized with ZIF-8/matrimid mixed-matrix membranes. *J. Membr. Sci.* **2010**, *361*, 28–37.
11. Adams, R.; Carson, C.; Ward, J. Metal organic framework mixed matrix membranes for gas separations. *Microporous Mesoporous Mater.* **2010**, *31*, 13–20. [CrossRef]
12. Lin, H.Q.; Freeman, B.D. Materials selection guidelines for membranes that remove CO_2 from gas mixtures. *J. Mol. Struct.* **2005**, *3*, 57–74. [CrossRef]
13. Murtaza, G. Ethylcellulose microparticles: A review. *Acta Pol. Pharm.* **2012**, *69*, 11–22. [PubMed]
14. Xing, Q.; Zhao, F.; Chen, S.M. Porous biocompatible three-dimensional scaffolds of cellulose microfiber/gelatin composites for cell culture. *Acta Biomater.* **2010**, *6*, 2132–2139. [CrossRef]
15. MoonR, J.; Martini, A.; Nairn, J. Cellulose nanomaterials review: Structure, properties and nanocomposites. *Chem. Soc. Rev.* **2011**, *40*, 3941–3994. [CrossRef]

16. Davidovich-Pinhas, M.; Barbut, S.; Marangoni, A.G. Physical structure and thermal behavior of ethylcellulose. *Cellulose* **2014**, *21*, 3243–3255. [CrossRef]
17. Wang, Y.; Ma, X.; Ghanem, B.S. Polymers of intrinsic microporosity for energy-intensive membrane-based gas separations. *Mater. Today Nano* **2018**, *3*, 69–95. [CrossRef]
18. Lu, H.; Wang, Q.; Li, G. Electrospun water-stable zein/ethyl cellulose composite nanofiber and its drug release properties. *Mat. Sci. Eng. C-Mater.* **2017**, *74*, 86–93. [CrossRef]
19. Khan, F.Z.; Sakaguchi, T.; Shiotsuki, M. Synthesis, characterization, and gas permeation properties of silylated derivatives of ethyl cellulose. *Macromolecules* **2006**, *39*, 6025–6030. [CrossRef]
20. Li, X.G.; Kresse, I. Jürgen springer, morphology and gas permselectivity of blend membranes of polyvinylpyridine with ethylcellulose. *Polymer* **2001**, *16*, 6859–6869. [CrossRef]
21. Zhang, Z.; Zhang, H.; Zhang, Q. Thermotropic liquid crystallinity, thermal decomposition behavior, and aggregated structure of poly (propylene carbonate)/ethyl cellulose blends. *J. Appl. Polym. Sci.* **2006**, *100*, 584–592. [CrossRef]
22. Hu, Q. Poly (amide-imide)/TiO_2 nano-composite gas separation membranes: Fabrication and characterization. *J. Membr. Sci.* **1997**, *135*, 65–79. [CrossRef]
23. Moaddeb, M.; Koros, W.J. Gas transport properties of thin polymeric membranes in the presence of silicon dioxide particles. *J. Membr. Sci.* **1997**, *125*, 143–163. [CrossRef]
24. Sablok, A.; Jindal, R. Preparation and applications of room temperature ionic liquids in organic synthesis: a review on recent efforts. *Curr. Green Chem.* **2015**, *2*, 135–155.
25. Andreatta, G.; Lee, L.T.; Lee, F.K. Gas permeability in polymer and surfactant-stabilized bubble films. *J. Phys. Chem. B* **2006**, *11*, 19537–19542. [CrossRef]
26. Nikolaeva, D. The performance of affordable and stable cellulose-based poly-ionic membranes in CO_2/N_2 and CO_2/CH_4 gas separation. *J. Membr. Sci.* **2018**, *564*, 552–561. [CrossRef]
27. Wang, J.; Zang, Y.; Yin, G. Facile synthesis of five 2D surface modifiers by highly selective photocyclic aromatization and efficient enhancement of oxygen permselectivities of three polymer membranes by surface modification using a small amount of the 2D surface modifiers. *Polymer* **2014**, *55*, 1384–1396. [CrossRef]
28. Fragaa, S.; Monteleoneb, M. A novel time lag method for the analysis of mixed gas diffusion in polymeric membranes by on-line mass spectrometry: Method development and validation. *J. Membr. Sci.* **2018**, *561*, 39–58. [CrossRef]
29. Beckman, I.; Shalygin, M.; Tepliakov, V. *Particularities of membrane gas separation under unsteady state conditions. Mass Transfer in Chemical Engineering Processes*; InTech: Rijeka, Croatia, 2011; Volume 1, pp. 205–232.
30. Li, Y.; Sun, Y.; Deng, X. Graft polymerization of acrylic acid onto polyphenylene sulfide nonwoven initiated by low temperature plasma. *J. Appl. Polym. Sci.* **2006**, *102*, 5884–5889. [CrossRef]
31. Shannon, M.S.; Bara, J.E. Reactive and reversible ionic liquids for CO_2 capture and acid gas removal. *Sep. Sci. Technol.* **2012**, *47*, 178–188. [CrossRef]
32. Jia, H.; Luo, J.; Aoki, T. Synthesis and oxygen permselectivity of copoly (substituted acetylene)s with bulky fused polycyclic aliphatic groups. *Polymer* **2016**, *99*, 95–703. [CrossRef]
33. Robeson, L.M. The upper bound revisited. *J. Membr. Sci.* **2008**, *320*, 390–400. [CrossRef]
34. Comesaña-Gándara, B. Redefining the robeson upper bounds for CO_2/CH_4 and CO_2/N_2 separations using a series of ultrapermeable benzotriptycene-based polymers of intrinsic microporosity. *Energ. Environ. Sci.* **2019**, *12*, 2733–2740. [CrossRef]
35. Swaidan, R.; Ghanem, B.; Pinnau, I. Fine-tuned Intrinsically ultramicroporous polymers redefine the permeability/selectivity upper bounds of membrane-based air and hydrogen separations. *ACS Macro Lett.* **2015**, *4*, 947–951. [CrossRef]
36. Pathak, B.; Samanta, D.; Ahuja, R. Borane derivatives: A new class of super and hyperhalogens. *ChemPhysChem* **2011**, *12*, 2423–2428. [CrossRef]
37. Li, B.; Xu, D.; Zhang, X. Rubbery polymer inorganic nanocomposite membranes: Free volume characteristics on separation property. *Ind. Eng. Chem. Res.* **2010**, *49*, 12444–12451. [CrossRef]

© 2019 by the authors. Licensee MDPI, Basel, Switzerland. This article is an open access article distributed under the terms and conditions of the Creative Commons Attribution (CC BY) license (http://creativecommons.org/licenses/by/4.0/).

polymers

MDPI

Article

Polyamide-Laccase Nanofiber Membrane for Degradation of Endocrine-Disrupting Bisphenol A, 17α-ethinylestradiol, and Triclosan

Milena Maryskova [1,2,*], Miroslava Rysova [1,2], Vit Novotny [1] and Alena Sevcu [1,2,*]

1 Institute for Nanomaterials, Advanced Technologies and Innovation, Technical University of Liberec, Bendlova 1409/7, 46117 Liberec, Czech Republic; miroslava.rysova@tul.cz (M.R.); vit.novotny@tul.cz (V.N.)
2 Faculty of Mechatronics, Informatics and Interdisciplinary Studies, Technical University of Liberec, Studentska 1402/2, 46117 Liberec, Czech Republic
* Correspondence: alena.sevcu@tul.cz (A.S.); milena.maryskova@tul.cz (M.M.)

Received: 27 August 2019; Accepted: 23 September 2019; Published: 25 September 2019

Abstract: Contamination of potable water by endocrine disrupting chemicals (EDCs) is a growing problem worldwide. One of the possible treatments is the utilization of laccase enzyme catalyzing oxidation of phenolic structures of EDC when anchored in a polymeric nanofiber membrane. Previous studies failed to develop a membrane with a sufficiently active enzyme, or the immobilization process was too complicated and time-consuming. Here, we established an elegant method for immobilizing *Trametes versicolor* laccase onto polyamide 6 nanofibers (PA6-laccase) via adsorption and glutaraldehyde crosslinking, promoting high enzyme activity and easier applicability in water treatment technology. This simple and inexpensive immobilization ensures both repeated use, with over 88% of initial activity retained after five ABTS catalytic cycles, and enhanced storage stability. PA6-laccase was highly effective in degrading a 50-μM EDC mixture, with only 7% of bisphenol A, 2% of 17α-ethinylestradiol, and 30% of triclosan remaining after a 24-h catalytic process. The PA6-laccase membrane can lead to the improvement of novel technologies for controlling of EDC contamination in potable water.

Keywords: laccase; polyamide 6; nanofibers; wastewater treatment; endocrine disrupting chemicals

1. Introduction

Over recent decades, the list of known environmental pollutants has been widened by chemical compounds occurring at very low concentrations. Most of these only became detectable following significant progress in available analytical methods. These emerging micropollutants represent a new and, as yet, insufficiently explored form of toxicity, not least due to their remarkable persistence in the aquatic environment and their ability to bioaccumulate. Many of these compounds are capable of short- and long-term toxicity, disruption to the endocrine system, or contribute to the antibiotic resistance of microorganisms [1].

Conventional wastewater treatment methods are insufficient for complete reduction of some pollutants, especially endocrine disrupting chemicals (EDCs). Wastewater treatment plants are only capable of removing or transforming a limited amount of these compounds, either through sorption onto activated sludge or common degradation processes [2]. While progressive technologies such as photocatalysis, UV oxidation, ozonation, Fenton's reagent, super-critical water oxidation, or ultrasound and ionizing radiation appear to be more effective in removing some EDCs [3,4], most of these approaches require high energy and reagent input. The future strategy of EDCs treatment in the EU, according to the Directive 2013/39/EU of the European Parliament, is based only on two alternative processes: ozonation and treatment with powdered activated carbon [5]. Ozonation is potentially hazardous due to toxicity associated with the formation of possible harmful by-products (e.g., suspected human carcinogen bromate when bromine appears in water) [6]. Activated carbon possesses high adsorption capacity of organic matter in a combination of small particle size and prolonged contact time. On the other hand, used carbon needs to be separated and sent for destruction and re-activation through incineration [7]. Alternative technologies involve nanofiltration, reverse osmosis, and enzymatic treatment. From this perspective, nanofibers represent the most promising material [8,9].

Laccase, an oxidoreductase, shows great promise for use in wastewater treatment due to its sufficient substrate specificity, short reaction time, and low energy consumption [10]. Pure laccase is costly, mainly due to the complicated purification process needed and low commercial demand. As such, it makes little sense to target the application of laccase in its soluble form at present. Instead, it may prove more cost-effective to develop methods for enzyme immobilization, thereby ensuring a long-term effect and enhanced durability in the aquatic environment.

Numerous previous studies have addressed enzyme immobilization, including immobilization of oxidoreductases for wastewater treatment. Most of these focus on laccase as the optimal candidate [11–14], with peroxidase [15–18] and fungal tyrosinase [19–21] less often chosen. More recent studies have also described immobilization of two enzymes synergistically, thereby combining their efficiencies [22–25]. Of the available immobilization techniques using different forms of the matrix (e.g., nanoparticles, beads, foams, nanofibers, mats), nanofibers appear to be the most promising for wastewater treatment as they can be used to form safe and easily handled macroscopic mats with a high specific surface area.

Cost-effectiveness and safety of the final nanofiber-laccase membrane are required to be applicable in the water treatment technology. Polyamide 6 (PA6) is probably the cheapest electrospun material with adequate mechanical properties, high stability, and safety. In our previous study, we attempted to develop nanofibers made of PA6/chitosan with bovine serum albumin or hexamethylenediamine as spacers and glutaraldehyde as a crosslinker to reach the highest possible laccase loading and activity [11]. Although the final material was promising due to high enzyme activity and the presence of popular chitosan, the immobilization process took a long time and required too many activation steps, which may prevent its commercialization.

Therefore, our next step was to develop an easier, less laborious method for laccase immobilization onto PA6 nanofibers, achieving the same or better properties as the PA6/chitosan-laccase. The resultant PA6-laccase membrane was tested for its applicability for EDC degradation using a mixture of three well-known EDCs, bisphenol A, 17α-ethinylestradiol, and triclosan.

2. Material and Methods

2.1. Reagents

Laccase from *Trametes versicolor* (>10 U/mg, powder), 2,2′-azino-bis(3-ethylbenzothiazoline-6-sulphonic acid) (ABTS; >98%), bisphenol A (BPA; ≥99%), 17α-ethynylestradiol (EE2; ≥98%), triclosan (TCS; analytical standard), and glutaraldehyde (GA; Grade II, 25% in H_2O) were all obtained from Sigma-Aldrich (St. Louis, MO, USA). Pelleted PA6 (B24, M_w 37000 g/mol) was obtained from BASF (Ludwigshafen, Germany). All other reagents used were of analytical grade.

2.2. Electrospinning of the Nanofiber Matrix

PA6 pellets were dissolved in a mixture of formic acid and acetic acid (2:1; *v*/*v*) to prepare a 12% wt. solution. The nanofibers were then prepared by electrospinning, using Nanospider™ NS 1WS500U equipment (Elmarco, Czech Republic) with a voltage of −20/60 kV and distance between electrodes set at 180 mm. The surface density of prepared nanofiber layers was controlled by adjusting speed of collecting material and by a number of electrospun layers. Specifically, the speed for fabrication nanofibers with 1.5 g/m^2 was 26 mm/min, and for 3 g/m^2 was 15 mm/min. Both nanofibers with 5 g/m^2

and 8 g/m^2 were prepared from two layers, and the speed was adjusted to 30 mm/min and 15 mm/min, respectively.

2.3. Nanofiber Characterization

Images of pristine PA6 nanofibers and nanofibers with immobilized laccase (PA6-laccase) were obtained using SC7620 sputter coater (Quorum Technologies, Lewes, UK) with 10 nm gold layers, Carl Zeiss ULTRA Plus (Zeiss, Oberkochen, Germany) and VEGA3 Tescan (Tescan, Czech Republic) scanning electron microscopes (SEM). Subsequently, the SEM images were analyzed using VEGA TC software for assessing average fiber diameter (obtained by averaging the values of 100 individual measurements).

2.4. Enzyme Immobilization

T. versicolor laccase was immobilized onto the PA6 nanofibers via adsorption followed by GA crosslinking. A range of parameters, including nanofiber matrix surface density, enzyme solution volume, buffer concentration, and pH, adsorption and crosslinking time, and GA concentration, were examined to establish the most effective immobilization method. Our preliminary experiments identified the optimal immobilization process temperature as 4 °C and the most convenient mode of agitation providing uniform enzyme solution distribution as orbital shaking at 150 rpm. After each immobilization process, the samples were washed with citrate-phosphate buffer (McIlvaine's buffer) at pH 3 until no laccase activity was detected in the washings, following which the activity of the immobilized laccase was determined.

2.4.1. Influence of Nanofiber Surface Density on the Immobilization Process

Four types of PA6 nanofiber sheets with different surface densities were prepared by adjusting the speed of the electrospinning process. The finest sheet had a surface density of 1.5 g/m^2 and an average fiber diameter of 79.3 ± 19.8 nm. Subsequent sheets had a surface density of ca. 3 g/m^2 and a fiber diameter of 87.9 ± 14.7 nm, ca. 5 g/m^2 and 109.4 ± 19.1 nm, and 8 g/m^2 and 100.4 ± 23.8 nm.

To evaluate the most suitable nanofiber sheet, circular PA6 samples (diameter 1.5 cm) were immersed into 500 μL of 2 mg/mL laccase solution in 50% McIlvaine's buffer at pH 3 and shaken for 15 h at 4 °C and 150 rpm. GA was then added to achieve a final concentration of 2.5% (*v*/*v*), and the mixture was shaken for a further five hours under the same conditions. Finally, the samples were washed thoroughly and enzyme activity measured as described in Sections 2.4 and 2.5.

2.4.2. Influence of Laccase Solution Volume on the Immobilization Process

PA6 samples were immersed into different volumes of 2 mg/mL laccase solution (250 μL, 300 μL, and 500 μL) and incubated according to the procedure described in Section 2.4.1. As the optimal volume appeared to depend on the geometry of the vessel used for immobilization, type of agitation, and sorption of the supporting material, we also examined the optimal vessel type and form of agitation.

2.4.3. Effect of Buffer Concentration on the Immobilization Process

Undiluted (100%) McIlvaine's buffer was prepared from a mixture of 200 mM disodium hydrogen phosphate and 100 mM citric acid at appropriate ratios that depended on the required pH [26]. Subsequently, the buffer was diluted with ultrapure water to produce a 20% and 50% solution. All three concentrations (20, 50, and 100%) were then used for laccase immobilization. All other parameters of the immobilization process were identical to those described in Section 2.4.2 (i.e., 300 μL of 2mg/mL laccase solution, pH 3).

2.4.4. Effect of Time, pH and GA Concentration on the Immobilization Process

Six different combinations of adsorption time (3, 5, 15, and 24 h) and crosslinking time (3, 5, and 8 h) were used to determine the optimal immobilization duration. The optimal pH for immobilization was evaluated using 20% McIlvaine's buffer with a pH of 3, 4, 5, 6, and 7. Finally, we tested the GA crosslinker at concentrations of 0.5, 1, 2.5, and 5% (*v*/*v*).

2.5. *Enzyme Activity Assay*

Laccase catalytic activity was measured at 420 nm (ε = 36 $mM^{-1} \cdot cm^{-1}$) at room temperature according to Hassani et al. [27] and Maryšková et al. [11] using a BioTech Synergy HTX microplate reader (BioTech Instruments Inc., Winooski, VT, USA). The reaction was performed in McIlvaine's buffer containing 20 μL of the stock enzyme solution (diluted with 160 μL of buffer) and the same volume of 0.5 mM ABTS. The activity of the immobilized enzyme was measured by adding the sample to 3.6 mL of McIlvaine's buffer (pH 3) and adding 400 μL of 0.5 mM ABTS. Samples were taken at selected time intervals and measured under the conditions described above.

2.6. *Storage Stability and Reusability of PA6-Laccase*

PA6-laccase was incubated in pH 6 20% buffer at 4 °C to assess storage stability. Two replicate samples were taken at selected time points (13, 20, and 30 days), and their enzyme activity was measured. Free laccase solution was stored under the same conditions for comparison with the immobilized laccase.

Reusability of the PA6-laccase nanofibers was determined by measuring enzyme activity over several catalytic cycles using ABTS as a substrate, as described in Section 2.5. The samples were removed from the ABTS mixture and thoroughly washed with fresh buffer at pH 3 after each catalytic cycle.

2.7. *Degradation of BPA, EE2, and TCS*

The degradation efficiency of free and immobilized laccase was determined by decreasing the concentration of micropollutants over time. One sample of PA6-laccase, PA6 (blank) or 25 μL of the enzyme stock solution (2 mg of enzyme per 1 mL of ultrapure water) were added to glass vials containing 5 mL of a mixture of BPA, EE2, and TCS (50 μM) in ultrapure water containing 30% methanol. Over the selected time intervals, 70 μL of the mixture supernatant was collected into vials containing 65 μL of deionized water and 5 μL of 10% sodium azide, thereby preventing further EDC degradation if some of the enzymes had been collected with the supernatant [28]. Each experiment was performed in duplicate, and the results presented as the mean value ± standard deviation.

EDC (BPA, EE2, TCS) degradation by laccase and PA6-laccase was measured using a Dionex Ultimate 3000 high-pressure liquid chromatograph (HPLC, Thermo Fisher Scientific, Waltham, MA, USA), with an LPG-3400SD quaternary gradient pump, a SR-3000 solvent rack, a WPS-3000TSC autosampler, a TCC-3000SD column compartment, a DAD-3000 detector, and a Phenomenex Kinetex F5 core-shell column (Torrance, CA, USA, length 150 mm, internal diameter 4.6 mm). The aqueous component of the mobile phase consisted of 10 mM phosphoric acid in 5% (*v*/*v*) acetonitrile, while the organic component consisted of 10 mM phosphoric acid in a mixture of 90% acetonitrile and 10% methanol (*v*/*v*). A set of linear gradients were produced, starting with the proportion of the organic component in the mobile phase at 10 %. At 0.7 min the proportion had increased to 15%, at 1.7 min it was 25%, at 3 min 35%, at 4.2 min 40%, at 5.9 min 50%, and at 7.6 min 70%. The proportion of the organic component reached 80% at 8.8 min and from this point up to 9.3 min, the composition of the mobile phase returned to the starting conditions. The chromatogram for each sample was recorded for 12.65 min, with an injection volume of 20 μL, a flow rate of 1.5 mL/min, and column temperature kept at 40 °C. Chromatograms were recorded at wavelengths of 200, 227, 278, and 285 nm, with a sampling rate of 2 Hz.

3. Results and Discussion

3.1. Nanofiber Matrix Morphology

Support material morphology is an important parameter affecting the enzyme immobilization process. Nanofiber materials are mainly characterized by average fiber diameter and surface density, the combination of these two parameters influencing specific surface area and the immobilization capacity of the matrix [29]. As all our PA6 nanofiber sheets were produced under similar electrospinning conditions, the average fiber diameter was similar for all the nanofiber samples at 79.3 to 109.4 nm (Figure 1a–d). In comparison, the nanofiber materials used in previous studies all consisted of thicker fibers, with typical diameters ranging from 150 to 500 nm [30–32]. As such, our PA6 nanofiber sheets had a greater specific surface area for laccase immobilization.

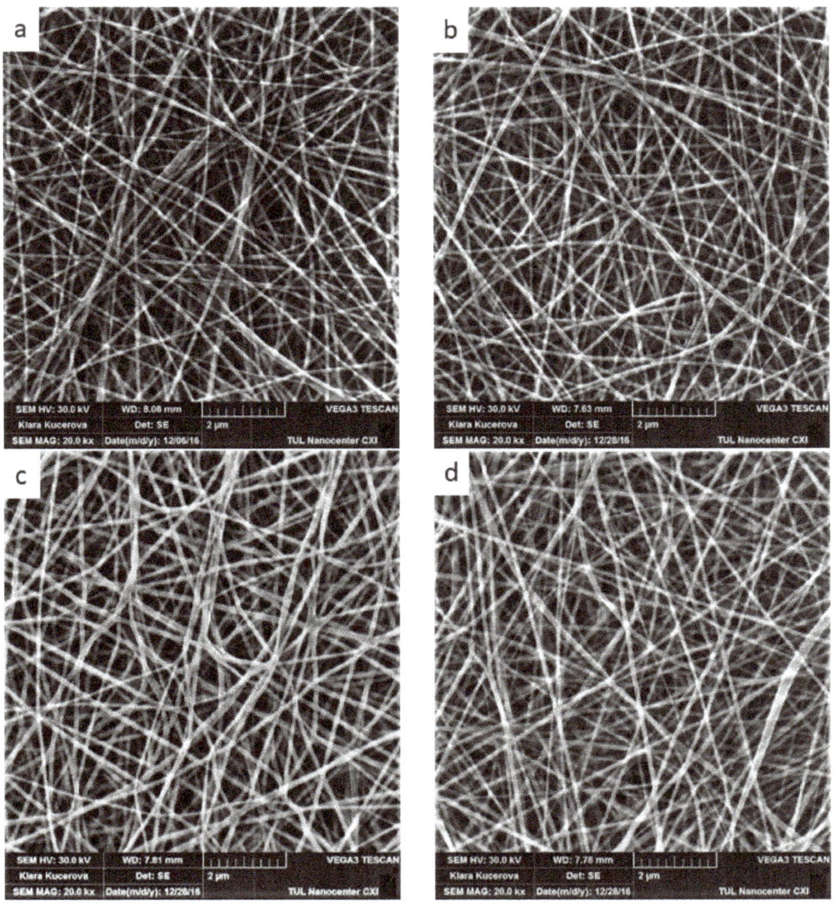

Figure 1. Pristine polyamide 6 nanofibers with surface densities of (**a**) 1.5 g/m^2, (**b**) 3 g/m^2, (**c**) 5 g/m^2, and (**d**) 8 g/m^2. Images were obtained using Carl Zeiss ULTRA Plus scanning electron microscope (Zeiss, Germany) at a magnification of 20k×.

Despite surface density being an important parameter influencing enzyme loading, substrate/product diffusion, and nanofiber sheet mechanical properties, most studies focus solely on the physico-chemical nature of the immobilization process and tend to ignore the "usability" of the product under actual water treatment conditions [31,33–35]. In our study, highest activity levels recorded for immobilized laccase, and the most effective reuse within three ABTS oxidation catalytic cycles, was achieved by PA6-laccase nanofibers with a density of 3 g/m^2 (Figure 2a). Unfortunately, these nanofibers were not easy to handle due to their poor mechanical properties, a tendency to tear, and deteriorative wettability compared to nanofibers of higher surface density. As such, PA6 nanofibers with a surface density of 5 g/m^2 were chosen for enzyme immobilization, despite displaying 13 % less enzymatic activity than 3 g/m^2 PA6 nanofibers.

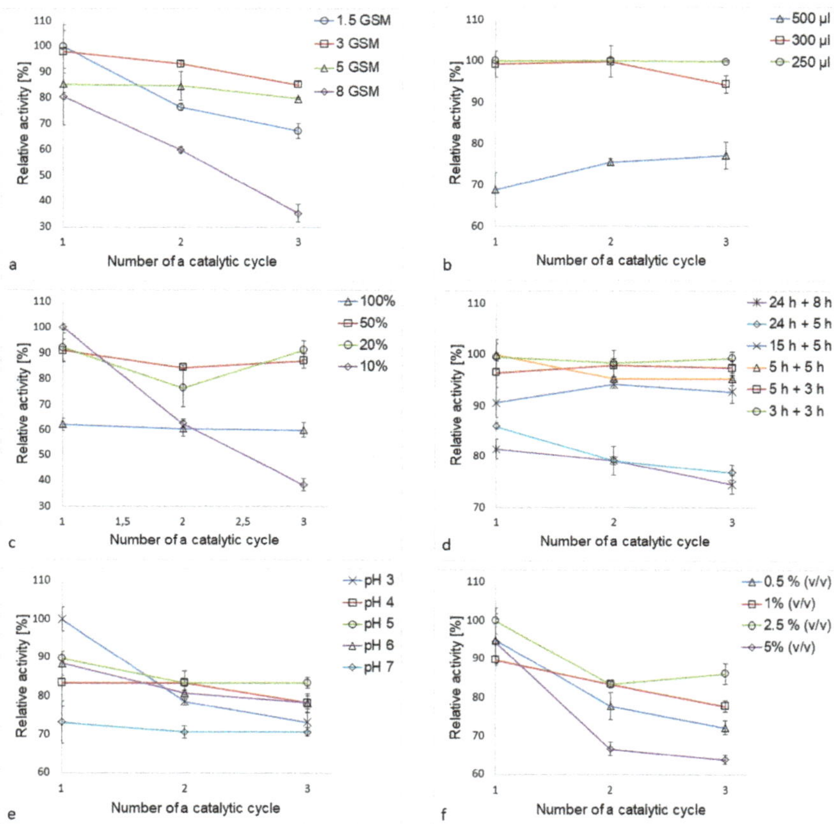

Figure 2. Parameters influencing laccase immobilization—(**a**) influence of PA6 nanofiber matrix surface density, (**b**) influence of enzyme solution volume, (**c**) buffer concentration, (**d**) adsorption and crosslinking time, (**e**) pH, and (**f**) GA concentration.

3.2. Optimal Volume of Laccase Solution

Were pristine PA6 nanofibers to be used for enzyme adsorption, it can be assumed that there would be insufficient functional groups to bind the enzyme selectively during the adsorption process. Thus, the volume of enzyme solution will have a strong effect on adsorption efficiency by forcing the liquid to fully adsorb into the nanofiber structure, and as such, the lowest volume ensuring full wetting of the carrier would be most desirable. With insufficient agitation, however, enzyme molecules tend to agglomerate and form sediment. Thus, the volume selected must enable satisfactory enzyme flux in the solution. The lowest volume of laccase solution tested required a very small vessel (0.5 mL vial), which required a higher agitation speed (around 220 rpm) to ensure sufficient motion in the solution. In comparison, a 24-well non-treated polystyrene microplate was able to accommodate PA6 nanofiber samples with a diameter of 1.5 cm, meaning that the samples remained flat throughout the immobilization process. Further, though a slightly higher liquid volume was required (>300 µL), such a vessel allowed easy monitoring of nanofiber immersion level and required a lower agitation speed (<150 rpm). While the best results were obtained using 250 µL of enzyme solution (Figure 2b), such a low volume required a special vessel in ensure complete immersion of the nanofiber sample. Hence, we recommend using 300 µL when using a 24-well plate for immobilization of 1.5 cm nanofiber samples as this encourages full adsorption of the enzyme solution into the nanofibers and ensures that most of the enzyme molecules are in contact with the matrix.

3.3. Optimal Buffer Concentration

McIlvaine's buffer has been used for laccase immobilization and activity assays in numerous previous studies [14,36–38]. However, none of these have addressed the optimization of the buffer concentration. Ionic strength is an important parameter influencing both enzyme solubility and the charge of the free functional groups on the enzyme molecules and nanofiber matrix, which in turn enables the development of electrostatic protein–protein and protein–matrix interactions. Generally speaking, lyophilized enzyme powder requires a buffer of sufficient ionic strength to fully solubilize and separate clusters formed by enzyme oligomers [39].

In this study, the clearest influence of ionic strength on the immobilization process was observed between 100% and 10% buffer (Figure 2c). While low ionic strength buffer increased the amount of laccase immobilized, it did not allow for a sufficiently strong interaction between the carrier, enzyme, and GA molecules formed during adsorption and crosslinking, resulting in immobilized laccase of poor stability. In comparison, the highest buffer concentration promoted the formation of strong interactions, resulting in increased enzyme stability. However, possibly due to conformational changes caused by the formation of strong bonds, this concentration yielded the lowest activity of all the samples tested. As a compromise, we were able to achieve both reasonable activity and stability when using 20% and 50% McIlvaine's buffer, though, from an economic point of view, the less concentrated buffer would be preferable.

3.4. Optimal Time for Adsorption and Crosslinking

Time taken for laccase adsorption and GA crosslinking has a clear influence on the activity and stability of immobilized laccase. Surprisingly, the longest time (24 + 8 h) for both adsorption and crosslinking gave the worst results, with lowest immobilized enzyme activity and stability (Figure 2d), probably as the catalyst was damaged through long agitation. Enzyme activity and reuse increased at shorter times. Optimal results were obtained after three hours adsorption and three hours GA crosslinking. This total of six hours compares well with previous studies, where the most common time for laccase adsorption was up to two hours, though adsorption usually took place overnight [31,40].

3.5. Optimal pH and GA Concentration

One of the most important parameters influencing immobilization is pH. In our study, the most suitable pH for immobilization of laccase onto PA6 nanofibers was evaluated as pH 5, although the highest stability over three ABTS oxidation catalytic cycles was recorded at pH 7 (Figure 2e). In previous studies [30–32,40,41], a similar pH (4–5.5) was also considered as optimal, probably as laccase is best able to maintain its functional conformation when attached to a matrix.

Of the four GA concentrations tested (0.5, 1, 2.5, and 5% *v*/*v*), the highest operational stability was observed at the highest GA concentration (5% *v*/*v*). Surprisingly, the optimum concentration turned out to be 2.5% *v*/*v*, providing both the highest activity and best reuse of immobilized laccase. We hypothesize that, at this concentration, larger enzyme clusters were formed outside the matrix area, which decreased the number of bonds between the enzyme and the support (Figure 2f).

GA has long been favored as a crosslinking agent for enzyme immobilization. In a number of cases, it has been used for matrix stabilization [42,43] or functionalization of the supporting material [30,44–46], the optimal GA concentration in such cases varying between 1–4% *v*/*v*. In most cases, however, GA has been used as a bifunctional agent for the introduction of aldehyde groups to the matrix surface prior to enzyme attachment [47–50]. However, this type of functionalization requires the presence of free primary amino groups on the matrix. In comparison, our method does not require the matrix to have such a chemical composition and, as such, provides much greater freedom in the selection of materials for enzyme immobilization.

3.6. Summary of the Optimal Immobilization Process

Overall, our results suggest an optimal average PA6 nanofiber diameter of 105 ± 19.1 nm, with a surface density of 5 g/m^2 (Figure 3). Under these conditions, the nanofibers display excellent mechanical properties, ease of handling, repeatability, low cost, and homogenous surface density, thereby providing perfect conditions for laccase immobilization [51]. We immobilized *T. versicolor* laccase onto PA6 nanofibers via adsorption followed by GA crosslinking. The optimal immobilization process required the PA6 nanofiber samples (~1 mg each) to be submerged separately into 300 μL of 2 mg/mL laccase solution in 20% buffer at pH 5 and shaken at 4 °C in an orbital shaker at 150 rpm for three hours. GA was then added to achieve a final concentration of 2.5 % *v*/*v*, and the samples were shaken in the immobilization solution for a further three hours. Finally, the samples were washed with McIlvaine's buffer at pH 3. Pristine PA6 nanofibers display a homogenous structure and smooth surface (Figure 3a), while PA6-laccase displays a grainy surface formed by cross-linked laccase clusters strongly attached to the nanofibers, even after thorough washing (Figure 3b).

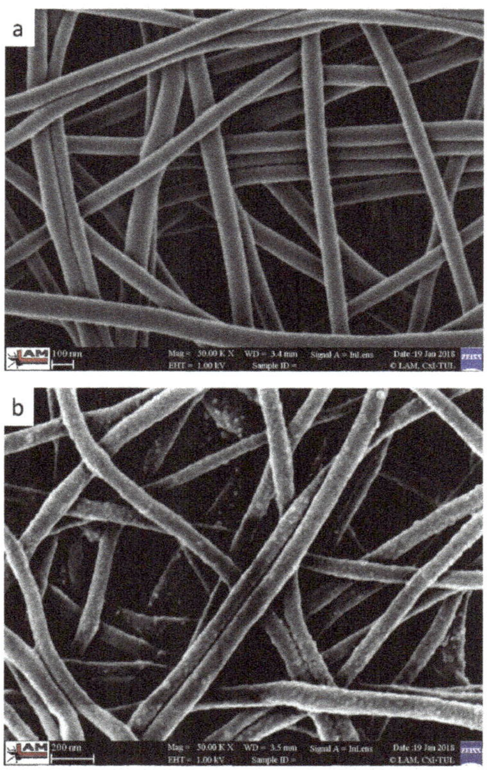

Figure 3. Structure of (**a**) pristine PA6, and (**b**) nanofibers with immobilized laccase (PA6–laccase). Images of pristine PA6 nanofibers and nanofibers with immobilized laccase were obtained using a Carl Zeiss ULTRA Plus scanning electron microscope (Zeiss, Germany) at a magnification of 50k×.

In comparison with our previous study [11], the immobilization technique using PA6 as a matrix is more straightforward. While before, we aimed for the highest possible enzyme loading onto PA6/chitosan material, here we focused on the least expensive and time-consuming method. First, PA6 nanofibers are more easily manufactured and much cheaper compared to PA6/chitosan blend. Second, we succeeded in reducing and simplifying the crosslinking process to a single reaction with GA, instead of double GA crosslinking and single bovine serum albumin or hexamethylendiamine activation. The whole procedure was in total shortened by 21 h. Besides that, the new laccase immobilization step required approximately three times lower bath volume. Although the PA6-laccase membrane reached slightly lower catalytic activity than PA6/chitosan-laccase, this drawback was balanced by 50% higher reuse activity.

3.7. Storage Stability and Reuse

The storage stability of free and immobilized laccase was tested by assessing activity after 13, 20, and 30 days of storage at 4 °C. Use of pH 6 McIlvaine's buffer and a temperature of 4 °C provided optimal conditions for the preservation of enzyme activity, with laccase retaining 42% of its initial activity. PA6-laccase displayed similar storage stability to the free enzyme, with 50% of activity retained after 30 days (Figure 4).

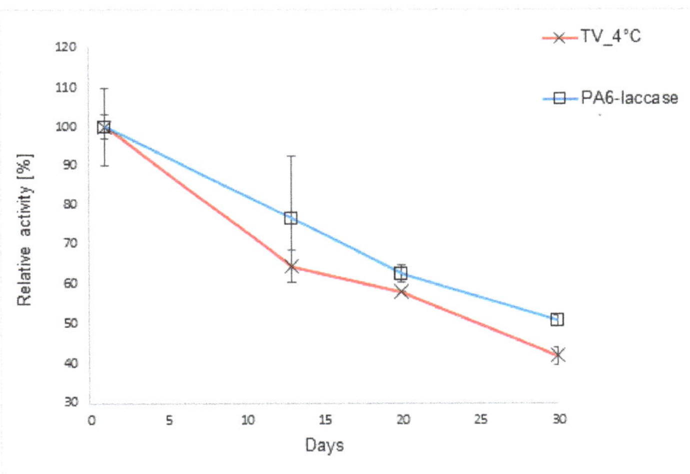

Figure 4. Storage stability of free and immobilized laccase. PA6 nanofibers loaded with immobilized laccase were incubated in 20% buffer (pH 6) at 4 °C. Free laccase solution was stored under the same conditions for comparison with the immobilized laccase.

In previous studies, laccase immobilized onto fibrous polymer-grafted polypropylene chloride film preserved around 57% of initial activity after 30 days storage in a pH 5.5 buffer at 4 °C [52], while laccase immobilized onto a PVDF membrane retained 43% of initial activity after 36 days of storage in pH 4 McIlvaine's buffer at 4 °C [14]. In an exceptional case, laccase immobilized onto carbon nanotubes retained 80% of initial activity under similar storage conditions, while free laccase retained a similar activity level as that in our study [48]. Thus, it appears that particle or nanoparticle carriers are better in preserving the enzyme activity, probably due to their favorable pore size distribution compared to fibrous or nanofibrous structures. However, these nanomaterials cannot compete with the PA6 nanofiber matrix regarding safety, ease of handling, and applicability under actual wastewater treatment conditions. PA6-laccase retained 88% of its initial activity after five ABTS oxidation catalytic cycles; a very promising result compared with the study of Xu et al. [53], where laccase covalently immobilized onto carbon nanotubes retained ca. 80% activity after five ABTS transformation cycles, and [54], where *T. versicolor* laccase immobilized onto magnetic bimodal mesoporous carbon retained 70% activity after five cycles.

3.8. Degradation of BPA, EE2, and TCS

We tested the degradation activity of laccase and PA6-laccase against a 50 µM BPA, EE2, and TCS micropollutant mixture. Our PA6-laccase displayed activity of 0.03 U, similar to that of free laccase in 25 µL of the enzyme solution. PA6-laccase removal efficiency of BPA was lower than that of free laccase (22% remaining after 24 h), though the removal profile was similar to that of free laccase with EE2. PA6-laccase was notably more efficient in TCS reduction within 24 h of incubation (ca. 70% decrease compared with 38% for free laccase; Figure 5). In comparison, a blank sample containing only PA6 nanofibers did not adsorb any EDCs over the 24 h.

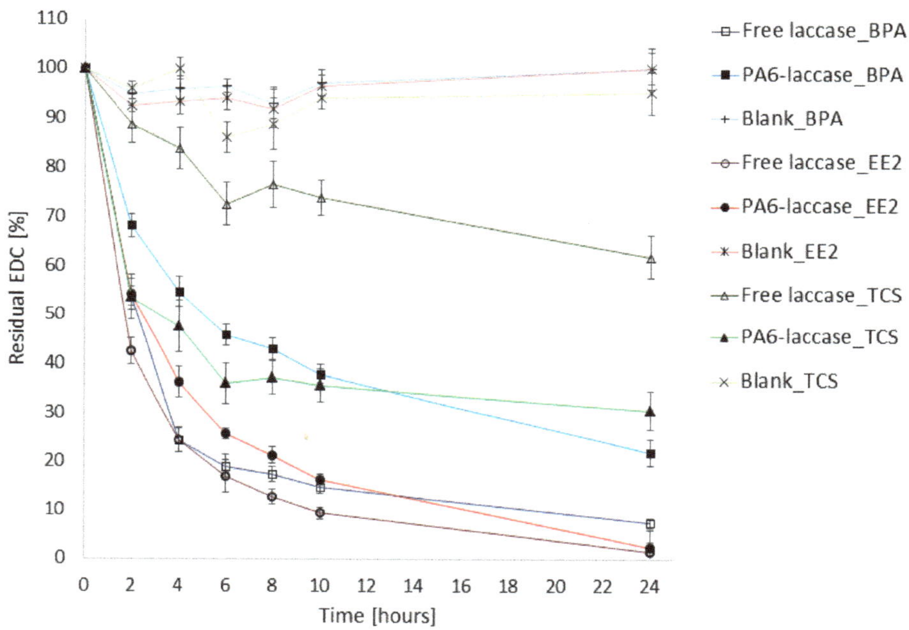

Figure 5. Comparison of degradation efficiency of free and PA6 immobilized laccase against BPA, EE2, and TCS.

Degradation of EDCs using immobilized laccase has been the subject of a number of studies [55,56]. *Cerrena unicolor* laccase (12 U/5 mL) captured on porous silica beads, for example, has been shown to eliminate around 90% of BPA (50 µM) after 60 min of incubation [57]. Application of nanoparticles for enzyme immobilization has recently been criticized due to issues connected with their commercial application, such as their tendency to agglomerate and the need to separate them from treated water [58,59]. The mechanical properties of PA6, on the other hand, showed no signs of damage over 30 days of storage. Hence, PA6 has a major advantage over particle- or nanoparticle-matrices in that their macroscopic and compact form allows them to be handled as textiles, providing great potential for their use as stable filters in water treatment processes.

4. Conclusions

The effectiveness of PA6-laccase for EDC degradation, along with its capacity for repeated use and enhanced storage stability, predetermine this material for real application in advanced wastewater treatment technology. Currently, the cost-effectiveness of immobilized enzymes for water treatment depends on the price of commercially available products and the immobilization method used. Recently, the most popular materials used as enzyme carriers were alginate and similar gels, nanocellulose, zeolites, or bentonite. Compared to these materials, pristine PA6 nanofibers do not require any further chemical modifications, are low-cost, easily manufactured, safe to handle, and possess exceptional mechanical properties beside other electrospun polymers. In this study, we developed a simple and reproducible technique for laccase immobilization onto PA6 nanofibers using only GA as a reagent. Our methodology proved simple and inexpensive thanks to short adsorption and crosslinking time and the use of easily fabricated support and readily available GA. All these features can facilitate inclusion into already existing filtration units without a risk that the matrix-enzyme system will leak into the effluent. Furthermore, we tested degradation efficiency toward the elimination of highly concentrated solution of environmentally hazardous endocrine disruptors BPA, EE2 and TCS. The PA6-laccase system was successful, and the immobilized laccase reached very similar results compared to the free soluble form. If higher demand increases commercial production of laccase and thus reduces its market price, PA6-laccase can enable a new cost-effective and efficient technology for the elimination of EDCs in potable water.

Author Contributions: M.M. performed most of the enzymatic measurements, laccase immobilization, and wrote the manuscript. M.R. prepared nanofiber layers, and V.N. contributed with HPLC analysis. A.S. coordinated the research and corrected the manuscript.

Funding: This research was supported through the Technology Agency of the Czech Republic (project NanoBioWat TE01020218 and project TH02030858) and by the MŠMT of the Czech Republic and the EU—European Structural and Investment Funds in the frames of Operational Programme Research, Development and Education—project Hybrid Materials for Hierarchical Structures "HyHi" (CZ.02.1.01/0.0/0.0/16_019/ 0000843). The authors would like to acknowledge the assistance provided under Research Infrastructure NanoEnviCz, supported by the MŠMT of the Czech Republic (project LM2015073). The research of M. M. was supported by the MŠMT of the Czech Republic through the SGS project no. 21176/115 of the Technical University of Liberec.

Conflicts of Interest: The authors declare no conflict of interest.

References

1. Fent, K.; Weston, A.A.; Caminada, D. Ecotoxicology of human pharmaceuticals. *Aquat. Toxicol.* **2006**, *76*, 122–159. [CrossRef] [PubMed]
2. Luo, Y.; Guo, W.; Ngo, H.H.; Nghiem, L.D.; Hai, F.I.; Zhang, J.; Liang, S.; Wang, X.C. A review on the occurrence of micropollutants in the aquatic environment and their fate and removal during wastewater treatment. *Sci. Total Environ.* **2014**, *473–474*, 619–641. [CrossRef] [PubMed]
3. Gültekin, I.; Ince, N.H. Synthetic endocrine disruptors in the environment and water remediation by advanced oxidation processes. *J. Environ. Manag.* **2007**, *85*, 816–832. [CrossRef] [PubMed]
4. Gmurek, M.; Olak-Kucharczyk, M.; Ledakowicz, S. Photochemical decomposition of endocrine disrupting compounds–A review. *Chem. Eng. J.* **2017**, *310*, 437–456. [CrossRef]
5. EUR-Lex-32013L0039-EN-EUR-Lex. Available online: https://eur-lex.europa.eu/eli/dir/2013/39/oj (accessed on 20 September 2019).
6. Lin, T.; Wu, S.; Chen, W. Formation potentials of bromate and brominated disinfection by-products in bromide-containing water by ozonation. *Environ. Sci. Pollut. Res. Int.* **2014**, *21*, 13987–14003. [CrossRef] [PubMed]
7. Matthis, A. Activated Carbon Options for Wastewater Treatment and Removal of Contaminants. Available online: https://www.watertechonline.com/activated-carbon-options-0517/ (accessed on 19 September 2019).
8. Ge, J.C.; Choi, N.J. Performance of electrospun nanofibrous membranes for trapping of BTX aromatic hydrocarbons and heavy metal ions: Mechanisms, isotherms and kinetics. *J. Clean. Prod.* **2019**, *217*, 388–397. [CrossRef]
9. Figoli, A.; Ursino, C.; Ramirez, D.O.S.; Carletto, R.A.; Tonetti, C.; Varesano, A.; Santo, M.P.D.; Cassano, A.; Vineis, C. Fabrication of electrospun keratin nanofiber membranes for air and water treatment. *Polym. Eng. Sci.* **2019**, *59*, 1472–1478. [CrossRef]

10. Viswanath, B.; Rajesh, B.; Janardhan, A.; Kumar, A.P.; Narasimha, G. Fungal laccases and their applications in bioremediation. *Enzym. Res.* **2014**, *2014*, 163242. [CrossRef]
11. Maryšková, M.; Ardao, I.; García-González, C.A.; Martinová, L.; Rotková, J.; Ševců, A. Polyamide 6/chitosan nanofibers as support for the immobilization of Trametes versicolor laccase for the elimination of endocrine disrupting chemicals. *Enzym. Microb. Technol.* **2016**, *89*, 31–38. [CrossRef]
12. Nguyen, L.N.; Hai, F.I.; Price, W.E.; Leusch, F.D.L.; Roddick, F.; McAdam, E.J.; Magram, S.F.; Nghiem, L.D. Continuous biotransformation of bisphenol A and diclofenac by laccase in an enzymatic membrane reactor. *Int. Biodeterior. Biodegrad.* **2014**, *95*, 25–32. [CrossRef]
13. Xu, R.; Cui, J.; Tang, R.; Li, F.; Zhang, B. Removal of 2,4,6-trichlorophenol by laccase immobilized on nano-copper incorporated electrospun fibrous membrane-high efficiency, stability and reusability. *Chem. Eng. J.* **2017**, *326*, 647–655. [CrossRef]
14. Jahangiri, E.; Thomas, I.; Schulze, A.; Seiwert, B.; Cabana, H.; Schlosser, D. Characterisation of electron beam irradiation-immobilised laccase for application in wastewater treatment. *Sci. Total Environ.* **2018**, *624*, 309–322. [CrossRef] [PubMed]
15. Bilal, M.; Rasheed, T.; Iqbal, H.M.N.; Hu, H.; Wang, W.; Zhang, X. Novel characteristics of horseradish peroxidase immobilized onto the polyvinyl alcohol-alginate beads and its methyl orange degradation potential. *Int. J. Biol. Macromol.* **2017**, *105*, 328–335. [CrossRef] [PubMed]
16. Pandey, V.P.; Rani, J.; Jaiswal, N.; Singh, S.; Awasthi, M.; Shasany, A.K.; Tiwari, S.; Dwivedi, U.N. Chitosan immobilized novel peroxidase from Azadirachta indica: Characterization and application. *Int. J. Biol. Macromol.* **2017**, *104*, 1713–1720. [CrossRef] [PubMed]
17. Peng, Y.; Ling-Ling, H.; Yu-Zhi, D.; Yong-Juan, X.; Hua-Gang, N.; Cong, C.; Xiao-Lin, L.; Xiao-Jun, H. Hooking horseradish peroxidase by using the affinity Langmuir-Blodgett technique for an oriented immobilization. *Appl. Surf. Sci.* **2017**, *403*, 89–94. [CrossRef]
18. Mohamed, S.A.; Al-Harbi, M.H.; Almulaiky, Y.Q.; Ibrahim, I.H.; El-Shishtawy, R.M. Immobilization of horseradish peroxidase on Fe3O4 magnetic nanoparticles. *Electron. J. Biotechnol.* **2017**, *27*, 84–90. [CrossRef]
19. Bayramoglu, G.; Akbulut, A.; Yakup Arica, M. Immobilization of tyrosinase on modified diatom biosilica: Enzymatic removal of phenolic compounds from aqueous solution. *J. Hazard. Mater.* **2013**, *244*, 528–536. [CrossRef]
20. Marín-Zamora, M.E.; Rojas-Melgarejo, F.; García-Cánovas, F.; García-Ruiz, P.A. Production of o-diphenols by immobilized mushroom tyrosinase. *J. Biotechnol.* **2009**, *139*, 163–168. [CrossRef]
21. Marín-Zamora, M.E.; Rojas-Melgarejo, F.; García-Cánovas, F.; García-Ruiz, P.A. Effects of the immobilization supports on the catalytic properties of immobilized mushroom tyrosinase: A comparative study using several substrates. *J. Biotechnol.* **2007**, *131*, 388–396. [CrossRef]
22. Taboada-Puig, R.; Junghanns, C.; Demarche, P.; Moreira, M.T.; Feijoo, G.; Lema, J.M.; Agathos, S.N. Combined cross-linked enzyme aggregates from versatile peroxidase and glucose oxidase: Production, partial characterization and application for the elimination of endocrine disruptors. *Bioresour. Technol.* **2011**, *102*, 6593–6599. [CrossRef]
23. Kochana, J.; Nowak, P.; Jarosz-Wilkołazka, A.; Bieroń, M. Tyrosinase/laccase bienzyme biosensor for amperometric determination of phenolic compounds. *Microchem. J.* **2008**, *89*, 171–174. [CrossRef]
24. Ohtsu, T.; Shigenari, S.; Yoshimoto, M.; Umakoshi, H. Reactive bienzyme systems fabricated through immobilization of biotinylated glucose oxidase and peroxidase molecules onto neutralized avidin-conjugated liposomes. *BioChem. Eng. J.* **2017**, *125*, 81–87. [CrossRef]
25. Ruan, C.; Li, Y. Detection of zeptomolar concentrations of alkaline phosphatase based on a tyrosinase and horse-radish peroxidase bienzyme biosensor. *Talanta* **2001**, *54*, 1095–1103. [CrossRef]
26. McIlvaine, T.C. A Buffer Solution for Colorimetric Comparison. *J. Biol. Chem.* **1921**, *49*, 183–186.
27. Hassani, T.; Ba, S.; Cabana, H. Formation of enzyme polymer engineered structure for laccase and cross-linked laccase aggregates stabilization. *Bioresour. Technol.* **2013**, *128*, 640–645. [CrossRef] [PubMed]
28. Ardao, I.; Magnin, D.; Agathos, S.N. Bioinspired production of magnetic laccase-biotitania particles for the removal of endocrine disrupting chemicals. *Biotechnol. Bioeng.* **2015**, *112*, 1986–1996. [CrossRef]
29. Kim, J.; Jia, H.; Wang, P. Challenges in biocatalysis for enzyme-based biofuel cells. *Biotechnol. Adv.* **2006**, *24*, 296–308. [CrossRef]
30. Fatarella, E.; Spinelli, D.; Ruzzante, M.; Pogni, R. Nylon 6 film and nanofiber carriers: Preparation and laccase immobilization performance. *J. Mol. Catal. B Enzym.* **2014**, *102*, 41–47. [CrossRef]

31. Sathishkumar, P.; Kamala-Kannan, S.; Cho, M.; Kim, J.S.; Hadibarata, T.; Salim, M.R.; Oh, B.-T. Laccase immobilization on cellulose nanofiber: The catalytic efficiency and recyclic application for simulated dye effluent treatment. *J. Mol. Catal. B Enzym.* **2014**, *100*, 111–120. [CrossRef]
32. Xu, R.; Si, Y.; Wu, X.; Li, F.; Zhang, B. Triclosan removal by laccase immobilized on mesoporous nanofibers: Strong adsorption and efficient degradation. *Chem. Eng. J.* **2014**, *255*, 63–70. [CrossRef]
33. Sathishkumar, P.; Chae, J.-C.; Unnithan, A.R.; Palvannan, T.; Kim, H.Y.; Lee, K.-J.; Cho, M.; Kamala-Kannan, S.; Oh, B.-T. Laccase-poly (lactic-co-glycolic acid) (PLGA) nanofiber: Highly stable, reusable, and efficacious for the transformation of diclofenac. *Enzym. Microb. Technol.* **2012**, *51*, 113–118. [CrossRef] [PubMed]
34. Bansal, M.; Kumar, D.; Chauhan, G.S.; Kaushik, A. Preparation, characterization and trifluralin degradation of laccase-modified cellulose nanofibers. *Mater. Sci. Energy Technol.* **2018**, *1*, 29–37. [CrossRef]
35. Wu, D.; Feng, Q.; Xu, T.; Wei, A.; Fong, H. Electrospun blend nanofiber membrane consisting of polyurethane, amidoxime polyarcylonitrile, and β-cyclodextrin as high-performance carrier/support for efficient and reusable immobilization of laccase. *Chem. Eng. J.* **2018**, *331*, 517–526. [CrossRef]
36. Zhang, X.; Wang, M.; Lin, L.; Xiao, G.; Tang, Z.; Zhu, X. Synthesis of novel laccase-biotitania biocatalysts for malachite green decolorization. *J. Biosci. Bioeng.* **2018**, *126*, 69–77. [CrossRef] [PubMed]
37. Barrios-Estrada, C.; Rostro-Alanis, M.J.; Parra, A.L.; Belleville, M.-P.; Sanchez-Marcano, J.; Iqbal, H.M.N.; Parra-Saldívar, R. Potentialities of active membranes with immobilized laccase for Bisphenol A degradation. *Int. J. Biol. Macromol.* **2017**, *108*, 837–844. [CrossRef]
38. Traunsteiner, C.; Sek, S.; Huber, V.; Valero-Vidal, C.; Kunze-Liebhäuser, J. Laccase immobilized on a mixed thiol monolayer on Au (111)–structure-dependent activity towards oxygen reduction. *Electrochim. Acta* **2016**, *213*, 761–770. [CrossRef]
39. Reinikainen, T.; Teleman, O.; Teeri, T.T. Effects of pH and high ionic strength on the adsorption and activity of native and mutated cellobiohydrolase I from Trichoderma reesei. *Proteins Struct. Funct. Bioinform.* **2004**, *22*, 392–403. [CrossRef]
40. Kim, R.E.; Hong, S.-G.; Ha, S.; Kim, J. Enzyme adsorption, precipitation and crosslinking of glucose oxidase and laccase on polyaniline nanofibers for highly stable enzymatic biofuel cells. *Enzym. Microb. Technol.* **2014**, *66*, 35–41. [CrossRef]
41. Xu, R.; Zhou, Q.; Li, F.; Zhang, B. Laccase immobilization on chitosan/poly (vinyl alcohol) composite nanofibrous membranes for 2,4-dichlorophenol removal. *Chem. Eng. J.* **2013**, *222*, 321–329. [CrossRef]
42. Kadam, A.A.; Jang, J.; Jee, S.C.; Sung, J.-S.; Lee, D.S. Chitosan-functionalized supermagnetic halloysite nanotubes for covalent laccase immobilization. *Carbohydr. Polym.* **2018**, *194*, 208–216. [CrossRef]
43. Ghodake, G.S.; Yang, J.; Shinde, S.S.; Mistry, B.M.; Kim, D.-Y.; Sung, J.-S.; Kadam, A.A. Paper waste extracted α-cellulose fibers super-magnetized and chitosan-functionalized for covalent laccase immobilization. *Bioresour. Technol.* **2018**, *261*, 420–427. [CrossRef] [PubMed]
44. Drozd, R.; Rakoczy, R.; Wasak, A.; Junka, A.; Fijałkowski, K. The application of magnetically modified bacterial cellulose for immobilization of laccase. *Int. J. Biol. Macromol.* **2018**, *108*, 462–470. [CrossRef] [PubMed]
45. Skoronski, E.; Souza, D.H.; Ely, C.; Broilo, F.; Fernandes, M.; Fúrigo, A.; Ghislandi, M.G. Immobilization of laccase from Aspergillus oryzae on graphene nanosheets. *Int. J. Biol. Macromol.* **2017**, *99*, 121–127. [CrossRef] [PubMed]
46. García-Morales, R.; García-García, A.; Orona-Navar, C.; Osma, J.F.; Nigam, K.D.P.; Ornelas-Soto, N. Biotransformation of emerging pollutants in groundwater by laccase from P. sanguineus CS43 immobilized onto titania nanoparticles. *J. Environ. Chem. Eng.* **2018**, *6*, 710–717. [CrossRef]
47. Barbosa, O.; Ortiz, C.; Berenguer-Murcia, Á.; Torres, R.; Rodrigues, R.C.; Fernandez-Lafuente, R. Glutaraldehyde in bio-catalysts design: A useful crosslinker and a versatile tool in enzyme immobilization. *RSC Adv.* **2013**, *4*, 1583–1600. [CrossRef]
48. Lassouane, F.; Aït-Amar, H.; Amrani, S.; Rodriguez-Couto, S. A promising laccase immobilization approach for Bisphenol A removal from aqueous solutions. *Bioresour. Technol.* **2018**, *271*, 360–367. [CrossRef] [PubMed]
49. Lonappan, L.; Liu, Y.; Rouissi, T.; Pourcel, F.; Brar, S.K.; Verma, M.; Surampalli, R.Y. Covalent immobilization of laccase on citric acid functionalized micro-biochars derived from different feedstock and removal of diclofenac. *Chem. Eng. J.* **2018**, *351*, 985–994. [CrossRef]

50. López-Gallego, F.; Betancor, L.; Mateo, C.; Hidalgo, A.; Alonso-Morales, N.; Dellamora-Ortiz, G.; Guisán, J.M.; Fernández-Lafuente, R. Enzyme stabilization by glutaraldehyde crosslinking of adsorbed proteins on aminated supports. *J. Biotechnol.* **2005**, *119*, 70–75. [CrossRef] [PubMed]
51. Guibo, Y.; Qing, Z.; Yahong, Z.; Yin, Y.; Yumin, Y. The electrospun polyamide 6 nanofiber membranes used as high efficiency filter materials: Filtration potential, thermal treatment, and their continuous production. *J. Appl. Polym. Sci.* **2013**, *128*, 1061–1069. [CrossRef]
52. Arica, M.Y.; Salih, B.; Celikbicak, O.; Bayramoglu, G. Immobilization of laccase on the fibrous polymer-grafted film and study of textile dye degradation by MALDI–ToF-MS. *Chem. Eng. Res. Des.* **2017**, *128*, 107–119. [CrossRef]
53. Xu, R.; Tang, R.; Zhou, Q.; Li, F.; Zhang, B. Enhancement of catalytic activity of immobilized laccase for diclofenac biodegradation by carbon nanotubes. *Chem. Eng. J.* **2015**, *262*, 88–95. [CrossRef]
54. Liu, Y.; Zeng, Z.; Zeng, G.; Tang, L.; Pang, Y.; Li, Z.; Liu, C.; Lei, X.; Wu, M.; Ren, P.; et al. Immobilization of laccase on magnetic bimodal mesoporous carbon and the application in the removal of phenolic compounds. *Bioresour. Technol.* **2012**, *115*, 21–26. [CrossRef] [PubMed]
55. Nair, R.R.; Demarche, P.; Agathos, S.N. Formulation and characterization of an immobilized laccase biocatalyst and its application to eliminate organic micropollutants in wastewater. *New Biotechnol.* **2013**, *30*, 814–823. [CrossRef] [PubMed]
56. Zdarta, J.; Jankowska, K.; Wyszowska, M.; Kijeńska-Gawrońska, E.; Zgoła-Grześkowiak, A.; Pinelo, M.; Meyer, A.S.; Moszyński, D.; Jesionowski, T. Robust biodegradation of naproxen and diclofenac by laccase immobilized using electrospun nanofibers with enhanced stability and reusability. *Mater. Sci. Eng. C* **2019**, *103*, 109789. [CrossRef] [PubMed]
57. Songulashvili, G.; Jimenéz-Tobón, G.A.; Jaspers, C.; Penninckx, M.J. Immobilized laccase of Cerrena unicolor for elimination of endocrine disruptor micropollutants. *Fungal Biol.* **2012**, *116*, 883–889. [CrossRef]
58. Grieger, K.D.; Baun, A.; Owen, R. Redefining risk research priorities for nanomaterials. *J. Nanopart. Res.* **2010**, *12*, 383–392. [CrossRef] [PubMed]
59. Westerhoff, P.; Song, G.; Hristovski, K.; Kiser, M.A. Occurrence and removal of titanium at full scale wastewater treatment plants: Implications for TiO2 nanomaterials. *J. Environ. Monit.* **2011**, *13*, 1195–1203. [CrossRef]

© 2019 by the authors. Licensee MDPI, Basel, Switzerland. This article is an open access article distributed under the terms and conditions of the Creative Commons Attribution (CC BY) license (http://creativecommons.org/licenses/by/4.0/).

MDPI
St. Alban-Anlage 66
4052 Basel
Switzerland
Tel. +41 61 683 77 34
Fax +41 61 302 89 18
www.mdpi.com

Polymers Editorial Office
E-mail: polymers@mdpi.com
www.mdpi.com/journal/polymers

www.ingramcontent.com/pod-product-compliance
Lightning Source LLC
Chambersburg PA
CBHW040833250926
51850CB00007BA/1306
* 9 7 8 3 0 3 6 5 3 4 7 1 8 *